REEDS MARINE ENGINEERING AND TECHNOLOGY

# NAVAL ARCHITECTURE
# FOR MARINE ENGINEERS

## REEDS MARINE ENGINEERING AND TECHNOLOGY SERIES

# 4

REEDS MARINE ENGINEERING AND TECHNOLOGY

# NAVAL ARCHITECTURE
## FOR MARINE ENGINEERS

Revised by Richard Pemberton

EA Stokoe

# REEDS
LONDON • OXFORD • NEW YORK • NEW DELHI • SYDNEY

REEDS
Bloomsbury Publishing Plc
50 Bedford Square, London, WC1B 3DP, UK
29 Earlsfort Terrace, Dublin 2, Ireland

BLOOMSBURY, REEDS, and the Reeds logo are trademarks of Bloomsbury Publishing Plc

First published in Great Britain 1963
Second edition 1967
Third edition 1973
Reprinted 1975, 1977, 1982
Fourth edition 1991
Reprinted 1997, 1998, 2000, 2001, 2003, 2007 (twice), 2009, 2010, 2011, 2012, 2013, 2015 (twice), 2016 (twice)
Fifth edition 2018
Sixth edition 2024

Bloomsbury Publishing Plc does not have any control over, or responsibility for, any third-party websites
referred to or in this book. All internet addresses given in this book were correct at the time of going to press.
The author and publisher regret any inconvenience caused if addresses have changed or sites have
ceased to exist, but can accept no responsibility for any such changes

A catalogue record for this book is available from the British Library

Library of Congress Cataloguing-in-Publication data has been applied for.

ISBN: PB: 978-1-3994-1012-0; ePub: 978-1-3994-1011-3; ePDF: 978-1-3994-1013-7

2 4 6 8 10 9 7 5 3 1

Typeset by Newgen KnowledgeWorks Pvt. Ltd., Chennai, India
Printed and bound in Great Britain by CPI Group (UK) Ltd, Croydon CR0 4YY

To find out more about our authors and books visit www.bloomsbury.com
and sign up for our newsletters

# CONTENTS

# PREFACE

The previous version of Reed's Marine Engineering Series Volume 4 was edited to update the original version by Stokoe. Having first been published in 1963, it has covered the Naval Architecture syllabus of the early stage examinations for those looking to become a Marine Engineering Officer under the exams of the UK Maritime and Coastguard Agency. Sixty years after that first publication, the world is a very different place, and shipping is undergoing the largest technological change since either sail gave way to steam, or steam gave way to oil/diesel.

Any Marine Engineering Officer going through training today will need an awareness of a wider range of technologies than was previously the case as the vessel owners and operators look to meet the International Maritime Organization's (IMO) target of having net-zero Greenhouse Gas (GHG) emissions from international shipping by or close to 2050. Whilst knowledge of these technologies is not a requirement of the current syllabus, it was felt important to support new cadets in their future careers, by making them aware of those technologies. With this intention, the new chapter has been included.

The examples presented within this book aim to show all relevant stages of calculation, and it is recommended that this approach is followed by candidates. First, should a mistake be made, it is possible for an examiner to identify it and award marks for knowledge of the correct method, rather than just basing the mark on obtaining the correct answer. Second, the process is likely to be applicable to a wider range of problems than methods that have shortcuts.

It should be noted that a large proportion of the worked solutions include diagrams and it is suggested that the students follow this practice. Students should note that the solutions given here are correct to the decimal precision quoted at each stage of a calculation. It is quite likely that differences will occur, if multiple stages are conducted within the calculator's memory, but these differences are most likely small, and should not be of great concern.

The original examples have been reviewed, and a set of current typical exam questions are reproduced with kind permission of Alan Sleigh, MCA Contract Manager with the Scottish Qualifications Authority. While there have been changes to this volume, the sentiment expressed at the end of the preface remains the same. An engineer who works systematically through this volume will find that their time is amply repaid when attending a course of study at a college and their chance of success in the examination will be greatly increased.

# 1

# UNDERLYING MATHS AND PHYSICS

While the concepts of ship stability can be described without resorting to formulae, very quickly we require a good knowledge of certain mathematical topics to be able to operate at the required standard. This chapter is included primarily for purposes of revision, as any naval architecture student will be more comfortable with technical topics if their underlying mathematical foundation is solid.

## Forces and Moments

Fundamental to the understanding of ship stability are the concepts of forces and moments. A force is defined as an interaction that when unopposed will cause an object to move. An example of this would be a book lying on a table. If the book is pushed, a force is applied, and it will move. How the book moves is dependent on three factors:

1. The amount, or magnitude, of the applied force (measured in newtons, N).
2. The direction in which the force is applied.
3. The location of where the force was applied.

These three factors should always be considered when determining what forces are acting on a body. The movement of a body may be a linear motion in a direction, due to the applied forces, or it might be a rotation. The rotation will be due to the total moments of the forces. Moments are defined as being the product of a force and a perpendicular distance between the line of action of a force and a reference point.

If an object is at rest, then all of the forces and moments acting on it must be in equilibrium, or cancelling the effect of one another.

# Units

Engineers and scientists are always dealing with quantities and properties of items. In much of the world, the system followed is the Système International d'Unités (SI), or the metric system of measurement. This system comprises the following six basic units of measurement:

| Quantity | Unit | Symbol |
|---|---|---|
| Length | Metre | m |
| Mass | Kilogram | kg |
| Time | Second | s |
| Temperature | Kelvin | K |
| Electric current | Ampere | A |
| Luminous intensity | Candela | cd |

## Derived units

It is possible to obtain derived units from these basic units. The system has been designed in such a way that the basic units are used without numerical multipliers to obtain the fundamental derived units. The system is therefore said to be *coherent*.

$$
\begin{aligned}
\text{unit area} &= m^2 \\
\text{unit volume} &= m^3 \\
\text{unit velocity} &= m/s \\
\text{unit acceleration} &= m/s^2
\end{aligned}
$$

The unit of *force* is the *newton* N.

Now,

$$\text{Force} = \text{mass} \times \text{acceleration}$$
$$1\,\text{N} = 1\,\text{kg} \times 1\,\text{m/s}^2$$

Hence

$$\text{N} = \text{kg m/s}^2$$

It is worth highlighting at this stage the difference between mass and weight. When discussing objects, we often use weight and mass interchangeably, but there is a distinct difference. The mass of an object is a measure of how much matter it contains, and this will be given in kg. The weight of an object is how much force that object imparts on its support due to gravitational attraction, and this will be in Newtons.

The unit of *work* is the *joule* J

$$\text{work done} = \text{force} \times \text{distance}$$
$$1\,\text{joule} = 1\,\text{N} \times 1\,\text{m}$$
$$\text{J} = \text{Nm}$$

The unit of *power* is the *watt* W

$$\text{Power} = \text{work done per unit time}$$
$$1\,\text{watt} = 1\,\text{J} \div 1\,\text{s}$$
$$\text{W} = \text{J/s}$$

The unit of *pressure* is the *pascal* Pa

$$\text{pressure} = \text{force per unit area}$$
$$1\,\text{Pa} = 1\,\text{N} \div 1\,\text{m}^2$$
$$\text{Pa} = \text{N/m}^2$$

Solving engineering problems, be they in an exam or experienced during your career, requires the ability to interpret the available information. It is not always the case that information is presented in a manner that allows comparison with a straight textbook equation, therefore an engineer should be happy with knowing how fundamental quantities (such as forces, work done and power) are derived.

This will allow solutions to problems where the necessary information is not in the format you might expect.

It is important to be aware of the other main unit system, the imperial system, which is currently in use in North America. Some units from it are still in regular usage, such as knots, but the UK's transition from this to the SI system took place in the late 1960s. In terms of maintaining accuracy in your calculations, it is often advisable to work only in one unit system, which would typically be the SI system. It is important to be aware, however, that some empirical formulae, such as those used in calculating the resistance of a ship, have coefficients associated with them that require the input values to the equation to be in imperial rather than SI units. A check should always be made on the required format of the inputs to an equation with coefficients in it.

Familiarity with units and derived units is a key skill for any engineer, and can help in terms of checking your calculations. By keeping track of the units in a formula, you are able to identify whether a mistake has been made, because if the derived units are not correct, then the numerical value is not what you are expecting either.

# Indices

Indices are the mathematical style for abbreviating multiplication of a number or symbol by itself. For example, 5 x 5 is written as $5^2$ and 3 x 3 x 3 x 3 as $3^4$. The quantities are multiplied, and so the indices are added. This technique is often used for units, in particular the derived units described above. An area of a square with sides 4 m would be 4 m x 4 m equals 16 $m^{1+1}$ or 16 $m^2$ (you may also see 16 sq.m, but it is preferable to keep to the scientific notation). If the quantities are divided, then the indices are subtracted.

## Multiples and submultiples

In order to keep the number of names of units to a minimum, multiples and submultiples of the fundamental units are used. In each case, powers of ten are found to be most convenient and are represented by prefixes, which are combined with the symbol of the unit.

| Multiplication factor | | Standard form | Prefix | Symbol |
|---|---|---|---|---|
| 1 000 000 000 000 | | $10^{12}$ | tera | T |
| 1 000 000 000 | | $10^{9}$ | giga | G |
| 1 000 000 | | $10^{6}$ | mega | M |
| 1 000 | | $10^{3}$ | kilo | k |
| 100 | | $10^{2}$ | hecto | h |
| 10 | | $10^{1}$ | deca | da |
| | 0.1 | $10^{-1}$ | deci | d |
| | 0.01 | $10^{-2}$ | centi | C |
| | 0.001 | $10^{-3}$ | milli | m |
| | 0.000 001 | $10^{-6}$ | micro | μ |
| | 0.000 000 001 | $10^{-9}$ | nano | n |
| | 0.000 000 000 001 | $10^{-12}$ | pico | p |

Only one prefix may be used with each symbol. Thus a thousand kilograms would be expressed as a Mg and not kkg. When a prefix is attached to a unit, it becomes a new unit symbol on its own account and this can be raised to positive or negative powers of ten.

Multiples of $10^{3}$ are recommended but others are recognised because of convenient sizes and established usage and custom. A good example of this convenient usage lies in the calculation of volumes. If only metres or millimetres are used for the basic dimensions, the volume is expressed in $m^{3}$ or $mm^{3}$.

Now,

$$1 \, m^{3} = 10^{9} \, mm^{3}$$

ie the gap is too large to be convenient. If, on the other hand, the basic dimensions may be expressed in decimetres or centimetres in addition to metres and millimetres, the units of volume change in $10^{3}$ intervals.

$$ie \quad 1 \, m^{3} = 1000 \, dm^{3}$$
$$1 \, dm^{3} = 1000 \, cm^{3}$$
$$1 \, cm^{3} = 1000 \, mm^{3}$$

Several special units are introduced, again because of their convenience. A megagramme, for instance, is termed a *tonne*, which is approximately equal to an imperial ton mass.

Pressure may be expressed in *bars* (b) of value $10^5 N/m^2$. A bar is approximately equal to one atmosphere. Stresses may be expressed in *Megapascals* ($10^6 N/m^2$).

It is unwise, however, to consider comparisons between imperial and SI units and it is probable that the pressure and stress units will revert to the basic unit and its multiples. Keeping track of both the unit and its multiples is an important habit to have when performing calculations, as it will help you work towards the correct answer.

# Rearranging Formulae/Equations

Exam questions will often require the candidate to rearrange a formula in order to answer the question, and it is therefore recommended that you should be proficient in this. Equations are similar to a see-saw, in that for them to remain balanced, whatever happens to one side of the equation must happen to the other – whether that be an addition, subtraction, multiplication or division, the same operation must happen to both sides.

The process of rearranging a formula should start by identifying the variable you want, and looking to perform various mathematical operations, until it is only that variable on one side.

For example, in the equation below, if we are given the volume, width and length of a rectangular box, we can find the height by rearranging the following formula:

$$\text{Volume (m}^3) = \text{Length (m)} \times \text{Width (m)} \times \text{Height (m)}$$

$$\frac{\text{Volume (m}^3)}{\text{Length (m)}} = \text{Width (m)} \times \text{Height (m)}$$

$$\frac{\text{Volume (m}^3)}{\text{Length (m)} \times \text{Width (m)}} = \text{Height (m)}$$

# Trigonometry

Trigonometry is the branch of maths that covers the relationships between angles and lengths of a right-angled triangle, and when performing stability calculations, there can be the requirement to use trigonometry.

Figure 1.1 shows the relationships between the various lengths and angles, and it is important to remember these relationships, and how they and the inverse functions can be used within your calculator.

The angle of interest is typically referred to by the Greek letter θ (pronounced th-ee-ta), with the shortest side next to it being called the *adjacent*, and the *opposite* side being at 90° to the adjacent. The longest side of the triangle is known as the *hypotenuse*, and the lengths of these sides are related to one another by Pythagorus' theorem:

$$\text{Hypotenuse}^2 = \text{Adjacent}^2 + \text{Opposite}^2$$

A method for remembering the trigonometric relationships is often by saying the sounds made by the first letters of the terms in the formula: SOH, CAH, TOA, which represent sine, cosine and tangent respectively.

Figure 1.1 shows the relationships between the functions and their inverse. The inverse trigonometric functions allow the angle of interest to be calculated, but there are a number of different naming conventions for these. They can be named with the prefix arc, such as arcsin, arccos or arctan.

When using a computer spreadsheet program, this convention forms the basis for the function names, the main functions being SIN(*angle*), COS(*angle*) and TAN(*angle*), with their inverses being ASIN(*opp/hyp*), ACOS(*adj/hyp*) and ATAN(*opp/adj*). It is important to realise that within a spreadsheet package these functions will either take radians as an input or give it as an output; it is the user who must remember to convert to and from degrees.

When using a calculator, as is typical within the exam setting, the inverse trigonometric functions are found under either the second function or the inverse key, and are often shown with a −1 index, such as SIN$^{-1}$, in an alternative colour to the main functions.

$$\sin \theta = \frac{\text{Opposite}}{\text{Hypotenuse}}$$

$$\cos \theta = \frac{\text{Adjacent}}{\text{Hypotenuse}}$$

$$\tan \theta = \frac{\text{Opposite}}{\text{Adjacent}}$$

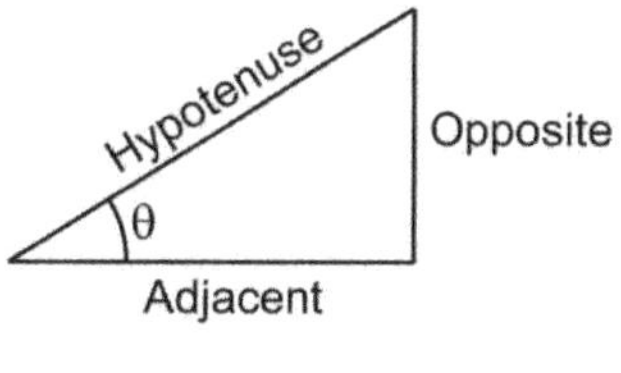

$$\theta = \arcsin \frac{\text{Opposite}}{\text{Hypotenuse}} \qquad \theta = \sin^{-1} \frac{\text{Opposite}}{\text{Hypotenuse}}$$

$$\theta = \arccos \frac{\text{Adjacent}}{\text{Hypotenuse}} \qquad \theta = \cos^{-1} \frac{\text{Adjacent}}{\text{Hypotenuse}}$$

$$\theta = \arctan \frac{\text{Opposite}}{\text{Adjacent}} \qquad \theta = \tan^{-1} \frac{\text{Opposite}}{\text{Adjacent}}$$

▲ **Figure 1.1** *Trigonometric relationships*

In addition to trigonometric functions, when answering questions it can often be useful to remember the rule of similar triangles, shown in figure 1.2 and given in the equation below.

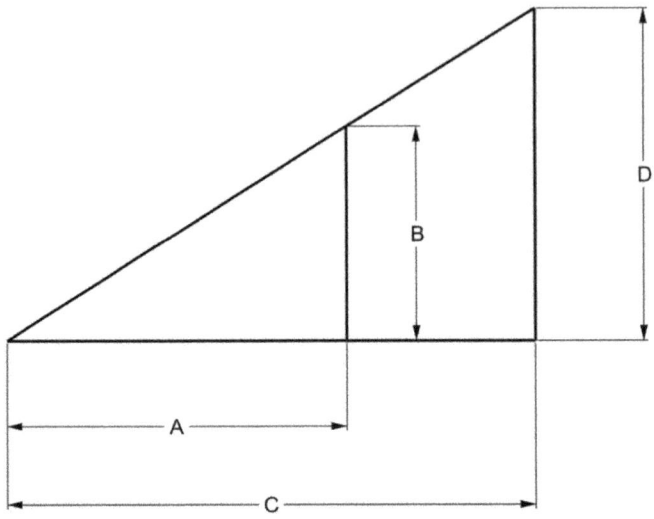

▲ **Figure 1.2** *Similar triangles*

$$\frac{A}{B} = \frac{C}{D}$$

Therefore, if you have knowledge of three of the lengths, you can determine the other, without calculating the angle of the triangle.

When using trigonometry to define lengths, it is always advised to draw a diagram, and always check which sides or angles you are dealing with. In particular, check that you are working with a right-angled triangle.

# Using a Calculator

As mentioned previously, you will be using a calculator to calculate your answers for the exams you take. There are a number of points to remember when using calculators.

- Make sure you are familiar with your calculator and how it functions. The order in which you have to enter numbers and perform operations is particularly important.
- Some calculators will return answers in fractions, whereas for the exams you will be expected to give the answer as a decimal. Make sure you have this setting on your calculator, or can get a decimal answer from it.

- As well as all the usual functions (ie trigonometric), it is important that you are able to perform the inverse of a function, and this varies from calculator to calculator.

- While we are used to working with degrees as the unit of angles, you should also be aware that there are others as your calculator will probably be capable of degrees (DEG), gradians (GRA) and radians (RAD). There are 360 degrees, 400 gradians and $2\pi$ radians in a circle. It is important to be aware which unit your calculator is working with, and how to change it should you wish. For reference, gradians are never used within naval architecture, and the majority of the questions will require you to work in degrees. Where area under a curve is calculated, radians should be used.

The importance of writing down each stage of your calculation and the values you get along the way cannot be emphasised enough. This will make it easier for you to spot errors and easier for the examiner to mark, giving you marks for your process even if you should make a mistake. It is important to get into this habit while you work through this book, as it will prepare you for your future exams.

# Calculation of Area, Volume, First and Second Moments

## Simpson's First Rule

Whether it relates to the stability and flotation of a vessel or the structural integrity of part of the ship, engineers need to know the area and volume properties of complex objects, and this is typically determined by the integration of certain quantities.

While modern computer programs have reduced the amount of hand calculations that take place, specialist software is not always available, and an engineer should not rely on guessing. It is important that engineers are capable of performing their own area and volume calculations for arbitrary shapes without resorting to a computer program.

The simplest method used to calculate an area is known as the *trapezium rule*. This is normally not used in the exam calculations, but is worth being aware of as an introduction to more complex methods of integration.

The trapezium rule states that the area beneath a curve can be approximated by the summation of the areas of a series of trapezia beneath the curve, as shown in figure 1.3.

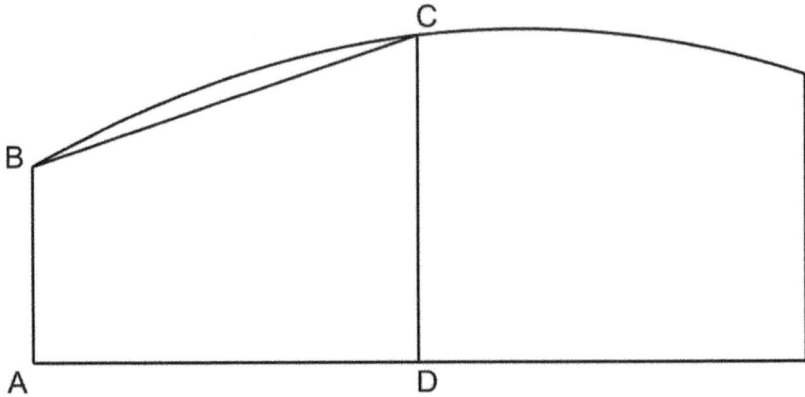

▲ **Figure 1.3** *The trapezium rule*

$$Area = AD \times \frac{1}{2}(AB + CD)$$

The area of the trapezium is not exactly the same size as the area under the curve, and this can lead to there being a cumulative error between the calculated value and the actual value. This error increases as the length *AD* increases and the discrepancy between the straight edge of the trapezium and the curve is greater. It can also increase if there are large changes of curvature in the curve relative to the spacing *AD*.

A more accurate set of methods for establishing the area beneath a curve are Simpson's rules. Simpson's First Rule is based on the assumption that the curved portion of a figure forms part of a parabola ($y = ax^2 + bx + c$), and gives the area contained between *three* consecutive, equally spaced ordinates.

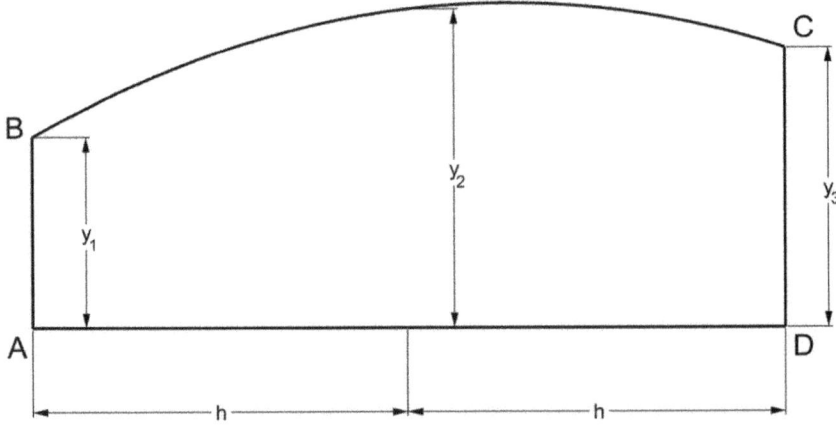

▲ **Figure 1.4** *Simpson's First Rule*

$$\text{Area ABCD} = \frac{h}{3}(y_1 + 4y_2 + y_3)$$

The area beneath the curve is then given by the equation above, and this rule may be applied repeatedly to determine the area of a larger plane such as EFGH, figure 1.5.

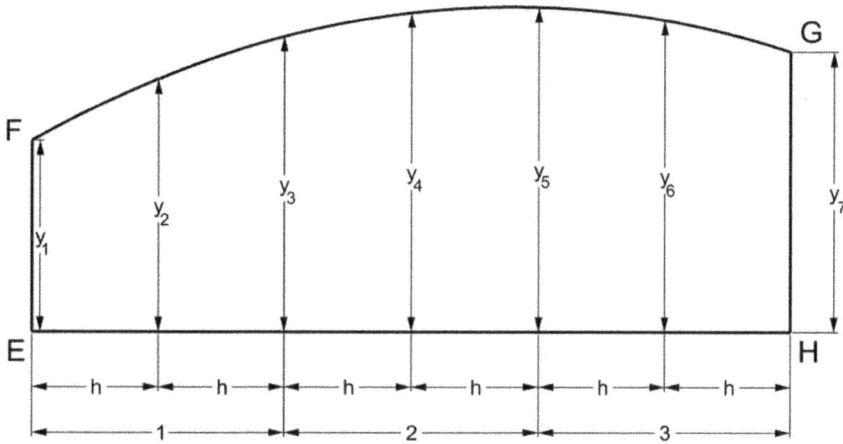

▲ **Figure 1.5** *Simpson's Rule with multiple ordinates*

$$\text{Area 1} = \frac{h}{3}(y_1 + 4y_2 + y_3)$$

$$\text{Area 2} = \frac{h}{3}(y_3 + 4y_4 + y_5)$$

$$\text{Area 3} = \frac{h}{3}(y_5 + 4y_6 + y_7)$$

$$\text{Area EFGH} = \text{Area 1} + \text{Area 2} + \text{Area 3}$$

$$\text{Area EFGH} = \frac{h}{3}\big[(y_1 + 4y_2 + y_3) + (y_3 + 4y_4 + y_5) + (y_5 + 4y_6 + y_7)\big]$$

$$\text{Area EFGH} = \frac{h}{3}\big[y_1 + 4y_2 + 2y_3 + 4y_4 + 2y_5 + 4y_6 + y_7\big]$$

It should be noted at this stage that it is necessary to apply the whole rule and thus an *odd* number of equally spaced ordinates is necessary. Greater speed and accuracy is obtained if this rule is applied in the form of a table. The distance $h$ is termed the common interval and the numbers 1, 4, 2, 4, etc are termed Simpson's multipliers.

EXAMPLE: The $y$ values for a curve against $x$ between 0 and 39 m are 1.8, 3.6, 5.3, 6.8, 7.1, 5.7 and 4.0 m respectively.

Calculate the area under the curve.

Since there are seven ordinates there will be six spaces, therefore:

$$\text{Common interval} = \frac{39}{6} = 6.5\,\text{m}$$

Construct the table:

| Position (x) | Offset (y) | Simpson's multiplier | Product for area |
|---|---|---|---|
| 0 | 1.8 | 1 | 1 x 1.8 = 1.8 |
| 6.5 | 3.6 | 4 | 4 x 3.6 = 14.4 |
| 13 | 5.3 | 2 | 2 x 5.3 = 10.6 |
| 19.5 | 6.8 | 4 | 4 x 6.8 = 27.2 |
| 26 | 7.1 | 2 | 2 x 7.1 = 14.2 |
| 32.5 | 5.7 | 4 | 4 x 5.7 = 22.8 |
| 39 | 4.0 | 1 | 1 x 4.0 = 4.0 |
| | | | $\Sigma_A = 95.0$ |

$$\text{Area} = \frac{h}{3}\sum_A = \frac{6.5}{3} \times 95.0$$
$$= 205.83\,\text{m}^2$$

## Application to volumes

Simpson's Rule is a mathematical rule that will give the area under any continuous curve, no matter what the ordinates represent. It can be used in order to calculate volumes, as a curve of cross-sectional area against length for an object, when integrated, is equal to the volume.

Consider the case of a typical barrel form, which has circular cross sections, but is narrower at its top and base than in the middle. The volume of the barrel could be found by measuring the diameter at evenly spaced intervals from the base of the barrel.

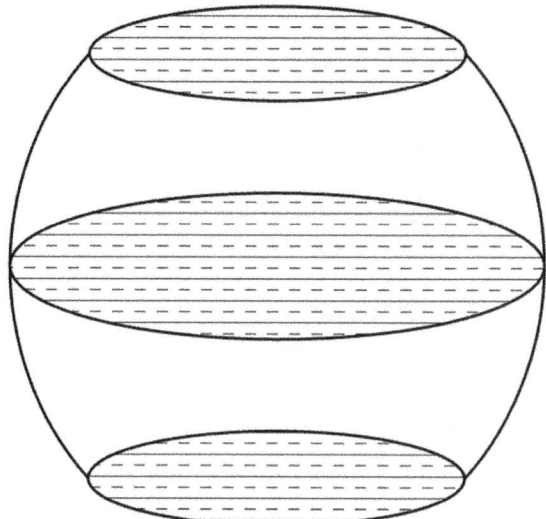

▲ **Figure 1.6** *Volume of barrel*

| Position (m) | Diameter (m) | Cross Section Area (m²) | Simpson's Multiplier | Product for Area (m²) |
|---|---|---|---|---|
| Base | 0.4 | $=\pi \times 0.4^2 \div 4 = 0.126$ | 1 | $1 \times 0.126 = 0.126$ |
| 0.2 | 0.5 | $=\pi \times 0.5^2 \div 4 = 0.196$ | 4 | $4 \times 0.196 = 0.784$ |
| 0.4 | 0.6 | $=\pi \times 0.6^2 \div 4 = 0.283$ | 2 | $2 \times 0.283 = 0.566$ |
| 0.6 | 0.5 | $=\pi \times 0.5^2 \div 4 = 0.196$ | 4 | $4 \times 0.196 = 0.784$ |
| 0.8 | 0.4 | $=\pi \times 0.4^2 \div 4 = 0.126$ | 1 | $1 \times 0.126 = 0.126$ |
| | | | | $\Sigma_v = 2.387$ |

$$\text{Barrel volume} = \frac{h}{3}\Sigma_v$$
$$= \frac{0.2}{3} \times 2.388$$
$$= 0.159 \text{ m}^3$$

While the example of a barrel has been given, provided we have information regarding the cross-sectional area of a shape, this technique can be used to calculate the volume of any shape – in our case, a ship's hull.

## First and second moments of area

In addition to the area, we are often interested in the moment of that area about a reference line. The centre of area of a shape is often used as a point about which we can assume its properties act, and the centre of area is found by dividing the first moment of area by the area:

$$\text{First moment of area} = \int_A y\, dA$$

$$\text{Centre of area} = \frac{\text{First moment of area}}{\text{Area}}$$

$$= \frac{\int_A y\, dA}{A}$$

Another property of area that is of great use to engineers is the second moment of area. We have seen how the first moment of area can be used for establishing properties of a shape, but it can often be difficult to visualise what the second moment of area relates to. As a quantity, it is a measure of the difficulty of rotating that shape about an axis. The larger the second moment of area, the harder it is for it to rotate.

The formula for second moment of area is given in the equation below, and it is of particular use when calculating the deflection of a beam, or in some of the stability calculations we use in following chapters.

$$\text{Second moment of area} = \int_A y^2\, dA$$

The equation above refers to the second moment of area about an axis through the centroid of the cross section (the neutral axis), and is often referred to as $I_{NA}$. It may often be necessary to calculate the second moment about an axis 0-0 parallel to the neutral axis and distance $H$ from it. This is done using the *theorem of parallel axes* and is given by:

$$I_{OO} = I_{NA} + AH^2$$

where $A$ is the area of the cross section.

As will be discussed in Chapters 2 and 3, it is often found necessary to determine the centroid of a curved plane such as a waterplane and the second moment of area of a waterplane.

Consider a plane ABCD, figure 1.7.

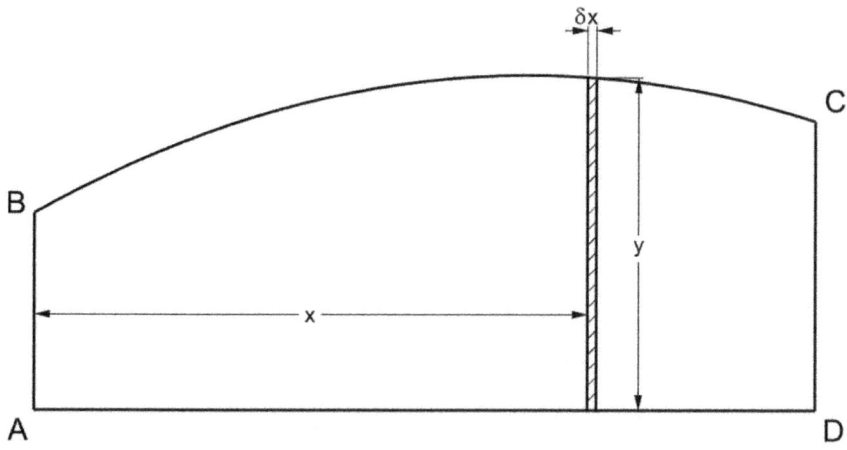

Divide the plane into thin strips of length $\delta x$, with one such strip, distance $x$ from AB, having an ordinate $y$.

$$\text{Area of strip} = y \times \delta x$$
$$\text{Total area of plane} = (y_1 + y_2 + y_3 + \ldots)\delta x$$
$$= \sum y\delta x$$

The area of the plane may be found by putting the ordinates $y$ through Simpson's Rule.

$$\text{First moment of area of strip about AB} = x \times y\delta x$$
$$= xy\delta x$$
$$\text{First moment of area of plane about AB} = (x_1 y_1 + x_2 y_2 + x_3 y_3 + \ldots)\delta x$$
$$= \sum y\delta x$$

Now, it was mentioned earlier that Simpson's Rule may be used to find the area under any continuous curve, no matter what the ordinates represent. Such a curve may be drawn on a base equal to bc, with ordinates of $xy$, and the area under this curve may be found by putting the values of $xy$ through Simpson's Rule:

$$\text{Second moment of area of strip about AB} = I_{NA} + Ax^2$$
$$= \frac{y}{12}(\delta x)^3 + x^2 y\delta x$$

This may be reduced to $(x^2 y\,\delta x)$ since $\delta x$ is very small and when cubed, becomes negligible.

Second moment of area of plane about $AB = (x_1^2 y_1 + x_2^2 y_2 + x_3^2 y_3 + \ldots)\delta x$

$$= \sum x^2 y \delta x$$

This may be found by putting the values of $x^2 y$ through Simpson's Rule.

First moment of area of strip about $BC = \dfrac{y}{2} \times y \delta x$

$$= \dfrac{y^2}{2}\delta x$$

First moment of area of plane about $BC = (y_1^2 + y_2^2 + y_3^2 + \ldots)\dfrac{1}{2}\delta x$

$$= \sum \dfrac{1}{2}y^2 \delta x$$

This may be found by putting ½ $y^2$ through Simpson's Rule.

Second moment of area of strip about $BC = \dfrac{y^3}{12}\delta x + \left(\dfrac{y^2}{2}\right)y\delta x$

$$= \dfrac{y^3}{3}\delta x$$

Second moment of area of plane about $BC = (y_1^3 + y_2^3 + y_3^3 + \ldots)\dfrac{1}{3}\delta x$

$$= \sum \dfrac{1}{3}y^3 \delta x$$

This may be found by putting ⅓ $y^3$ through Simpson's Rule.

# Summary

This chapter is intended to serve as a brief refresher to candidates as to what mathematics and physics will be applied throughout the remainder of the text and is not intended to cover the topics in as much depth as specific mathematical and mechanics textbooks. The following chapters will show how these principles are applied.

# 2

# HYDROSTATICS
# AND FLOTATION

This chapter deals with the fundamental principles of flotation and the way in which the underwater volume of a ship is described.

## Density

The density of a substance is the mass of a unit volume of that substance and may be expressed in grams per millilitre (g/ml), kilograms per cubic metre (kg/m³) or tonnes per cubic metre (t/m³). The numerical values of g/ml are the same as t/m³. The density of fresh water may be taken as $1.000$ t/m³ or $1000$ kg/m³ and the density of sea water is typically $1.025$ t/m³ or $1025$ kg/m³. Due to the presence of dissolved salts in sea water, its density is greater than fresh water, and its value can change around the world dependent on salinity and temperature.

The relative density or specific gravity of a substance is the density of the substance divided by the density of fresh water, ie the ratio of the mass of any volume of the substance to the mass of the same volume of fresh water. Thus the relative density (rd) of fresh water is $1.000$ while the relative density of sea water is $1025 \div 1000$ or $1.025$. Often the Greek letter $\rho$ is used to refer to density, and subscripts will be used to define which density (eg $\rho_F$ refers to the density of fresh water and $\rho_S$ refers to the density of sea water).

It is useful to know that the density of a substance expressed in t/m³ is numerically the same as the relative density. If a substance has a relative density of $x$, then one cubic metre of the substance will have a mass of $x$ tonnes. $V$ cubic metres will have a mass of $Vx$ tonnes or $1000Vx$ kilograms.

Thus:

$$\text{mass of substance} = \text{volume of substance} \times \text{density of substance}$$

and

$$\text{Relative density} = \frac{\text{mass of body}}{\text{mass of equal volume of fresh water}}$$

EXAMPLE: If the relative density of lead is 11.2, find

a) its density
b) the mass of 0.25 m³ of lead.

$$\text{Density of lead} = \text{relative density of lead} \times \text{density of fresh water}$$
$$= 11.2 \times 1.0 \text{ tonne/m}^3$$
$$= 11.2 \text{ tonne/m}^3$$
$$\text{Mass of lead} = 0.25 \times 11.2$$
$$= 2.8 \text{ tonne}$$

EXAMPLE: A plank 6 m long, 0.3 m wide and 50 mm thick has a mass of 60 kg. Calculate the density of the wood.

$$\text{Volume of wood} = 6.0 \times 0.3 \times 0.05$$
$$= 90 \times 10^{-3} \text{ m}^3$$
$$\text{Density of wood} = \frac{\text{mass}}{\text{volume}}$$
$$= \frac{60}{90 \times 10^{-3}} \frac{\text{kg}}{\text{m}^3}$$
$$= 667 \text{ kg/m}^3$$

Typical abbreviations you will see relating to this are: FW Fresh Water, SW Salt Water, DW Dock Water, rd Relative Density or sg Specific Gravity.

# Archimedes' Principle

Archimedes' principle and the law of flotation can be experienced readily with this simple experiment. Take an empty bucket, and keeping the open end above the water, try to push the bucket into the water. It requires a reasonable amount of effort to do this, and the more you push the bucket down, the harder it becomes. The bucket is experiencing an upthrust, which Archimedes defined as:

'The upthrust experienced on a body immersed in a liquid is equal to the weight of the liquid displaced.'

This upthrust is experienced by any body immersed in a liquid, and can be used to determine the relative density of an object.

In calculating relative density, it can be seen below that either mass or weight may be used as long as the units of the numerator are the same as those of the denominator. This is acceptable, since the value of $g$ used to obtain the weight in air is the same as the value of $g$ used to obtain the upthrust in water.

$$\text{Relative density} = \frac{\text{mass of body}}{\text{mass of equal volume of fresh water}}$$
$$= \frac{\text{weight of body in air}}{\text{weight in air} - \text{weight in fresh water}}$$
$$= \frac{\text{weight of body in air}}{\text{upthrust in fresh water}}$$
$$= \frac{\text{mass of body}}{\text{apparent loss of mass in fresh water}}$$

EXAMPLE: A solid block of cast iron has a mass of 500 kg. When it is completely immersed in fresh water, the mass appears to be reduced to 430 kg. Calculate the relative density of cast iron.

$$\text{Mass of cast iron} = 500 \text{ kg}$$
$$\text{Apparent loss in mass in freshwater} = 500 - 430 \text{ kg}$$
$$= 70 \text{ kg}$$
$$\text{Relative density} = \frac{500}{70}$$
$$= 7.143$$

EXAMPLE: A piece of brass (rd 8.4) 0.06 m³ in volume is suspended in oil of rd 0.8. Calculate the apparent mass of the brass.

$$\text{Mass of brass} = 1000 \times 8.4 \times 0.06$$
$$= 504 \text{ kg}$$
$$\text{Mass of equal volume of oil} = 1000 \times 0.8 \times 0.06$$
$$= 48 \text{ kg}$$
$$\text{Apparent mass in oil} = 504 - 48$$
$$= 456 \text{ kg}$$

# Floating Bodies and the Law of Flotation

Archimedes' principle can be taken further, in that if a body can displace its own weight of liquid, it will float. The upthrust experienced by the body is the same as the weight of the body and it is in equilibrium. This also means that the relative density of the body is less than 1.

The percentage of volume of a floating body that is immersed depends upon the relative density of the body and the relative density of the liquid; eg a body of rd 0.8 will float in fresh water with 80% of its volume immersed, and in sea water with 1.000 ÷ 1.025 x 80% of its volume immersed.

EXAMPLE: A block of wood 4 m long, 0.3 m wide and 0.25 m thick floats at a draught of 0.15 m in sea water. Calculate the mass of the wood and its relative density.

$$\begin{aligned}
\text{mass of wood} &= \text{mass of water displaced} \\
&= 1025 \times 4 \times 0.3 \times 0.15 \\
&= 184.5 \text{ kg} \\
\text{mass of equal volume of fresh water} &= 1000 \times 4 \times 0.3 \times 0.25 \\
&= 300 \text{ kg} \\
\text{relative density of wood} &= \frac{184.5}{300} \\
&= 0.615
\end{aligned}$$

EXAMPLE: A box barge 40 m long and 9 m wide floats in sea water at a draught of 3.5 m. Calculate the mass of the barge.

$$\begin{aligned}
\text{mass of barge} &= \text{mass of water displaced} \\
&= 1025 \times 40 \times 9 \times 3.5 \\
&= 1292 \times 10^3 \text{ kg} \\
&= 1292 \text{ tonne}
\end{aligned}$$

# Marine Hydrometer

A hydrometer is a device that is used to measure the density of a liquid. It can be made of metal or glass, and typically consists of a weighted bulb with a long graduated stem.

The level on the stem at which the hydrometer will float varies depending on the density of the water. The denser the water, the higher the hydrometer will float, and conversely in less dense water the hydrometer will sink down.

A marine hydrometer is a glass or metal device that is specifically calibrated for the measurement of sea water density. The hydrometer has a weighted bulb, to ensure it floats with its stem vertical. The hydrometer should be immersed in the liquid, and then pushed down below its equilibrium point, so that it can return to equilibrium. Once the hydrometer has been checked to ensure there are no bubbles attached to it, which might adversely affect the result, the value of density can be read from the waterline on the stem scale. The reader's eye level should be in line with the flotation level, rather than at an angle, and the horizontal part of the fluid is used as the reference, not the edges of the meniscus. When using it to measure the density of water near a ship, it is important that samples of water are taken from forward, aft and either side of the ship. This accounts for any variations in density that might occur, possibly due to disturbed sediment (the author has seen a variation in density from one side of a harbour to the other). The marine hydrometer is of particular importance in conducting an inclining experiment (detailed in Chapter 4).

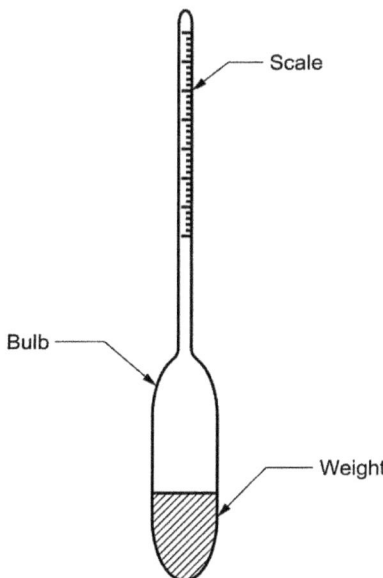

▲ **Figure 2.1** *A marine hydrometer*

# Hydrostatics

## Displacement

When a ship is floating freely at rest, the mass of the ship is equal to the mass of the volume of water displaced by the ship and is therefore known as the *displacement* of the ship. Thus if the volume of the underwater portion of the ship is known, together with the density of the water, it is possible to obtain the displacement of the ship.

It is usual to assume that a ship floats in sea water of density 1025 kg/m³ or 1.025t/m³.

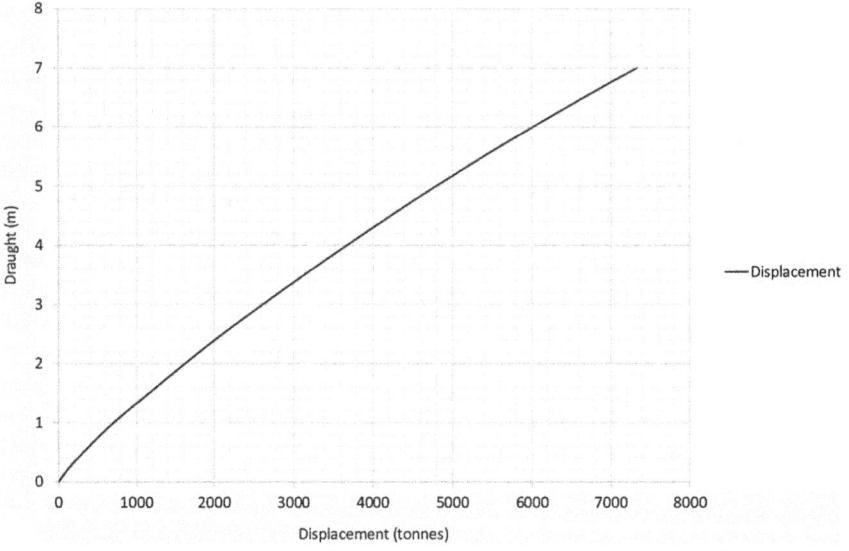

▲ **Figure 2.2** *Displacement draught curve*

Corrections may then be made if the vessel floats in water of any other density. Since the volume of water displaced depends upon the draught, it is useful to calculate values of displacement for a range of draughts. These values may then be plotted to form a *displacement curve*, from which the displacement may be obtained at any intermediate draught.

The following symbols will be used throughout the text:

$$\Delta = \text{displacement in tonnes}$$
$$\nabla = \text{volume of displacement in m}^3$$

Thus for sea water:

$$\Delta = \nabla \times 1.025$$

Some confusion exists between the *mass* of the ship and the *weight* of the ship. This confusion may be reduced if the displacement is always regarded as a mass. The gravitational force acting on this mass – the weight of the ship – will then be the product of the displacement $\Delta$ and the acceleration due to gravity $g$.

Hence:

$$\text{mass of ship (displacement)} = \Delta \text{ tonne}$$
$$\text{weight of ship} = \Delta g \text{ kN}$$

EXAMPLE: A ship displaces 12240 m³ of sea water at a particular draught.

**a)** Calculate the displacement of the ship.

**b)** How many tonnes of cargo would have to be discharged for the vessel to float at the same draught in fresh water?

a)

$$\text{Displacement in sea water} = 12240 \times 1.025$$
$$= 12546 \text{ tonne}$$

b)

$$\text{Displacement at same draught in fresh water} = 12240 \times 1.000$$
$$= 12240 \text{ tonne}$$
$$\text{Cargo to be discharged} = 12546 - 12240$$
$$= 306 \text{ tonne}$$

## Buoyancy

Buoyancy is the term given to the upthrust exerted by the water on the ship. If a ship floats freely, the buoyancy is equal to the weight of the ship.

The force of buoyancy acts at the *centre of buoyancy,* which is the centre of gravity of the underwater volume of the ship.

The *longitudinal* position of the centre of buoyancy (LCB) is usually given as a distance forward or aft of midships and is represented by the longitudinal centroid of the curve of immersed cross-sectional areas.

The *vertical* position of the centre of buoyancy (VCB) is usually given as a distance above the keel. This distance is denoted by *KB* and is represented by the vertical centroid of the waterplane area curve. The distance from the waterline to the VCB may be found by two other methods:

a) from the displacement curve, figure 2.3.

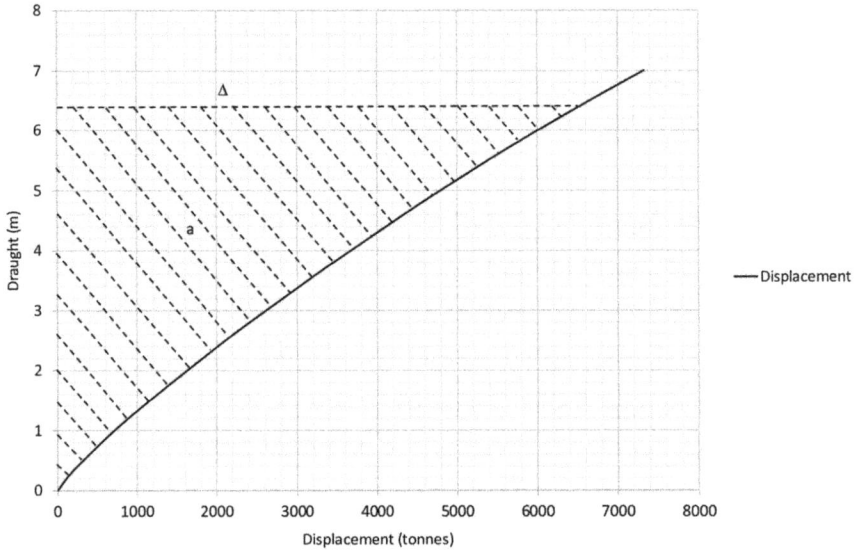

▲ **Figure 2.3** *Centre of buoyancy from displacement draught curve*

$$\text{VCB below waterline} = \frac{\text{area between curve and draught axis}}{\text{displacement}}$$

$$= \frac{a}{\Delta}$$

**(b)** by Morrish's approximate formula

$$\text{VCB below waterline} = \frac{1}{3}\left(\frac{d}{2} + \frac{\nabla}{A_w}\right)$$

where $d$ = draught in m

$\nabla$ = volume of displacement in m³

$A_w$ = waterplane area in m²

# Tonne Per Centimetre Immersion

The tonne per centimetre immersion (TPC) of a ship at any given draught is the mass required to increase the mean draught by 1 cm.

Consider a ship floating in water of density $\rho$ t/m³.

If the mean draught is increased by 1 cm, then:

$$\text{Increase in volume of displacement} = \frac{1}{100} \times \text{waterplane area}$$

$$= \frac{A_w}{100} \text{ m}^3$$

$$\text{Increase in displacement} = \frac{A_w}{100} \times \rho \text{ tonne}$$

$$\text{TPC} = \frac{A_w \times \rho}{100}$$

$$\text{TPC}_{sw} = 0.01025 A_w$$

At different draughts, variations in waterplane area cause variations in TPC. Values of TPC may be calculated for a range of draughts and plotted to form a TPC curve, from which values of TPC may be obtained at intermediate draughts.

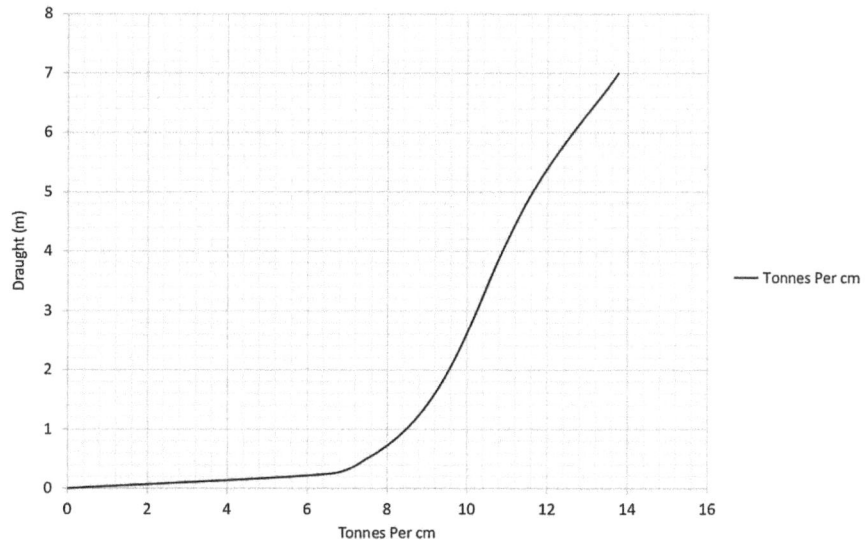

▲ **Figure 2.4** *Tonne Per Centimetre curve*

The area between the TPC curve and the draught axis to any given draught represents the displacement of the ship at that draught, while its centroid represents the vertical position of the centre of buoyancy.

It may be assumed for small alterations in draught that the ship is wall-sided and therefore TPC remains constant. If the change in draught exceeds about 0.5 m, then a mean TPC value should be used. If the change in draught is excessive, however, it is more accurate to use the area of the relevant part of the TPC curve.

EXAMPLE: The waterplane area of a ship is 1730 m². Calculate the TPC and the increase in draught if a mass of 270 tonne is added to the ship.

$$TPC = 0.01025 \times 1730$$
$$= 17.73$$
$$\text{Increase in draught} = \frac{\text{mass added}}{TPC}$$
$$= \frac{270}{17.73}$$
$$= 15.23 \text{ cm}$$

# Coefficients of Form

Coefficients of form have been devised to show the relation between the form of the ship and the dimensions of the ship.

**WATERPLANE AREA COEFFICIENT** $C_w$ is the ratio of the area of the waterplane to the product of the length and breadth of the ship, figure 2.5.

$$C_w = \frac{\text{waterplane area}}{\text{length} \times \text{beam}}$$
$$= \frac{A_w}{L \times B}$$

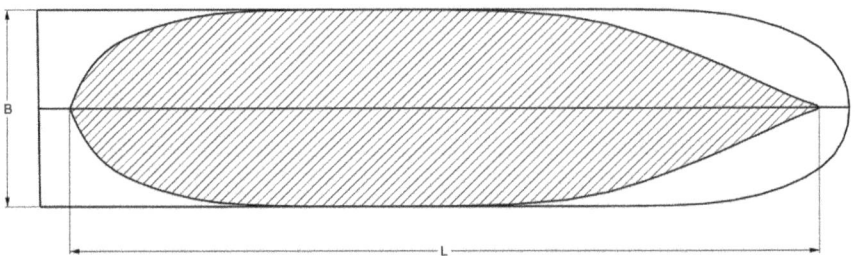

▲ **Figure 2.5** *Waterplane area*

**MIDSHIP SECTION AREA COEFFICIENT** $C_m$ is the ratio of the area of the immersed portion of the midship section to the product of the breadth and the draught, figure 2.6.

$$C_m = \frac{\text{area of immersed midship section}}{\text{draught} \times \text{beam}}$$

$$= \frac{A_m}{T \times B}$$

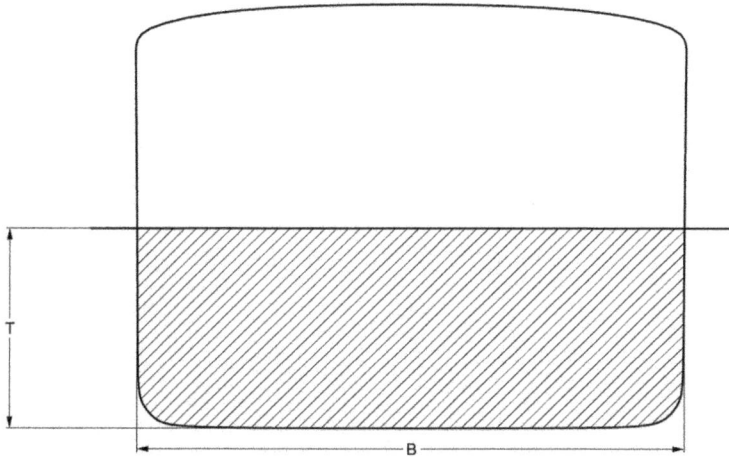

▲ **Figure 2.6** *Midship area coefficient*

**BLOCK COEFFICIENT OR COEFFICIENT OF FINENESS** $C_b$ is the ratio of the volume of displacement to the product of the length, breadth and draught, figure 2.7.

$$C_b = \frac{\text{volume of displacement}}{\text{length} \times \text{draught} \times \text{beam}}$$

$$= \frac{\nabla}{L \times T \times B}$$

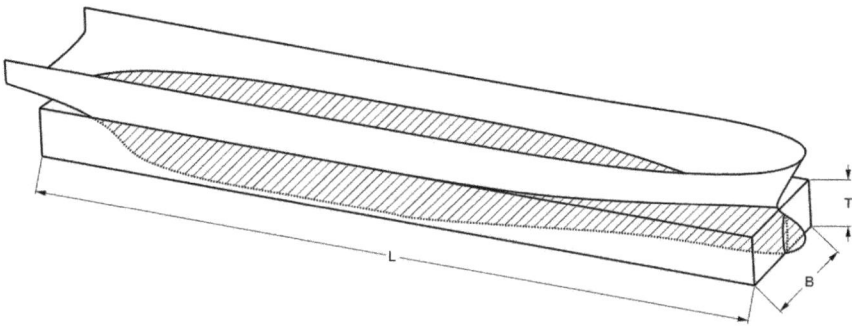

▲ **Figure 2.7** *Block coefficient*

**PRISMATIC COEFFICIENT** $C_p$ is the ratio of the volume of displacement to the product of the length and the area of the immersed portion of the midship section, figure 2.8.

$$C_p = \frac{\text{volume of displacement}}{\text{length} \times \text{area of immersed midship section}}$$

$$= \frac{\nabla}{L \times A_m}$$

But $\nabla = C_b \times L \times B \times d$

and $A_m = C_m \times B \times d$

Substituting these in the expression for $C_p$:

$$C_p = \frac{C_b \times L \times B \times T}{L \times C_m \times B \times T}$$

$$C_p = \frac{C_b}{C_m}$$

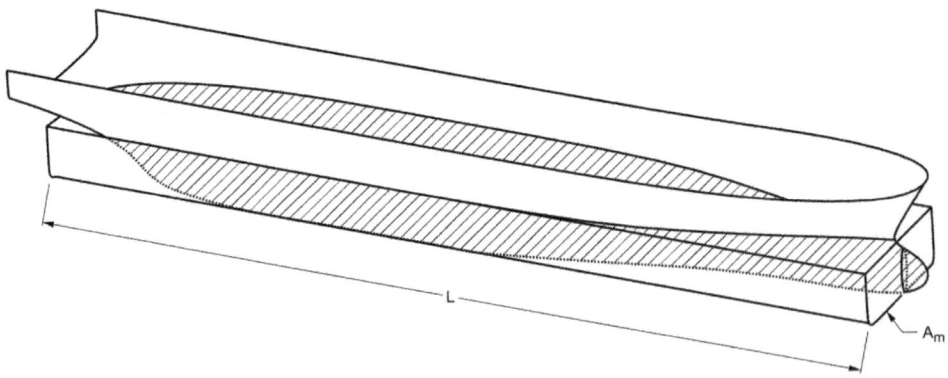

▲ **Figure 2.8** *Prismatic coefficient*

EXAMPLE: A ship 135 m long, 18 m beam and 7.6 m draught has a displacement of 14000 tonne. The area of the load waterplane is 1925 m² and the area of the immersed midship section 130 m².

Calculate:

a) $C_w$

b) $C_m$

c) $C_b$

d) $C_p$.

$$C_w = \frac{1925}{135 \times 18}$$
$$= 0.792$$

$$C_m = \frac{130}{18 \times 7.6}$$
$$= 0.950$$

$$\nabla = \frac{14000}{1.025}$$
$$= 13658 \text{ m}^3$$

$$C_b = \frac{13658}{135 \times 18 \times 7.6}$$
$$= 0.740$$

$$C_p = \frac{13658}{135 \times 130}$$
$$= 0.778$$

Alternatively,

$$C_p = \frac{0.740}{0.950}$$
$$= 0.778$$

# Wetted Surface Area

The wetted surface area of a ship is the area of the ship's hull that is in contact with the water. This area may be found by putting the transverse girths of the ship, from waterline to waterline, through Simpson's Rule and adding about ½% to allow for the longitudinal curvature of the shell. To this area should be added the wetted surface area of appendages such as cruiser stern, rudder and bilge keels. The importance of wetted area with regards to ship powering will be discussed in Chapter 7.

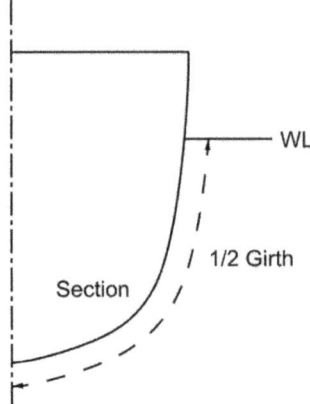

**▲ Figure 2.9** *Girth measurement*

Several approximate formulae for wetted surface area are available, two of which are:

Denny

$$S = 1.7LT + \frac{\nabla}{T}$$

Taylor

$$S = c\sqrt{\Delta L}$$

where $S$ = wetted surface area in m²
$L$ = length of ship in m
$T$ = draught in m
$\nabla$ = volume of displacement in m³
$\Delta$ = displacement in tonne
$c$ = a coefficient of about 2.6, depending upon the shape of the ship.

EXAMPLE: A ship of 5000 tonne displacement, 95 m long, floats at a draught of 5.5 m. Calculate the wetted surface area of the ship:

a) Using Denny's formula
b) Using Taylor's formula with c = 2.6

a)

$$S = 1.7LT + \frac{\nabla}{T}$$

$$S = 1.7 \times 95 \times 5.5 + \frac{5000}{1.025 \times 5.5}$$

$$= 888.3 + 886.9$$

$$= 1775.2 \text{ m}^2$$

b)

$$S = c\sqrt{\Delta L}$$

$$S = 2.6\sqrt{5000 \times 95}$$

$$= 1793 \text{ m}^2$$

# Similar Figures

Two planes or bodies are said to be similar when their linear dimensions are in the same ratio. This principle may be seen in a projector where a small image is projected from a slide on to a screen. The height of the image depends upon the distance of the screen from the light source, but the *proportions and shape* of the image remain the same as the image on the slide. Thus the image on the screen is a scaled-up version of the image on the slide.

The *areas* of similar figures vary as the *square* of their corresponding dimensions. This may be shown by comparing two circles having diameters $d$ and $D$ respectively.

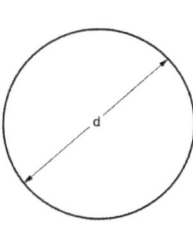 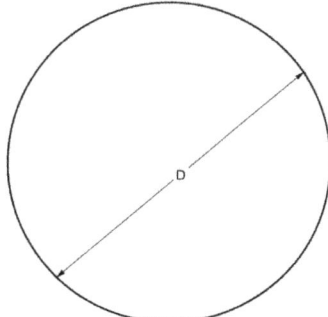

▲ **Figure 2.10** *Similar circles*

$$\text{Area of small circle } = \frac{\pi}{4}d^2$$

$$\text{Area of large circle } = \frac{\pi}{4}D^2$$

Since π/4 is constant:

$$\text{ratio of areas } = \frac{D^2}{d^2}$$

Thus if the diameter D is *twice* diameter d, the area of the former is *four times* the area of the latter.

$$\frac{A_1}{A_2} = \left(\frac{L_1}{L_2}\right)^2 = \left(\frac{B_1}{B_2}\right)^2$$

The *volumes* of similar bodies vary as the *cube* of their corresponding dimensions. This may be shown by comparing two spheres of diameters D and d respectively.

$$\text{Volume of large sphere } = \frac{\pi}{6}D^3$$

$$\text{Volume of small sphere } = \frac{\pi}{6}d^3$$

Since π/6 is constant:

$$\text{ratio of volumes } = \frac{D^3}{d^3}$$

Thus if diameter D is *twice* diameter d, the volume of the former is *eight times* the volume of the latter.

$$\frac{V_1}{V_2} = \left(\frac{L_1}{L_2}\right)^3 = \left(\frac{B_1}{B_2}\right)^3 = \left(\frac{T_1}{T_2}\right)^3$$

These rules may be applied to any similar bodies no matter what their shape, and in practice are applied to ships.

Thus if

L = length of ship

S = wetted surface area

Δ = displacement,

Then

$$S \propto L^2$$

Or

$$S^{\frac{1}{2}} \propto L$$

And

$$\Delta \propto L^3$$

Or

$$\Delta^{\frac{1}{3}} \propto L$$
$$S^{\frac{1}{2}} \propto \Delta^{\frac{1}{3}}$$

And

$$S \propto \Delta^{\frac{2}{3}}$$

Or

$$\Delta \propto S^{\frac{3}{2}}$$

EXAMPLE: A ship 110 m long displaces 9000 tonne and has a wetted surface area of 2205 m². Calculate the displacement and wetted surface area of a 6 m model of the ship.

$$\left(\frac{\Delta_1}{\Delta_2}\right) = \left(\frac{L_1}{L_2}\right)^3$$

$$\Delta_2 = 9000\left(\frac{6}{110}\right)^3$$

Displacement of model $= 1.46$ tonne

$$\frac{S_1}{S_2} = \left(\frac{L_1}{L_2}\right)^2$$

$$S_2 = \left(\frac{6}{110}\right)^2$$

Wetted surface area of model $= 6.56$ m²

# Applying Simpson's Rule to Ship Hulls

As was shown in the previous chapter, Simpson's Rule can be used to calculate areas and volumes. Its application to ship's hulls is shown here.

When calculating the area of a waterplane, it is usual to divide the length of the ship into about ten equal parts, giving eleven sections. These sections are numbered from 0 at the after end to 10 at the fore end. Thus, amidships will be section number 5. It is convenient to measure distances from the centreline to the ship side, giving half ordinates. These half ordinates are used in conjunction with Simpson's Rule and the answer multiplied by 2.

EXAMPLE: The equally spaced half ordinates of a waterplane 27 m long are 1.1, 2.7, 4.0, 5.1, 6.1, 6.9 and 7.7 m respectively.

Calculate the area of the waterplane.

| ½ Ordinate | Simpson's Multipliers | Product for Area |
|---|---|---|
| 1.1 | 1 | 1.1 |
| 2.7 | 4 | 10.8 |
| 4.0 | 2 | 8.0 |
| 5.1 | 4 | 20.4 |
| 6.1 | 2 | 12.2 |
| 6.9 | 4 | 27.6 |
| 7.7 | 1 | 7.7 |
| | | $87.8 = \Sigma_A$ |

Since there are seven ordinates, there will be six spaces.

$$\text{Common interval} = \frac{27}{6} = 4.5 \text{ m}$$

$$\text{Area} = \frac{h}{3} \Sigma_A \times 2 = \frac{4.5}{3} \times 87.8 \times 2$$

$$= 263.4 \text{ m}^2$$

If the immersed cross-sectional areas of a ship at a number of positions along the length of the ship are plotted on a base representing the ship's length, figure 2.11, the areas under the resulting curve will represent the volume of water displaced by the ship and may be found by putting the cross-sectional areas through Simpson's Rule. Hence the displacement of the ship at any given draught may be calculated. The longitudinal centroid of this figure represents the longitudinal centre of buoyancy of the ship.

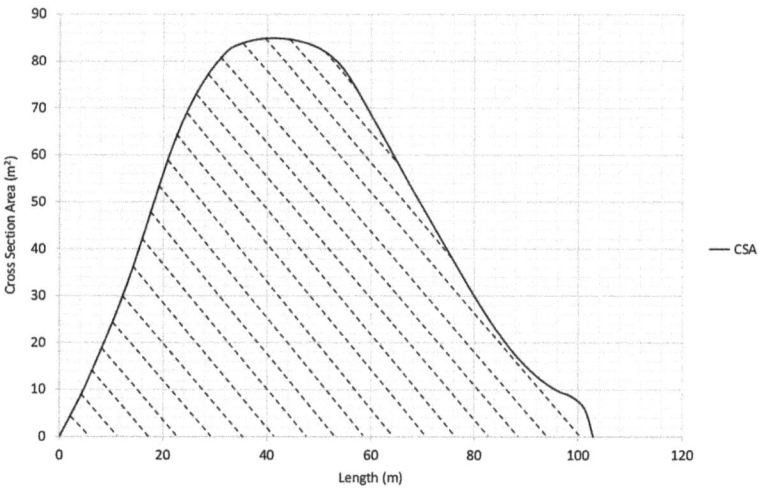

▲ **Figure 2.11** *Cross-sectional area curve*

It is also possible to calculate the displacement by using ordinates of waterplane area or tonne per centimetre immersion, with a common interval of draught, figure 2.12. The vertical centroids of these two curves represent the vertical centre of buoyancy of the ship.

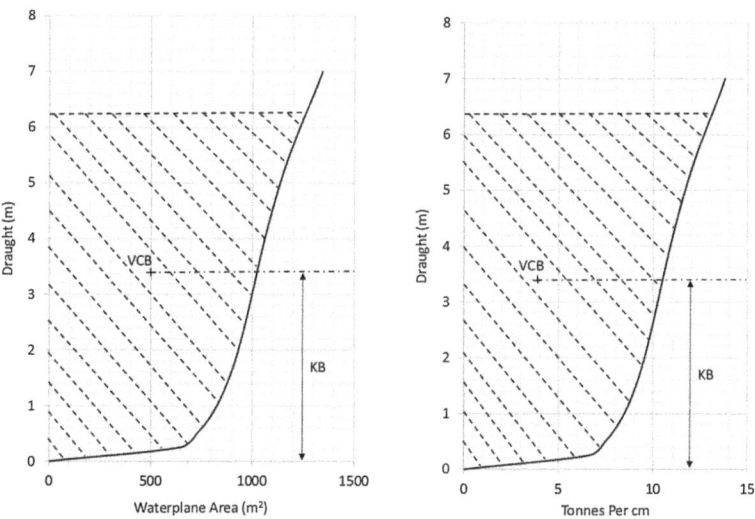

▲ **Figure 2.12** *KB from TPC or waterplane area curve*

Similar methods are used to determine hold and tank capacities.

EXAMPLE: The immersed cross-sectional areas through a ship 180 m long, at equal intervals, are 5, 118, 233, 291, 303, 304, 304, 302, 283, 171, and 0 m² respectively. Calculate the displacement of the ship in sea water of 1.025 tonne/m³.

| CSA | SM | Product for volume |
|-----|-----|-----|
| 5 | 1 | 5 |
| 118 | 4 | 472 |
| 233 | 2 | 466 |
| 291 | 4 | 1164 |
| 303 | 2 | 606 |
| 304 | 4 | 1216 |
| 304 | 2 | 608 |
| 302 | 4 | 1208 |
| 283 | 2 | 566 |
| 171 | 4 | 684 |
| 0 | 1 | 0 |
| | | $6995 = \Sigma_{_\nabla}$ |

$$\text{Common interval} = \frac{180}{10} = 18 \text{ m}$$

$$\text{Volume of displacement} = \frac{h}{3}\Sigma_{_\nabla}$$

$$= \frac{18}{3} \times 6995$$

$$= 41970 \text{ m}^3$$

Displacement = volume of displacement × density

Area from 0 to 4 =

$$\frac{2}{3}h\left[\left(\frac{1}{4}y_0 + 1y_{\frac{1}{4}} + \frac{1}{2}y_{\frac{1}{2}} + 1y_{\frac{3}{4}} + \frac{1}{4}y_1\right) + \left(\frac{1}{2}y_1 + 2y_{1\frac{1}{2}} + \frac{1}{2}y_2\right) + \left(1y_2 + 4y_3 + 1y_4\right)\right]$$

$$= 41970 \times 1.025$$

$$= 43019 \text{ tonne}$$

EXAMPLE: The TPC values for a ship at 1.2 m intervals of draught, commencing at the keel, are 8.2, 16.5, 18.7, 19.4, 20.0, 20.5 and 21.1 respectively. Calculate the displacement at 7.2 m draught.

| Waterplane | TPC | SM | Product for displacement |
|---|---|---|---|
| 0 | 8.2 | 1 | 8.2 |
| 1.2 | 16.5 | 4 | 66.0 |
| 2.4 | 18.7 | 2 | 37.4 |
| 3.6 | 19.4 | 4 | 77.6 |
| 4.8 | 20.0 | 2 | 40.0 |
| 6.0 | 20.5 | 4 | 82.0 |
| 7.2 | 21.1 | 1 | 21.1 |
| | | | $332.3 = \Sigma_\Delta$ |

Common interval = 120 cm

$$\text{Displacement} = \frac{h}{3}\Sigma_\Delta$$

$$= \frac{120}{3} \times 332.3$$

$$= 13292 \text{ tonne}$$

*Note:* The common interval must be expressed in *centimetres* since the ordinates are tonne per *centimetre* immersion.

# Use of Intermediate Ordinates

At the ends of the ship, where the curvature of a waterplane is considerable, it is necessary to reduce the spacing of the ordinates to ensure an accurate result. Intermediate ordinates are introduced to reduce the spacing to half or quarter of the normal spacing. While it is possible to calculate the area of such a waterplane by dividing it into separate sections, this method is not considered advisable. The following method may be used.

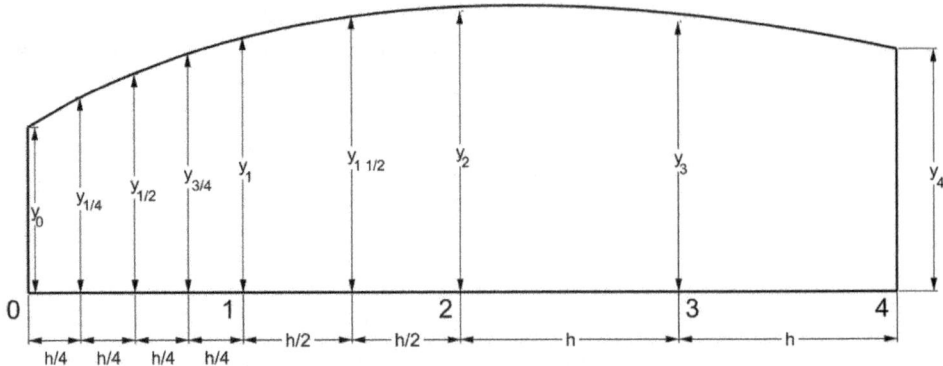

▲ **Figure 2.13** *Intermediate ordinates*

If the length of the ship is divided initially into ten equal parts, then:

$$\text{Common interval} = h = \frac{L}{10}$$

It is proposed to introduce intermediate ordinates at a spacing of $\frac{h}{4}$ h/4 from section 0 to section 1 and at a spacing of $h/2$ from section 1 to section 2. The ½ ordinates at sections AP, ¼, ½, ¾ etc will be denoted by $y_0$, $y_{¼}$, $y_{½}$, $y_{¾}$, etc respectively.

$$\text{Area from 0 to 1} = \frac{2}{3}\frac{h}{4}\left(1y_0 + 4y_{\frac{1}{4}} + 2y_{\frac{1}{2}} + 4y_{\frac{3}{4}} + 1y_1\right)$$

$$= \frac{2}{3}h\left(\frac{1}{4}y_0 + 1y_{\frac{1}{4}} + \frac{1}{2}y_{\frac{1}{2}} + 1y_{\frac{3}{4}} + \frac{1}{4}y_1\right)$$

$$\text{Area from 1 to 2} = \frac{2}{3}\frac{h}{2}\left(1y_1 + 4y_{1\frac{1}{2}} + 1y_2\right)$$

$$= \frac{2}{3}h\left(\frac{1}{2}y_1 + 2y_{1\frac{1}{2}} + \frac{1}{2}y_2\right)$$

$$\text{Area from 2 to 4} = \frac{2}{3}h\left(1y_2 + 4y_3 + 1y_4\right)$$

Thus:

$$\text{Area from 0 to 4} = \frac{2}{3}h\left[\left(\frac{1}{4}y_0 + 1y_{\frac{1}{4}} + \frac{1}{2}y_{\frac{1}{2}} + 1y_{\frac{3}{4}} + \frac{1}{4}y_1\right) + \left(\frac{1}{2}y_1 + 2y_{1\frac{1}{2}} + \frac{1}{2}y_2\right)\right.$$
$$\left. + \left(1y_2 + 4y_3 + 1y_4\right)\right]$$

$$= \frac{2}{3}h\left[\frac{1}{4}y_0 + 1y_{\frac{1}{4}} + \frac{1}{2}y_{\frac{1}{2}} + 1y_{\frac{3}{4}} + \frac{3}{4}y_1 + 2y_{1\frac{1}{2}} + 1\frac{1}{2}y_2 + 4y_3 + 1y_4\right]$$

When building up a system of multipliers, it is wise to ignore the ordinates and concentrate only on the spacing and the multipliers. The following example shows how these multipliers may be determined.

EXAMPLE: The half ordinates of a cross section through a ship are as follows:

| WL | keel | 0.25 | 0.50 | 0.75 | 1.0 | 1.5 | 2.0 | 2.5 | 3.0 | 4.0 | 5.0 | 6.0 | 7.0 m |
|---|---|---|---|---|---|---|---|---|---|---|---|---|---|
| ½ ord | 2.9 | 5.0 | 5.7 | 6.2 | 6.6 | 6.9 | 7.2 | 7.4 | 7.6 | 7.8 | 8.1 | 8.4 | 8.7 m |

Calculate the area of the cross section to the 7 m waterline.

Let the common interval $= h = 1$ m

The interval from the keel to 1 m waterline $= h/4$,

the interval from 1 m to 3 m waterline $= h/2$,

and the interval from 3 m to 7 m waterline $= h$.

Multipliers from keel to 1 m with common interval $h/4$

$= 1 : 4 : 2 : 4 : 1$

or with common interval of $h$

$= ¼ : 1 : ½ : 1 : ¼$

Multipliers from 1 m to 3 m with common interval $h/2$

$= 1 : 4 : 2 : 4 : 1$

or with common interval of $h$

$= \frac{1}{2} : 2 : 1 : 2 : \frac{1}{2}$

Multipliers from 3 m to 7 m with common interval $h$

$= 1 : 4 : 2 : 4 : 1$

Adding the respective multipliers, we have:

| keel | 0.25 | 0.50 | 0.75 | 1.0 | 1.5 | 2.0 | 2.5 | 3.0 | 4.0 | 5.0 | 6.0 | 7.0 m |
|------|------|------|------|-----|-----|-----|-----|------|-----|-----|-----|-------|
| ¼: | 1: | ½: | 1: | ¼ | | | | | | | | |
| | | | | ½: | 2: | 1: | 2: | ½ | | | | |
| | | | | | | | | 1: | 4: | 2: | 4: | 1 |
| ¼: | 1: | ½: | 1: | ¾ | 2: | 1: | 2: | 1½: | 4: | 2: | 4: | 1 |

These can be incorporated into a table to produce the area.

| Waterline | ½ ordinate | SM | Product for Area |
|-----------|-----------|-----|------------------|
| Keel | 2.9 | ¼ | 0.73 |
| 0.25 | 5.0 | 1 | 5.00 |
| 0.50 | 5.7 | ½ | 2.85 |
| 0.75 | 6.2 | 1 | 6.20 |
| 1.0 | 6.6 | ¾ | 4.95 |
| 1.5 | 6.9 | 2 | 13.80 |
| 2.0 | 7.2 | 1 | 7.20 |
| 2.5 | 7.4 | 2 | 14.80 |
| 3.0 | 7.6 | 1½ | 11.40 |
| 4.0 | 7.8 | 4 | 31.20 |
| 5.0 | 8.1 | 2 | 16.20 |
| 6.0 | 8.4 | 4 | 33.60 |
| 7.0 | 8.7 | 1 | 8.70 |
| | | | $\Sigma_A = 156.63$ |

$$\text{Area of cross section} = \frac{2}{3} \times 1.0 \times 156.63$$

$$= 104.42 \ m^2$$

It is usually necessary to calculate area and centroid when determining the second moment of area of a waterplane about a transverse axis. Since the centroid is near amidships, it is preferable to take moments about amidships. The following calculation shows the method used to determine area, centroid and second moment of area about the centroid for a waterplane having half ordinates of $y_0, y_1, y_2, \ldots y_{10}$ spaced $hm$ apart, commencing from aft.

The positive sign indicates an ordinate aft of midships.

The negative sign indicates an ordinate forward of midships.

| Section | $\frac{1}{2}$ ord | SM | Product for area | Lever | Product for 1st moment | Lever | Product for 2nd moment |
|---------|------|-----|------------------|-------|------------------------|-------|------------------------|
| AP | $y_0$ | 1 | $1y_0$ | $+5h$ | $+5y_0h$ | $+5h$ | $+25y_0h^2$ |
| 1 | $y_1$ | 4 | $4y_1$ | $+4h$ | $+16y_1h$ | $+4h$ | $+64y_1h^2$ |
| 2 | $y_2$ | 2 | $2y_2$ | $+3h$ | $+6y_2h$ | $+3h$ | $+18y_2h^2$ |
| 3 | $y_3$ | 4 | $4y_3$ | $+2h$ | $+8y_3h$ | $+2h$ | $+16y_3h^2$ |
| 4 | $y_4$ | 2 | $2y_4$ | $+1h$ | $+2y_4h$ | $+1h$ | $+2y_4h^2$ |
| 5 | $y_5$ | 4 | $4y_5$ | $0h$ | $\Sigma_{MA} \times h$ | $0h$ | – |
| 6 | $y_6$ | 2 | $2y_6$ | $-1h$ | $-2y_6h$ | $-1h$ | $+2y_6h^2$ |
| 7 | $y_7$ | 4 | $4y_7$ | $-2h$ | $-8y_7h$ | $-2h$ | $+16y_7h^2$ |
| 8 | $y_8$ | 2 | $2y_8$ | $-3h$ | $-6y_8h$ | $-3h$ | $+18y_8h^2$ |
| 9 | $y_9$ | 4 | $4y_9$ | $-4h$ | $-16y_9h$ | $-4h$ | $+64y_9h^2$ |
| FP | $y_{10}$ | 1 | $1y_{10}$ | $-5h$ | $-5y_{10}h$ | $-5h$ | $+25y_{10}h^2$ |
| | | | $\Sigma_A$ | | $\Sigma_{MF} \times h$ | | $\Sigma_I \times h^2$ |

Area of waterplane, $A$:

$$A = \frac{2}{3}h\Sigma_A$$

First moment of area of waterplane about amidships

$$= \frac{2}{3}h^2 \left( \sum_{MA} + \sum_{MF} \right)$$

$$\text{centroid from midships } x = \frac{\frac{2}{3}h^2 \left( \sum_{MA} + \sum_{MF} \right)}{\frac{2}{3}h \sum_A}$$

$$= h \frac{\left( \sum_{MA} + \sum_{MF} \right)}{\sum_A}$$

(NOTE: If $\sum_{MF}$ is greater than $\sum_{MA}$, the centroid will be forward of midships.)

Second moment of area of waterplane about amidships

$$I_M = \frac{2}{3}h \sum_I \times h^2$$

$$= \frac{2}{3}h^3 \sum_I$$

Second moment of area of waterplane about the centroid

$$I_F = I_M - Ax^2$$

It should be noted that $h$ remains in the table as a constant and left to the end of the calculation. It may be omitted from the table, as will be seen from the worked examples.

EXAMPLE: The half ordinates of a waterplane 180 m long are as follows:

section: AP, ½, 1, 2, 3, 4, 5, 6, 7, 8, 9, 9½, FP

½ ord(m): 0, 5.0, 8.0, 10.5, 12.5, 13.5, 13.5, 12.5, 11.0, 7.5, 3.0, 1.00

Calculate:

a) area of waterplane

b) distance of centroid from midships

c) second moment of area of waterplane about a transverse axis through the centroid.

| Section | ½ ordinate | SM | Product for Area | Lever | Product for 1st moment | Lever | Product for 2nd moment |
|---|---|---|---|---|---|---|---|
| AP | 0 | ½ | — | + 5 | — | + 5 | — |
| ½ | 5.0 | 2 | 10.0 | +4½ | + 45.0 | + 4½ | + 202.5 |
| 1 | 8.0 | 1½ | 12.0 | + 4 | + 48.0 | + 4 | + 192.0 |
| 2 | 10.5 | 4 | 42.0 | + 3 | + 126.0 | + 3 | + 378.0 |
| 3 | 12.5 | 2 | 25.0 | + 2 | + 50.0 | + 2 | + 100.0 |
| 4 | 13.5 | 4 | 54.0 | + 1 | + 54.0 | + 1 | + 54.0 |
| 5 | 13.5 | 2 | 27.0 | 0 | + 323.0 | 0 | — |
| 6 | 12.5 | 4 | 50.0 | − 1 | − 50.0 | − 1 | + 50.0 |
| 7 | 11.0 | 2 | 22.0 | − 2 | − 44.0 | − 2 | + 88.0 |
| 8 | 7.5 | 4 | 30.0 | − 3 | − 90.0 | − 3 | + 270.0 |
| 9 | 3.0 | 1½ | 4.5 | − 4 | − 18.0 | − 4 | + 72.0 |
| 9½ | 1.0 | 2 | 2.0 | − 4½ | − 9.0 | − 4½ | + 40.5 |
| FP | 0 | ½ | — | − 5 | — | − 5 | — |
| | | | 278.5 | | − 211.0 | | + 1447.0 |

$$\text{Common interval} = \frac{180}{10}$$
$$= 18 \text{ m}$$
$$\text{Area of waterplane} = \frac{2}{3} \times 18 \times 278.5$$
$$= 3342.0 \text{ m}^2$$
$$\text{Centroid from midships} = \frac{18(323 - 211)}{278.5}$$
$$= 7.238 \text{ m aft}$$
$$\text{Second moment of area of waterplane about midships} = \frac{2}{3} \times 18^3 \times 1447$$
$$= 5.626 \times 10^6 \text{ m}^4$$
$$= 5.626 \times 10^6 - 3342 \times 7.238^2$$
$$= 5.626 \times 10^6 - 0.175 \times 10^6$$
$$= 5.451 \times 10^6$$

To determine the second moment of area of the waterplane about the centreline of the ship, the half ordinates must be *cubed* and then put through Simpson's Rule.

| ½ ordinate | (½ ordinate)³ | SM | Product for 2nd moment |
|---|---|---|---|
| $y_0$ | $y_0^{\,3}$ | 1 | $1y_0^{\,3}$ |
| $y_1$ | $y_1^{\,3}$ | 4 | $4y_1^{\,3}$ |
| $y_2$ | $y_2^{\,3}$ | 2 | $2y_2^{\,3}$ |
| $y_3$ | $y_3^{\,3}$ | 4 | $4y_3^{\,3}$ |
| $y_4$ | $y_4^{\,3}$ | 2 | $2y_4^{\,3}$ |
| $y_5$ | $y_5^{\,3}$ | 4 | $4y_5^{\,3}$ |
| $y_6$ | $y_6^{\,3}$ | 2 | $2y_6^{\,3}$ |
| $y_7$ | $y_7^{\,3}$ | 4 | $4y_7^{\,3}$ |
| $y_8$ | $y_8^{\,3}$ | 2 | $2y_8^{\,3}$ |
| $y_9$ | $y_9^{\,3}$ | 4 | $4y_9^{\,3}$ |
| $y_{10}$ | $y_{10}^{\,3}$ | 1 | $1y_{10}^{\,3}$ |
| | | | $\Sigma I_{CL}$ |

Second moment of area of waterplane about the centreline.

$$= \frac{2}{3} \times h \times \sum I_{CL} \times \frac{1}{3}$$

$$= \frac{2}{9} \times h \times \sum I_{CL}$$

EXAMPLE: The half ordinates of a waterplane 180 m long are as follows:

Section: AP, ½, 1, 2, 3, 4, 5, 6, 7, 8, 9, 9½, FP

½ ord (m): 0, 5.0, 8.0, 10.5, 12.5, 13.5, 13.5, 12.5, 11.0, 7.5, 3.0, 1.00

Calculate the second moment of area of the waterplane about the centreline.

| Section | ½ ordinate | (½ ordinate)³ | SM | 2nd moment |
|---|---|---|---|---|
| AP | 0 | — | ½ | — |
| ½ | 5.0 | 125.0 | 2 | 250.0 |
| 1 | 8.0 | 512.0 | 1½ | 768.0 |
| 2 | 10.5 | 1157.6 | 4 | 4630.4 |

| Section | ½ ordinate | (½ ordinate)³ | SM | 2nd moment |
|---------|-----------|---------------|-----|------------|
| 3 | 12.5 | 1953.1 | 2 | 3906.2 |
| 4 | 13.5 | 2460.4 | 4 | 9841.6 |
| 5 | 13.5 | 2460.4 | 2 | 4920.8 |
| 6 | 12.5 | 1953.1 | 4 | 7812.4 |
| 7 | 11.0 | 1331.0 | 2 | 2662.0 |
| 8 | 7.5 | 421.9 | 4 | 1687.6 |
| 9 | 3.0 | 27.0 | 1½ | 40.5 |
| 9½ | 1.0 | 1.0 | 2 | 2.0 |
| FP | 0 | — | ½ | — |
| | | | | 36521.5 |

Common interval $= 18$ m

Second moment of area of waterplane about centreline

$$= \frac{2}{9} \times 18 \times 36521.5$$
$$= 146086 \text{ m}^4$$

It should be noted that the second moment of area about a transverse axis is considerably greater than the second moment about the centreline.

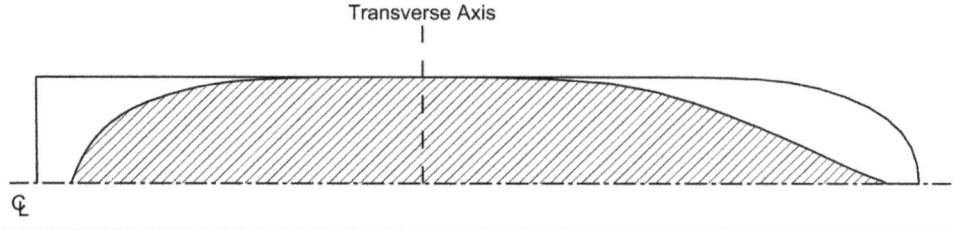

▲ **Figure 2.14** *Waterplane about centreline and transverse axis*

Since each waterplane is symmetrical, it is never necessary to calculate the first moment about the centreline. There are many occasions, however, on which the first moment of area of a tank surface must be calculated, as shown by the following example.

EXAMPLE: A double bottom tank extends from the centreline to the ship side. The widths of the tank surface, at regular intervals of $h$, are $y_1, y_2, y_3, y_4$ and $y_5$.

Calculate the second moment of area of the tank surface about a longitudinal axis through its centroid.

It is necessary in this calculation to determine the area, centroid from the centreline and the second moment of area.

| Width | SM | Product for Area | (Width)² | SM | Product for 1st moment | (Width)³ | SM | Product for 2nd moment |
|---|---|---|---|---|---|---|---|---|
| $y_1$ | 1 | $1y_1$ | $y_1^2$ | 1 | $1y_1^2$ | $y_1^3$ | 1 | $1y_1^3$ |
| $y_2$ | 4 | $4y_2$ | $y_2^2$ | 4 | $4y_2^2$ | $y_2^3$ | 4 | $4y_2^3$ |
| $y_3$ | 2 | $2y_3$ | $y_3^2$ | 2 | $2y_3^2$ | $y_3^3$ | 2 | $2y_3^3$ |
| $y_4$ | 4 | $4y_4$ | $y_4^2$ | 4 | $4y_4^2$ | $y_4^3$ | 4 | $4y_4^3$ |
| $y_5$ | 1 | $1y_5$ | $y_5^2$ | 1 | $1y_5^2$ | $y_5^3$ | 1 | $1y_5^3$ |
| | | $\Sigma_a$ | | | $\Sigma_m$ | | | $\Sigma_i$ |

$$\text{Area } a = \frac{h}{3}\Sigma_a$$

First moment of area about centreline

$$= \frac{h}{3}\times\Sigma_m\times\frac{1}{2}$$

$$= \frac{h}{6}\times\Sigma_m$$

$$\text{Centroid from centreline} = \frac{\dfrac{h}{6}\Sigma_m}{\dfrac{h}{3}\Sigma_a}$$

$$\bar{y} = \frac{\Sigma_m}{2\Sigma_a}$$

Second moment of area about centreline

$$i_{CL} = \frac{h}{9}\Sigma_i$$

Second moment of area about centroid

$$= i_{CL} - a\bar{y}^2$$

EXAMPLE: A double bottom tank 21 m long has a watertight centre girder. The widths of the tank top measured from the centreline to the ship's side are 10.0, 9.5, 9.0, 8.0, 6.5, 4.0 and 1.0 m respectively. Calculate the second moment of area of the tank surface about a longitudinal axis through its centroid, for one side of the ship only.

| Width | SM | Product for Area | (Width)² | SM | Product for 1st moment | (Width)³ | SM | Product for 2nd moment |
|-------|----|------------------|----------|----|------------------------|----------|----|------------------------|
| 10.0 | 1 | 10.0 | 100.00 | 1 | 100.00 | 1000.0 | 1 | 1000.0 |
| 9.5 | 4 | 38.0 | 90.25 | 4 | 361.00 | 857.4 | 4 | 3429.6 |
| 9.0 | 2 | 18.0 | 81.00 | 2 | 162.00 | 729.0 | 2 | 1458.0 |
| 8.0 | 4 | 32.0 | 64.00 | 4 | 256.00 | 512.0 | 4 | 2048.0 |
| 6.5 | 2 | 13.0 | 42.25 | 2 | 84.50 | 274.6 | 2 | 549.2 |
| 4.0 | 4 | 16.0 | 16.00 | 4 | 64.00 | 64.0 | 4 | 256.0 |
| 1.0 | 1 | 1.0 | 1.00 | 1 | 1.00 | 1.0 | 1 | 1.0 |
| | | 128.0 | | | 1028.50 | | | 8741.8 |

$$\text{Common interval} = \frac{21}{6} = 3.5 \text{ m}$$
$$\text{Area of tank surface} = \frac{3.5}{3} \times 128$$
$$= 149.33 \text{ m}^2$$
$$\text{Centroid from centreline} = \frac{1028.5}{2 \times 128}$$
$$= 4.018 \text{ m}$$

Second moment of area about centreline

$$= \frac{3.5}{9} \times 8741.8$$
$$= 3400.0 \text{ m}^4$$
$$= 3400.0 - 149.33 \times 4.018^2$$
$$= 3400.0 - 2410.8$$
$$= 989.2 \text{ m}^4$$

# Chapter 2 Test Examples

**Q. 2.1:** A piece of steel has a mass of 300 g and its volume is 42 cm³. Calculate:

    **a)** its density in kg/m³

    **b)** its relative density

    **c)** the mass of 100 cm³ of steel.

**Q. 2.2:** A piece of metal 250 cm³ in volume is attached to the bottom of a block of wood 3.5 dm³ in volume and having a relative density of 0.6. The system floats in fresh water with 100 cm³ projecting above the water. Calculate the relative density of the metal.

**Q. 2.3:** A raft 3 m long and 2 m wide is constructed of timber 0.25 m thick, having a relative density of 0.7. It floats in water of density 1018 kg/m³. Calculate the minimum mass that must be placed on top of the raft to sink it.

**Q. 2.4:** A box barge 65 m long and 12 m wide floats at a draught of 5.5 m in sea water. Calculate:

    **a)** the displacement of the barge

    **b)** its draught in fresh water.

**Q. 2.5:** A ship has a constant cross section in the form of a triangle, which floats apex down in sea water. The ship is 85 m long, 12 m wide at the deck and has a depth from keel to deck of 9 m. Draw the displacement curve using 1.25 m intervals of draught from the keel to the 7.5 m waterline. From this curve, obtain the displacement in fresh water at a draught of 6.5 m.

**Q. 2.6:** A cylinder 15 m long and 4 m outside diameter floats in sea water with its axis in the waterline. Calculate the mass of the cylinder.

**Q. 2.7:** Bilge keels of mass 36 tonne and having a volume of 22 m³ are added to a ship. If the TPC is 20, find the change in mean draught.

**Q. 2.8:** A vessel 40 m long has a constant cross section in the form of a trapezoid 10 m wide at the top, 6 m wide at the bottom and 5 m deep. It floats in sea water at a draught of 4 m. Calculate its displacement.

**Q. 2.9:** The waterplane areas of a ship at 1.25 m intervals of draught, commencing at the 7.5 m waterline, are 1845, 1690, 1535, 1355 and 1120 m². Draw the curve of tonne per centimetre immersion and determine the mass that must be added to increase the mean draught from 6.1 m to 6.3 m.

**Q. 2.10:** A ship 150 m long and 20.5 m beam floats at a draught of 8 m and displaces 19500 tonne. The TPC is 26.5 and midship section area coefficient 0.94. Calculate the block, prismatic and waterplane area coefficients.

**Q. 2.11:** A ship displaces 9450 tonne and has a block coefficient of 0.7. The area of immersed midship section is 106 m².

If beam = 0.13 × length = 2.1 × draught, calculate the length of the ship and the prismatic coefficient.

**Q. 2.12:** The length of a ship is 18 times the draught, while the breadth is 2.1 times the draught. At the load waterplane, the waterplane area coefficient is 0.83 and the difference between the TPC in sea water and the TPC in fresh water is 0.7. Determine the length of the ship and the TPC in fresh water.

**Q. 2.13:** The ½ girths of a ship 90 m long are as follows: 2.1, 6.6, 9.3, 10.5, 11.0, 11.0, 11.0, 9.9, 7.5, 3.9 and 0 m respectively. The wetted surface area of the appendages is 30 m² and ½% is to be added for longitudinal curvature. Calculate the wetted surface area of the ship.

**Q. 2.14:** A ship of 14000 tonne displacement, 130 m long, floats at a draught of 8 m. Calculate the wetted surface area of the ship using:

**a)** Denny's formula

**b)** Taylor's formula with c = 2.58.

**Q. 2.15:** A box barge is 75 m long, 9 m beam and 6 m deep. A similar barge having a volume of 3200 m³ is to be constructed. Calculate the length, breadth and depth of the new barge.

**Q. 2.16:** The wetted surface area of a ship is twice that of a similar ship. The displacement of the latter is 2000 tonne less than the former. Determine the displacement of the latter.

**Q. 2.17:** A ship 120 m long displaces 11000 tonne and has a wetted surface area of 2500 m². Calculate the displacement and wetted surface area of a 6 m model of the ship.

**Q. 2.18:** A ship 180 m long has ½ widths of waterplane of 1, 7.5, 12, 13.5, 14, 14, 14, 13.5, 12, 7 and 0 m respectively. Calculate:

**a)** waterplane area

**b)** TPC

**c)** waterplane area coefficient.

**Q. 2.19:** The waterplane areas of a ship at 1.5 m intervals of draught, commencing at the keel, are 865, 1735, 1965, 2040, 2100, 2145 and 2215 m² respectively. Calculate the displacement at 9 m draught.

**Q. 2.20:** A ship 140 m long and 18 m beam floats at a draught of 9 m. The immersed cross-sectional areas at equal intervals are 5, 60, 116, 145, 152, 153, 153, 151, 142, 85 and 0 m² respectively.

Calculate:

a) displacement

b) block coefficient

c) midship section area coefficient

d) prismatic coefficient.

**Q. 2.21:** The ½ ordinates of a waterplane 120 m long are as follows:

section: AP, ½, 1, 1½, 2, 3, 4, 5, 6, 7, 8, 8½, 9, 9½, FP

½ ord(m): 1.2, 3.5, 5.3, 6.8, 8.0, 8.3, 8.5, 8.5, 8.5, 8.4, 8.2, 7.9, 6.2, 3.5, 0

Calculate:

a) waterplane area

b) distance of centroid from midships.

**Q. 2.22:** The TPC values of a ship at 1.5 m intervals of draught, commencing at the keel, are 4.0, 6.1, 7.8, 9.1, 10.3, 11.4 and 12.0 m respectively. Calculate at a draught of 9 m:

a) displacement

b) KB.

**Q. 2.23:** The ½ breadths of the load waterplane of a ship 150 m long, commencing from aft, are 0.3, 3.8, 6.0, 7.7, 8.3, 9.0, 8.4, 7.8, 6.9, 4.7 and 0 m respectively.

Calculate:

a) area of waterplane

b) distance of centroid from midships

c) second moment of area about a transverse axis through the centroid.

**Q. 2.24:** The displacements of a ship at draughts of 0, 1, 2, 3 and 4 m are 0, 189, 430, 692 and 977 tonne. Calculate the distance of the centre of buoyancy above the keel when floating at a draught of 4 m, given:

VCB below waterline

= area between displacement curve and draught axis ÷ displacement

**Q. 2.25:** A ship 160 m long has ½ ordinates of waterplane of 1.6, 5.7, 8.8, 10.2, 10.5, 10.5, 10.5, 10.0, 8.0, 5.0 and 0 m respectively. Calculate the second moment of area of the waterplane about the centreline.

**Q. 2.26:** The immersed cross-sectional areas of a ship 120 m long, commencing from aft, are 2, 40, 79, 100, 103, 104, 104, 103, 97, 58 and 0 m². Calculate:

a) displacement

b) longitudinal position of the centre of buoyancy.

# 3

# CENTRE OF

# GRAVITY

The centre of gravity of an object is the point at which the whole weight of the object may be regarded as acting. If the object is suspended from this point, then it will remain balanced and will not tilt.

The distance of the centre of gravity from any axis is the total moment of *force* about that axis divided by the total *force*. If a body is composed of a number of different types of material, the force may be represented by the *weights* of the individual parts.

$$\text{Centre of gravity from axis} = \frac{\text{moment of weight about axis}}{\text{total weight}}$$

At any point on the earth's surface, the value of $g$ remains constant. Hence the weight may be represented by *mass,* and:

$$\text{Centre of gravity from axis} = \frac{\text{moment of mass about axis}}{\text{total mass}}$$

If the body is of the same material throughout, then the weight depends upon the *volume* and moments of *volume* may therefore be used.

$$\text{Centre of gravity from axis} = \frac{\text{moment of volume about axis}}{\text{total volume}}$$

The centre of gravity of a uniform lamina is midway through the thickness. Since both the thickness and the density are constant, moments of *area* may be used. This system may also be applied to determine the centre of gravity, or, more correctly, *centroid* of an area.

$$\text{Centroid from axis} = \frac{\text{moment of area about axis}}{\text{total area}}$$

The position of the centre of gravity of a ship may be found by taking moments of the individual masses. The actual calculation of the centre of gravity of a ship is a very lengthy process and, since many of the masses must be estimated, is not considered to be sufficiently accurate for stability calculations. Such a calculation is usually carried out for a passenger ship in the initial design stages, but the results are confirmed by an alternative method when the ship is completed, known as an *inclining experiment*, which is detailed in Chapter 4. Once the position of the centre of gravity of an empty ship is known, however, the centre of gravity of the ship in any loaded condition may be found.

It is usual to measure the *vertical* position of the centre of gravity (VCG) of the ship above the keel and this distance is denoted by *KG*. The height of the centre of gravity of an item on the ship above the keel is denoted by *Kg*. The *longitudinal* position of the centre of gravity (LCG) is usually given as a distance forward or aft of midships. If the ship is upright, the *transverse* centre of gravity lies on the centreline of the ship and it is not necessary to calculate its position.

EXAMPLE: A ship of 8500 tonne displacement is composed of masses of 2000, 3000, 1000, 2000 and 500 tonne at positions 2, 5, 8, 10 and 14 m above the keel. Determine the height of the centre of gravity of the ship above the keel.

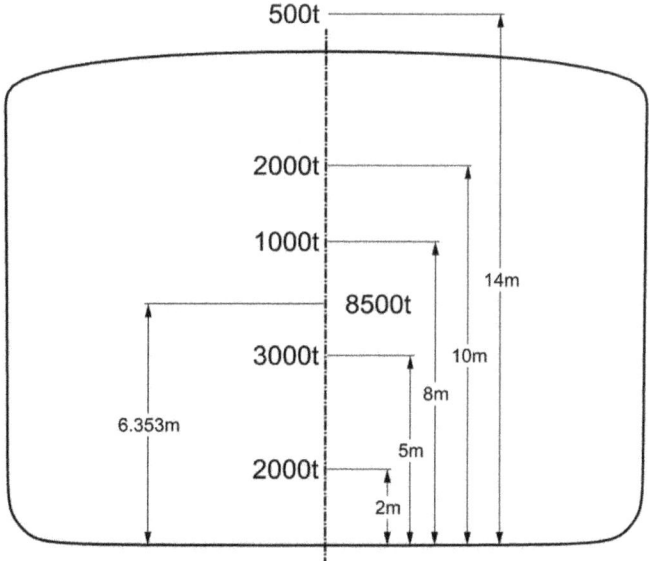

▲ **Figure 3.1** *Vertical mass distribution*

This example is preferably answered in table form.

| mass (tonne) | *Kg* (m) | Vertical moment (t m) | Calculations |
|---|---|---|---|
| 2000 | 2 | 4000 | 2 x 2000 = 4000 |
| 3000 | 5 | 15000 | 5 x 3000 = 15000 |
| 1000 | 8 | 8000 | 8 x 1000 = 8000 |
| 2000 | 10 | 20000 | 10 x 2000 = 20000 |
| 500 | 14 | 7000 | 14 x 500 = 7000 |
| TOTAL  8500 | | 54000 | 4000 + 15000 + 8000 + 20000 + 7000 = 54000 |

$$KG = \frac{\text{total moment}}{\text{total displacement}}$$

$$KG = \frac{54000}{8500}$$

$$KG = 6.353 \text{ m}$$

When answering exam questions, it is often a good idea to present work in a tabular manner, as it is easier for you to spot any errors should they occur. It is also easier for an examiner to mark, and give marks for process, even if the calculations include a small error.

EXAMPLE: A ship of 6000 tonne displacement is composed of masses of 300, 1200 and 2000 tonne at distances 60, 35 and 11 m aft of midships, and masses of 1000, 1000 and 500 tonne at distances 15, 30 and 50 m forward of midships. Calculate the distance of the centre of gravity of the ship from midships.

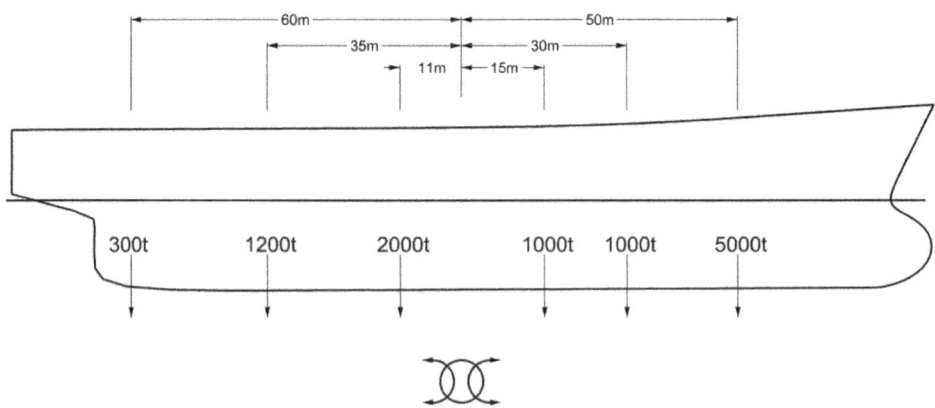

▲ **Figure 3.2** *Longitudinal mass distribution*

A table is again preferred.

| Mass (tonne) | Lcg from midships (m) | moment forward (t m) | moment aft (t m) |
|---|---|---|---|
| 300 | 60 aft | — | 18000 |
| 1200 | 35 aft | — | 42000 |
| 2000 | 11 aft | — | 22000 |
| 1000 | 15 forward | 15000 | — |
| 1000 | 30 forward | 30000 | — |
| 500 | 50 forward | 25000 | — |
| 6000 | | 70000 | 82000 |

The moment aft is greater than the moment forward and therefore the centre of gravity must lie aft of midships.

$$\text{Excess moment aft} = 82000 - 70000$$
$$= 12000 \text{ tonne m}$$
$$\text{Centre of gravity aft of midships} = \frac{\text{Excess moment aft}}{\text{total displacement}}$$
$$= \frac{12000}{6000}$$
$$= 2.00 \text{ m}$$

This example has been given with all distances given from midships. In these cases, it is important to keep track of which side of midships the masses are. There may well be examples or questions where the masses are given relative to either the forward or aft perpendicular. These do not change the overall result, it is just a different reference point. The most important point to remember is always make sure you know what reference point you are using, and be consistent with that through all your calculations.

# Shift in Centre of Gravity Due to Addition of Mass

When a mass is added to a ship, the centre of gravity of the ship moves towards the added mass. The distance moved by the ship's centre of gravity depends upon the magnitude of the added mass, the distance of the mass from the ship's centre of gravity and the displacement of the ship. If a mass is placed on the port side of the ship in the forecastle, the centre of gravity moves forward, upwards and to port. The actual distance and direction of this movement is seldom required but the separate components are most important, ie the longitudinal, vertical and transverse distances moved. When an item on a ship is removed, the centre of gravity moves away from the original position of that item.

EXAMPLE: A ship of 4000 tonne displacement has its centre of gravity 1.5 m aft of midships and 4 m above the keel. 200 tonne of cargo are now added 45 m forward of midships and 12 m above the keel. Calculate the new position of the centre of gravity.

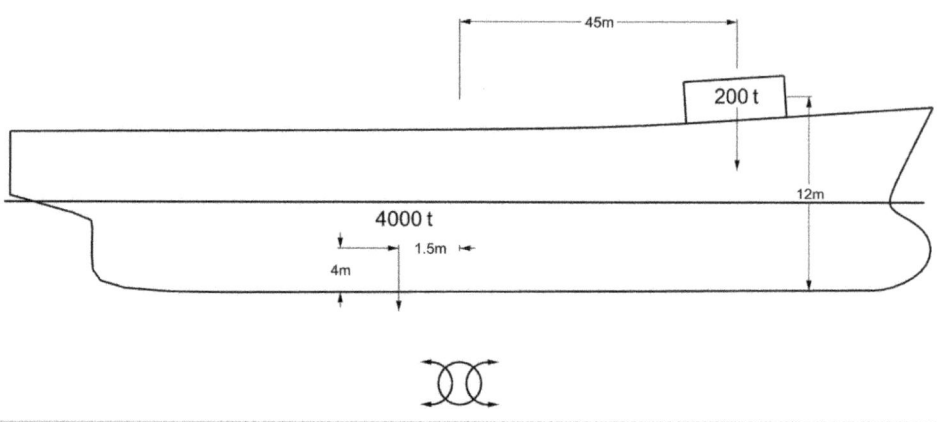

▲ **Figure 3.3** *Longitudinal movement in centre of gravity*

Taking moments about midships:

$$\text{Moment aft of midships} = 4000 \times 1.5$$
$$= 6000 \text{ tonne m}$$
$$\text{Moment forward of midships} = 200 \times 45$$
$$= 9000 \text{ tonne m}$$
$$\text{Excess moment forward} = 9000 - 6000$$
$$= 3000 \text{ tonne m}$$
$$\text{Centre of gravity forward of midships} = \frac{\text{Excess moment forward}}{\text{total displacement}}$$
$$= \frac{3000}{4000 + 200}$$
$$= 0.714 \text{ m forward}$$
$$\text{Centre of gravity from keel} = \frac{4000 \times 4 + 200 \times 12}{4000 + 200}$$
$$= \frac{16000 + 2400}{4200}$$
$$KG = 4.381 \text{ m}$$

Thus the centre of gravity rises 0.381 m.

The same answer may be obtained by taking moments about the original centre of gravity, thus:

$$\text{Moment of ship about centre of gravity} = 40000 \times 0$$
$$\text{Moment of added mass about centre of gravity} = 2000 \times (12 - 4)$$
$$= 1600 \text{ tonne m}$$

$$\text{Rise in centre of gravity} = \frac{\text{total moment}}{\text{total displacement}}$$
$$= \frac{1600}{4200}$$
$$= 0.381 \text{ m}$$

If the actual distance moved by the centre of gravity is required, it may be found from the longitudinal and vertical movements.

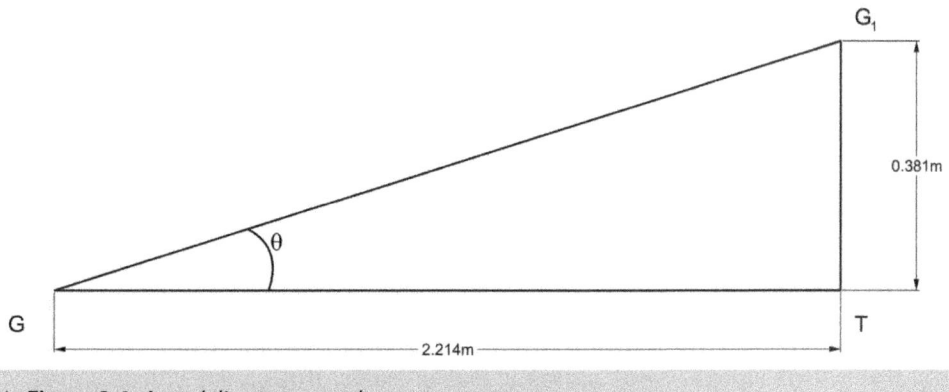

▲ **Figure 3.4** *Actual distance moved*

$$\text{Longitudinal shift in the centre of gravity } GT = 1.5 + 0.714$$
$$= 2.214 \text{ m}$$
$$GG_1 = \sqrt{GT^2 + TG_1^2}$$
$$= \sqrt{2.214^2 + 0.381^2}$$
$$= 2.247 \text{ m}$$

The angle $\theta$, which the centre of gravity moves relative to the horizontal, may be found from figure 3.4.

$$\tan \theta = \frac{0.381}{2.214}$$

$$\tan\theta = 0.172$$
$$\theta = \arctan(0.172)$$
$$\theta = \tan^{-1}(0.172)$$
$$\theta = 9.764°$$

# Shift in Centre of Gravity Due to Movement of Mass

When a mass that is already on board a ship is moved in any direction, there is a corresponding movement in the centre of gravity of the ship in the same direction.

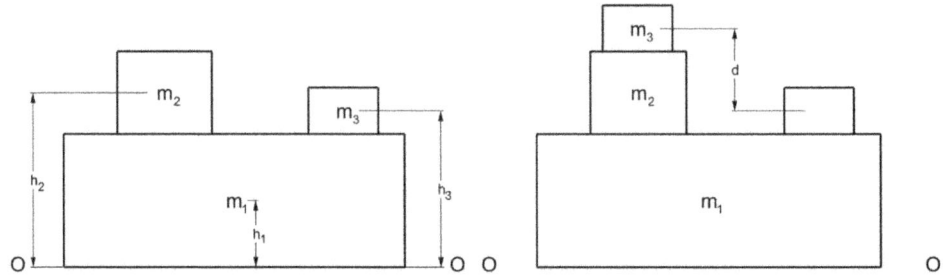

▲ **Figure 3.5** *Vertical movement in centre of gravity*

Consider a system composed of masses of $m_1$, $m_2$ and $m_3$ as shown in figure 3.5 (i), the centre of gravity of each being $h_1$, $h_2$ and $h_3$ respectively from the base O-O. The distance of the centre of gravity of the system from the base may be determined by dividing the total moment of mass about O-O by the total mass.

$$\text{Centre of gravity from } O-O = \frac{\text{total moment of mass}}{\text{total mass}}$$
$$= \frac{m_1 h_1 + m_2 h_2 + m_3 h_3}{m_1 + m_2 + m_3}$$
$$= y$$

If $m_3$ is now raised through a distance $d$ to the position shown in figure 3.5 (ii), the centre of gravity of the system is also raised.

$$\text{New centre of gravity from } O-O = \frac{m_1 h_1 + m_2 h_2 + m_3(h_3 + d)}{m_1 + m_2 + m_3}$$
$$= \frac{m_1 h_1 + m_2 h_2 + m_3 h_3}{m_1 + m_2 + m_3} + \frac{m_3 d}{m_1 + m_2 + m_3}$$
$$= y + \frac{m_3 d}{m_1 + m_2 + m_3}$$

Thus it may be seen that:

$$\text{Shift in centre of gravity} = \frac{m_3 d}{m_1 + m_2 + m_3}$$

or,

$$\text{Shift in centre of gravity} = \frac{\text{mass moved} \times \text{distance moved}}{\text{total mass}}$$

This is a very useful expression when making ship calculations and it is applied throughout stability and trim work. It allows you to account for the effect of a change in mass distribution in the ship, without recalculating the centre of gravity from the start. It should also be noted that it is not necessary to know either the position of the centre of gravity of the ship or the position of the mass relative to the centre of gravity of the ship. The rise in the centre of gravity is the same whether the mass is moved from the tank top to the deck or from the deck to the masthead as long as the distance moved is the same. The centre of gravity of the ship moves in the same direction as the centre of gravity of the mass. Thus if a mass is moved forwards and down, the centre of gravity of the ship also moves forwards and down.

EXAMPLE: A ship of 5000 tonne displacement has a mass of 200 tonne on the fore deck 55 m forward of midships. Calculate the shift in the centre of gravity of the ship if the mass is moved to a position 8 m forward of midships.

$$\text{Shift in centre of gravity} = \frac{\text{mass moved} \times \text{distance moved}}{\text{displacement}}$$
$$= \frac{200 \times (55 - 8)}{5000}$$
$$= 1.88 \text{ m aft}$$

# Effect of a Suspended Mass

When a ship loads or unloads cargo using its own gear or derricks, it is important to note that the mass must be considered to act at the derrick head, or point of suspension, rather than at the centre of gravity of the mass itself. The reason for this is that the mass is attached by a cable to the derrick head. It is the tension in that cable that exerts a force on the derrick, and it is at the derrick head that the ship feels this force.

In terms of movement of a suspended mass, there are three stages. For simplicity, we will consider the loading case:

**Stage 1:** Mass is on dock. At this point, the mass is not considered within the mass of ship.

**Stage 2:** Mass is hanging freely. As soon as the mass is clear of the ground, the mass is considered within the total mass of ship. The CoG of Mass acts at the derrick head and the overall CoG of Ship is $CG$ + Mass Correction using derrick head height.

**Stage 3:** Mass is on ship. Clearly, the mass is considered within total mass of ship, but as soon as the mass is no longer suspended, the CoG of the mass acts at its own position. The overall CoG of Ship is now $CG$ + Mass Correction using mass's own position as height.

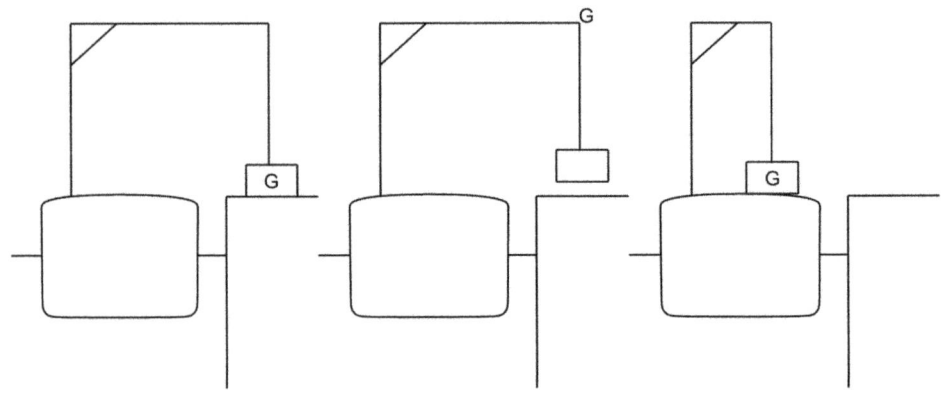

▲ **Figure 3.6** *Three stages of suspended masses*

The important point, which is often looked for in exams, is that as soon as the mass is clear of the tank top or the dock, the centre of gravity of the overall ship has risen. Operationally, it is important to note that when loading or discharging in a light condition, with tanks that are not pressed full or completely empty, the stability of the ship can be significantly less. This effect has resulted in the capsize of fishing vessels when unloading in port with minimal fuel.

Ships that are equipped to load heavy cargoes by means of heavy lift derricks must have a standard of stability that will prevent excessive heel when the cargo is suspended from the derrick. It may also be required to alter ballasting arrangements during loading and unloading, to ensure the ship's stability is not compromised. The effect of suspended masses must be considered when dealing with cargoes such as meat, which hangs within refrigerated containers. The centre of gravity must be taken as acting at the top of the container, where the hangers are attached, rather than at the middle.

EXAMPLE: A ship of 10000 tonne displacement has a mass of 60 tonne lying on the deck. A derrick, whose head is 7.5 m above the centre of gravity of the mass, is used to place the mass on the tank top 10.5 m below the deck. Calculate the shift in the vessel's centre of gravity when the mass is:

**a)** just clear of the deck

**b)** at the derrick head

**c)** in its final position.

**a)** When the mass is just clear of the deck, its centre of gravity is raised to the derrick head.

$$\text{Shift in centre of gravity} = \frac{\text{mass moved} \times \text{distance moved}}{\text{displacement}}$$

$$= \frac{60 \times 7.5}{10000}$$

$$= 0.045 \text{ m up}$$

**b)** When the mass is at the derrick head, there is no further movement of the centre of gravity of the ship.

$$\text{Shift in centre of gravity} = 0.045 \text{ m up}$$

**c)**

$$\text{Shift in centre of gravity} = \frac{60 \times 10.5}{10000}$$

$$= 0.063 \text{ m down}$$

# Chapter 3 Test Examples

**Q 3.1:** A ship of 4000 tonne displacement has its centre of gravity 6 m above the keel. Find the new displacement and position of the centre of gravity when masses of 1000, 200, 5000 and 3000 tonne are added at positions 0.8, 1.0, 5.0 and 9.5 m above the keel.

**Q 3.2:** The centre of gravity of a ship of 5000 tonne displacement is 6 m above the keel and 1.5 m forward of midships. Calculate the new position of the centre of gravity if 500 tonne of cargo are placed in the 'tween decks 10 m above the keel and 36 m aft of midships.

**Q 3.3:** A ship has 300 tonne of cargo in the hold, 24 m forward of midships. The displacement of the vessel is 6000 tonne and its centre of gravity is 1.2 m forward of midships. Find the new position of the centre of gravity if this cargo is moved to an after hold, 40 m from midships.

**Q 3.4:** An oil tanker of 17000 tonne displacement has its centre of gravity 1 m aft of midships and has 250 tonne of oil fuel in its forward deep tank, 75 m from midships. This fuel is transferred to the after oil fuel bunker, whose centre is 50 m from midships.

200 tonne of fuel from the after bunker is now burned.

Calculate the new position of the centre of gravity:
a) after the oil has been transferred
b) after the oil has been used.

**Q 3.5:** A ship of 3000 tonne displacement has 500 tonne of cargo on board. This cargo is lowered 3 m and an additional 500 tonne of cargo is taken on board 3 m vertically above the original position of the centre of gravity. Determine the alteration in position of the centre of gravity.

**Q 3.6:** A ship of 10000 tonne displacement has its centre of gravity 3 m above the keel. Masses of 2000, 300 and 50 tonne are removed from positions 1.5, 4.5 and 6 m above the keel. Find the new displacement and position of the centre of gravity.

**Q 3.7:** A vessel of 8000 tonne displacement has 75 tonne of cargo on the deck. It is lifted by a derrick whose head is 10.5 m above the centre of gravity of the cargo, and placed in the lower hold 9 m below the deck and 14 m forward of its original position. Calculate the shift in the vessel's centre of gravity from its original position when the cargo is:
a) just clear of the deck
b) at the derrick head
c) in its final position.

# STABILITY OF SHIPS

The term 'stability' can be applied to a number of different types of behaviour that a ship can exhibit, but in general it is used as a term to refer to the transverse static stability of a ship. This is the ship's ability to return to the upright (level heel) condition when external loads or forces are applied.

In the upright position, figure 4.1, the weight of the ship acts vertically down through the centre of gravity $G$, while the upthrust acts through the centre of buoyancy $B$. Since the weight is equal to the upthrust, and the centre of gravity and the centre of buoyancy are in the same vertical line, the ship is in equilibrium.

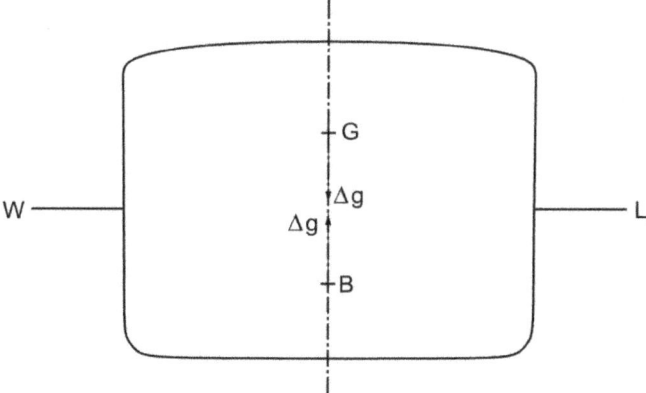

▲ **Figure 4.1** *Ship in equilibrium*

When the ship is inclined by an external force to an angle $\theta$, the centre of gravity remains in the same position but the centre of buoyancy moves from $B$ to $B_1$ (figure 4.2).

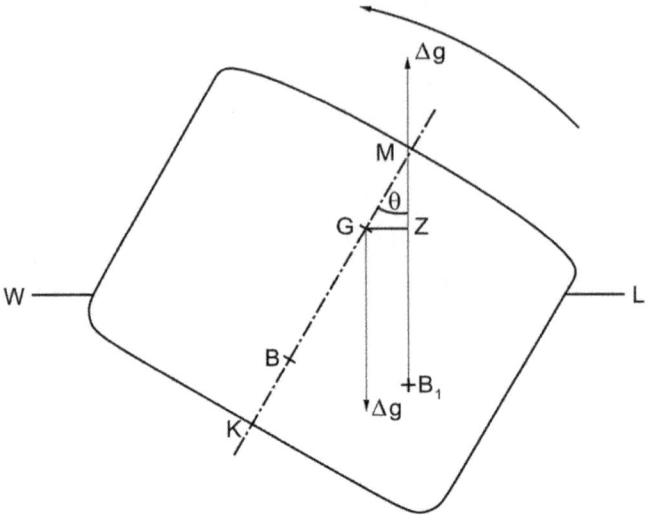

▲ **Figure 4.2** *Stable ship*

The buoyancy, therefore, acts up through $B_1$ while the weight still acts down through $G$, creating a moment of $\Delta g \times GZ$, which tends to return the ship to the upright. $\Delta g \times GZ$ is known as the *righting moment* and $GZ$ the *righting lever*. Since this moment tends to right the ship, the vessel is said to be *stable*.

For small angles of heel, up to about 10°, the vertical through the new centre of buoyancy $B_1$ intersects the centreline at $M$, the *transverse metacentre*. It may be seen from figure 4.2 that:

$$GZ = GM\sin\theta$$

Thus, for small angles of heel, $GZ$ is a function of $GM$, and since $GM$ is independent of $\theta$ while $GZ$ depends upon $\theta$, it is useful to express the initial stability of a ship in terms of $GM$, the *metacentric height*. $GM$ is said to be *positive* when $G$ lies below $M$ and the vessel is stable.

A ship with a small metacentric height will have a small righting lever at any angle and will roll easily; its roll will be a slow lollop. The ship is then said to be *tender*. A ship with a large metacentric height will have a large righting lever at any angle and will have a considerable resistance to rolling. The ship is then said to be *stiff*. A stiff ship will be very uncomfortable, having a very small rolling period, which produces a short and snappy roll. In extreme cases, this may result in structural damage.

If the centre of gravity lies above the transverse metacentre (Fig 4.3), the moment acts in the opposite direction, increasing the angle of heel. The vessel is then *unstable* and will not return to the upright, the metacentric height being regarded as *negative*.

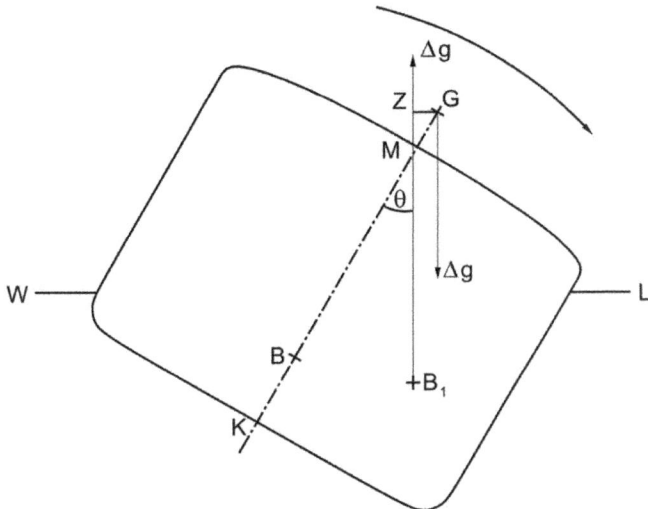

▲ **Figure 4.3** *Unstable ship*

When the centre of gravity and transverse metacentric coincide (Fig 4.4), there is no moment acting on the ship, which will therefore remain inclined to angle $\theta$. The vessel is then said to be in *neutral equilibrium*.

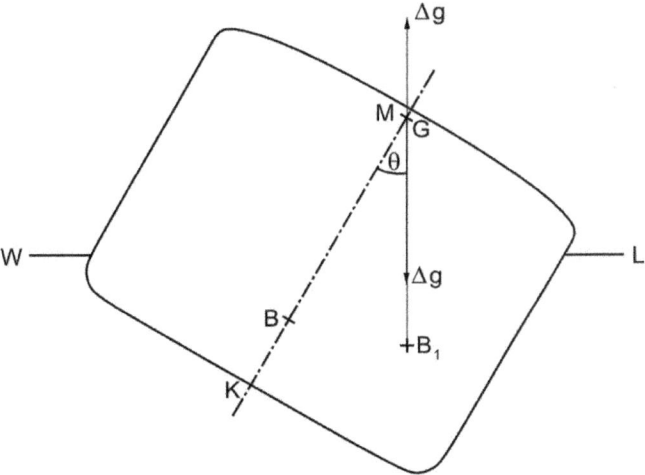

▲ **Figure 4.4** *Neutral equilibrium*

Since any reduction in the height of *G* will make the ship stable, and any rise in *G* will make the ship unstable, this condition is regarded as the point at which a ship *becomes* either stable or unstable.

# Position of the Metacentre, M

The distance of the transverse metacentre above the keel ($KM$) is given by:

$$KM = KB + BM$$

$KB$ is the distance of the centre of buoyancy above the keel and may be found by one of the methods shown previously.

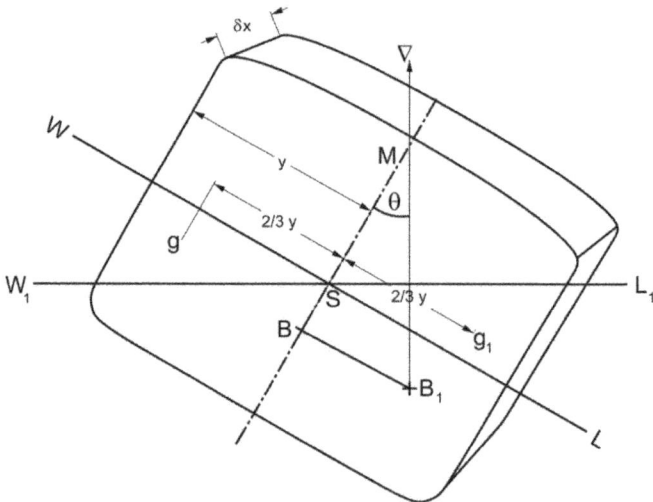

▲ **Figure 4.5** *Ship at small angle of heel*

$BM$ may be found as follows:

Consider a ship whose volume of displacement is $\nabla$, lying upright at waterline WL, the centre of buoyancy being on the centreline of the ship. If the ship is now inclined to a small angle $\theta$, it will lie at waterline $W_1L_1$, which intersects the original waterline at S (figure 4.5). Since $\theta$ is small, it may be assumed that S is on the centreline. A wedge of buoyancy $WSW_1$ has been moved across the ship to $L_1SL$, causing the centre of buoyancy to move from $B$ to $B_1$.

Let $v$ = volume of wedge

$gg_1$ = transverse shift in centre of gravity of wedge

Then

$$BB_1 = \frac{v \times gg_1}{\nabla}$$

But

$$BB_1 = BM \tan\theta$$

Therefore:

$$BM \tan\theta = \frac{v \times gg_1}{\nabla}$$

$$BM = \frac{v \times gg_1}{\nabla \tan\theta}$$

To determine the value of $v \times gg_1$, divide the ship into thin transverse strips of length $\delta x$, and let the half width of waterplane of the strip be $y$.

$$\text{Volume of strip of wedge} = \frac{1}{2} y \times y \tan\theta \delta x$$

$$\text{Moment of shift of strip of wedge} = \frac{4}{3} y \times \frac{1}{2} y^2 \tan\theta \delta x$$

$$= \frac{2}{3} y^3 \tan\theta \delta x$$

$$\text{Total moment of shift of wedge} = v \times gg_1$$

$$= \sum \frac{2}{3} y^3 \tan\theta \delta x$$

$$= \tan\theta \frac{2}{3} \sum y^3 \delta x$$

This equation can be simplified since the second moment of waterplane area about the centreline, $I$, is:

$$\frac{2}{3} \sum y^3 \delta x$$

Therefore:

$$v \times gg_1 = I \tan\theta$$

$$BM = \frac{I \tan\theta}{\nabla \tan\theta}$$

$$BM = \frac{I}{\nabla}$$

EXAMPLE: A box barge of length $L$ and breadth $B$ floats at a level keel draught $T$. Calculate the height of the transverse metacentre above the keel.

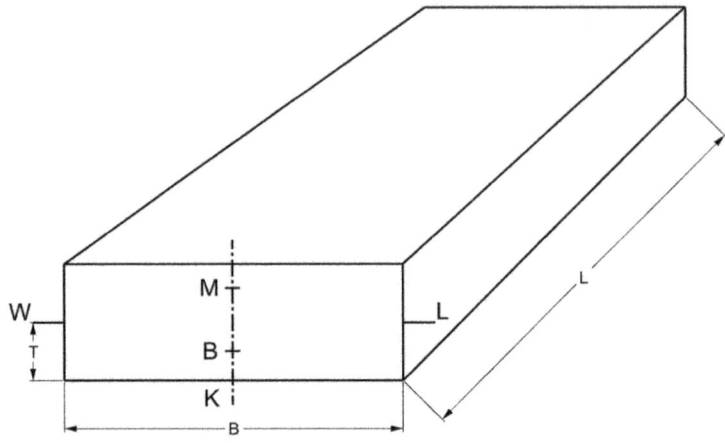

▲ **Figure 4.6** *Box barge*

$$KM = KB + BM$$

$$KB = \frac{T}{2}$$

$$BM = \frac{I}{\nabla}$$

$$I = \frac{1}{12}LB^3$$

$$\nabla = L \times B \times T$$

$$BM = \frac{LB^3}{12LBT}$$

$$= \frac{B^2}{12T}$$

$$KM = \frac{T}{2} + \frac{B^2}{12T}$$

It should be noted that while the above expression is applicable only to a box barge, similar expressions may be derived for vessels of constant triangular or circular cross sections. The waterplane in each case is in the form of a *rectangle,* the second moment of which is 1/12 × length × breadth³. As long as the length of a vessel having constant cross section exceeds the breadth, the length does not affect the transverse stability of the ship.

EXAMPLE: A vessel of constant triangular cross section has a depth of 12 m and a breadth at the deck of 15 m.

Calculate the draught at which the vessel will become unstable if the centre of gravity is 6.675 m above the keel.

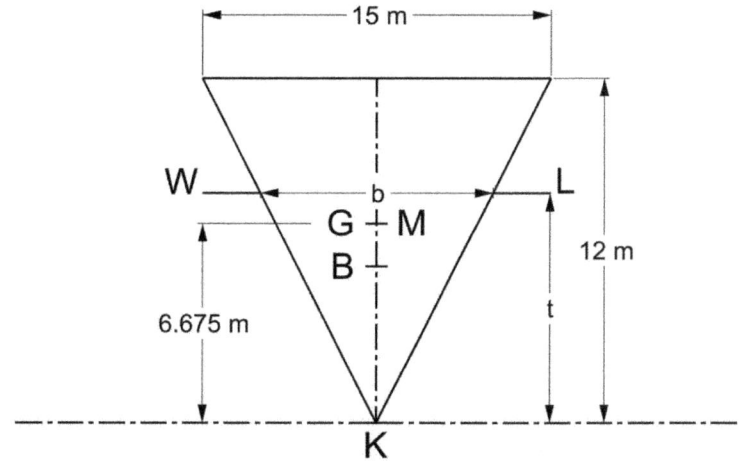

▲ **Figure 4.7** *Triangular cross section*

Let $t$ = draught

$b$ = breadth at waterline

By similar triangles:

$$\frac{b}{t} = \frac{B}{T}$$

$$b = \frac{15}{12}t$$

$$= \frac{5}{4}t$$

$$KB = \frac{2}{3}t$$

$$\nabla = \frac{1}{2}Lbt$$

$$I = \frac{1}{12}Lb^3$$

$$BM = \frac{I}{\nabla}$$

$$= \frac{1}{12}Lb^3 \div \frac{1}{2}Lbt$$

$$= \frac{b^2}{6t}$$

$$= \frac{1}{6t}\left(\frac{5}{4}t\right)^2$$

$$= \frac{25}{96}t$$

The vessel becomes unstable when $G$ and $M$ coincide.

$$KM = KG$$
$$= 6.675 \text{ m}$$
$$6.675 = \frac{2}{3}t + \frac{25}{96}t$$
$$= \frac{89}{96}t$$
$$t = 6.675 \times \frac{96}{89}$$
$$t = 7.2 \text{ m}$$

# Metacentric Diagram

Since both $KB$ and $BM$ depend upon draught, their values for any ship may be calculated for a number of different draughts, and plotted to form the *metacentric diagram* for the ship. The height of the transverse metacentre above the keel may then be found at any intermediate draught.

The metacentric diagram for a box barge is similar to that for a ship (figure 4.8), while the diagram for a vessel of constant triangular cross section is formed by two straight lines starting from the origin (figure 4.9).

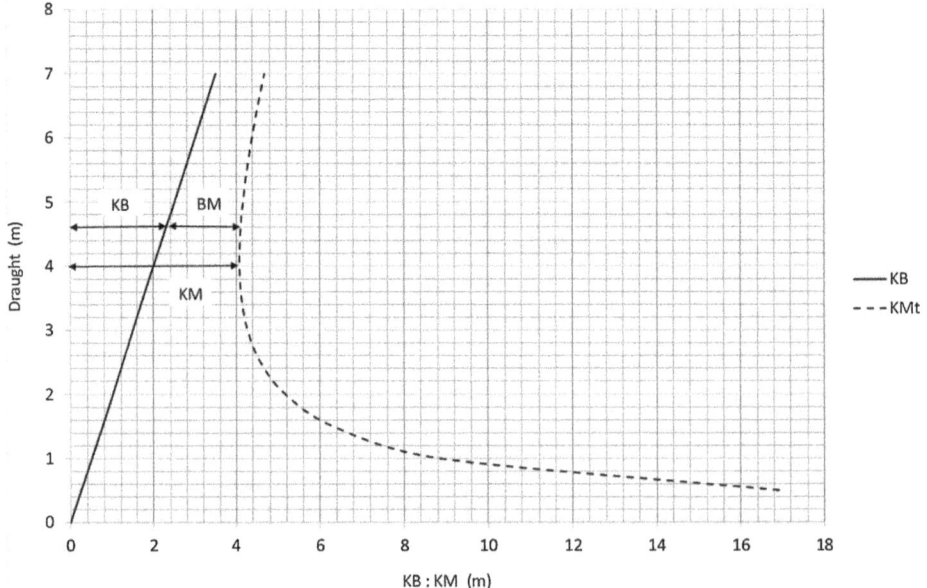

▲ Figure 4.8 *Box barge metacentric diagram*

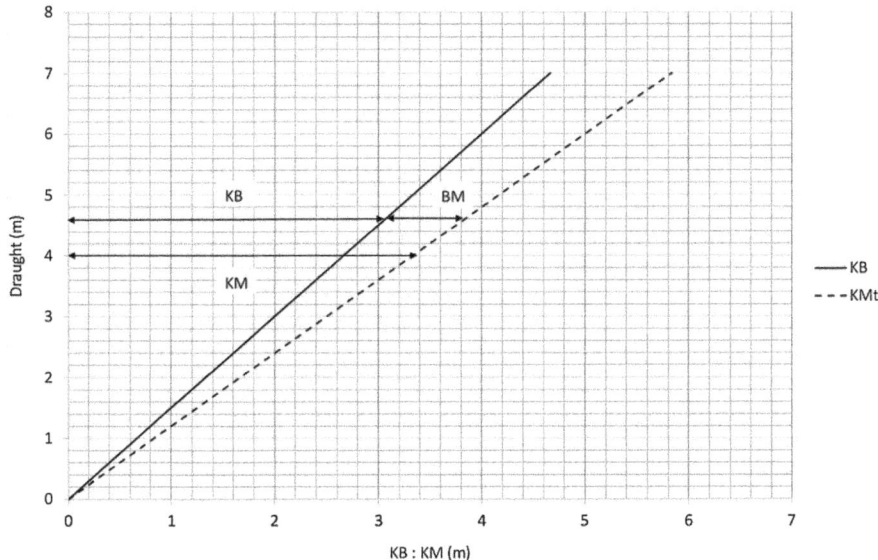

▲ **Figure 4.9** *Triangular cross section metacentric diagram*

EXAMPLE: A vessel of constant rectangular cross section is 12 m wide. Draw the metacentric diagram using 1.2 m intervals of draught up to the 7.2 m waterline.

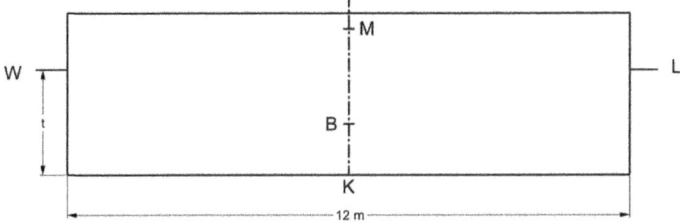

▲ **Figure 4.10** *Rectangular cross section*

It is useful in an example of this type to derive an expression for *KB* and *KM* in terms of the only variable – draught – and substitute the different draught values in tabular form.

$$KB = \frac{t}{2}$$

$$BM = \frac{B^2}{12t}$$

$$= \frac{12^2}{12t}$$

$$= \frac{12}{t}$$

$$KM = KB + BM$$

| t | KB | BM | KM |
|---|---|---|---|
| 0 | 0 | ∞ | ∞ |
| 1.2 | 0.6 | 10.00 | 10.60 |
| 2.4 | 1.2 | 5.00 | 6.20 |
| 3.6 | 1.8 | 3.33 | 5.13 |
| 4.8 | 2.4 | 2.50 | 4.90 |
| 6.0 | 3.0 | 2.00 | 5.00 |
| 7.2 | 3.6 | 1.67 | 5.27 |

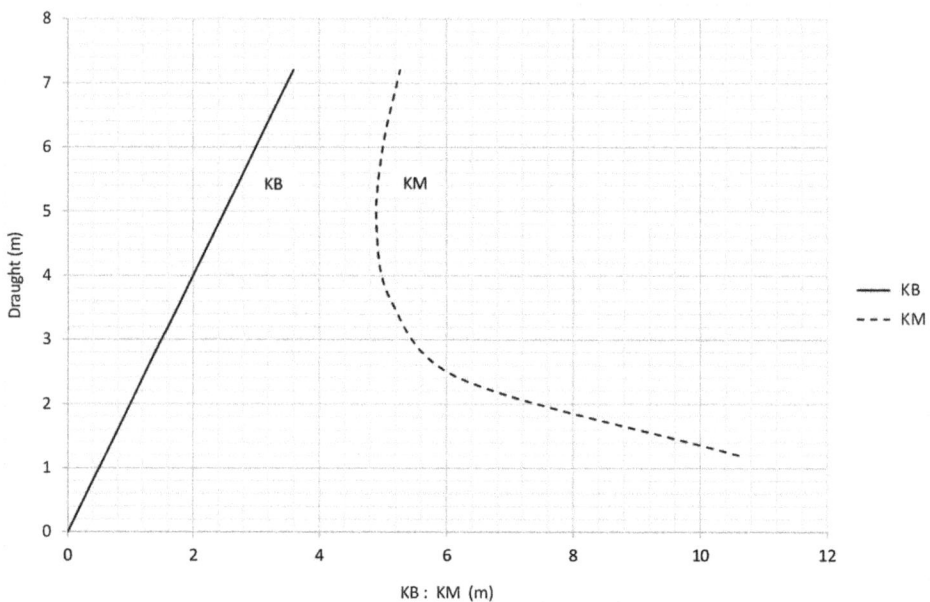

▲ **Figure 4.11** *Metacentric diagram*

# Stability at Large Angles of Heel

When a ship heels to an angle greater than about 10°, the principles on which the initial stability is based are no longer true. The proof of the formula for *BM* was based on the assumption that the two waterplanes intersect at the centreline and that the wedges are right-angled triangles. Neither of these assumptions may be made for large angles of heel, and the stability of the ship must be determined from first principles. In practice, this is calculated by computer programs for the ship by the naval architect, but the principles that stability software uses are as described below.

As a ship loads and unloads cargo, or burns fuel, the possible combinations of displacement and centre of gravity are infinite. Any method for calculating the stability of the vessel must take this into account.

The righting lever, $GZ$, is the perpendicular distance from a vertical axis through the centre of gravity $G$ to the centre of buoyancy $B_1$. As $G$ could have any number of locations, it is easier to calculate the perpendicular distance of the line of action for $B_1$ from the keel, for a range of displacements. This distance is called $KN$.

From figure 4.12 it can be seen that:

$$KN = KG\sin\theta + GZ$$

Which can be rearranged:

$$GZ = KN - KG\sin\theta$$

Therefore, if $KG$ and $KN$ are known, $GZ$ can be calculated. $KN$ is calculated by a computer program splitting the hull into sections along the ship's length. Each section will be inclined through a range of angles, say 15°, 30°, 45°, 60°, 75° and 90° for different displacements. The overall values of $KN$ for the hull are then calculated from the values at each of the sections, and summated using Simpson's Rule.

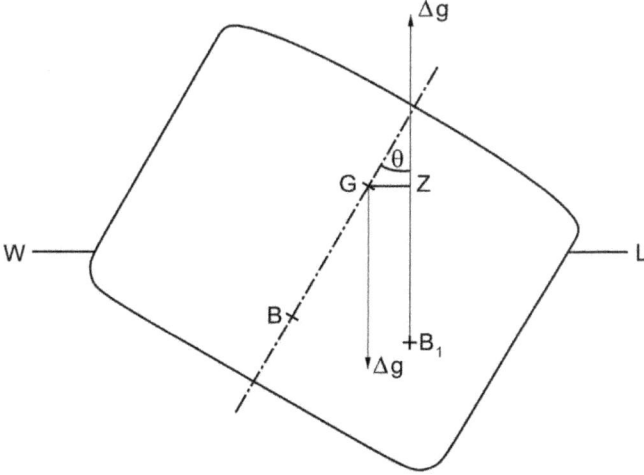

▲ **Figure 4.12** *Cross section of heeled ship*

Each ship will have either tables or graphs of *KN* for different displacements. These curves are known as the *cross curves of stability* and can be found in the ship's stability book (figure 4.13).

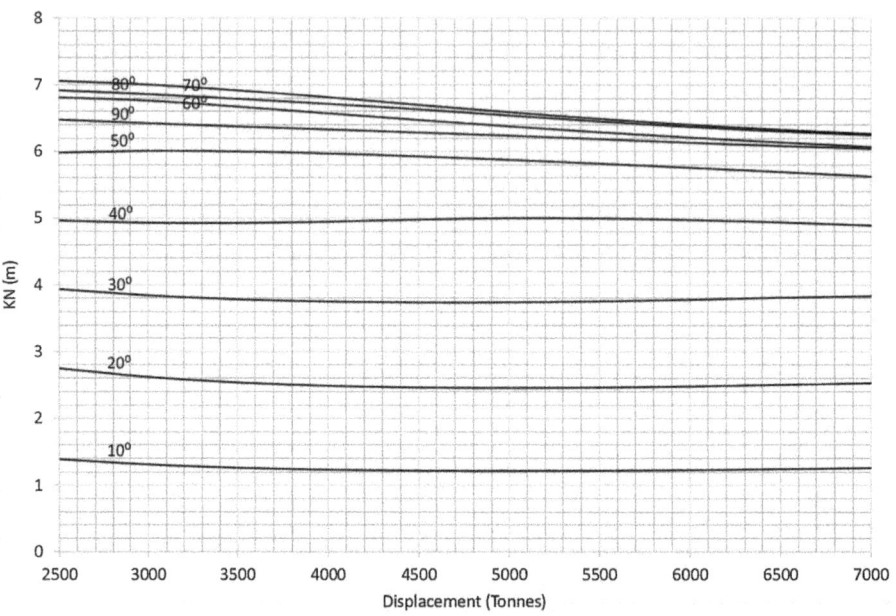

▲ **Figure 4.13** *Cross curves of stability KN*

The displacement, height of centre of gravity and metacentric height of a vessel may be calculated for any loaded condition. At this displacement, the righting levers may be obtained at the respective angles for the assumed position of the centre of gravity.

The righting levers are plotted on a base of angle of heel to form the *Curve of Statical Stability* for the ship in this condition of loading. The initial slope of the curve lies along a line drawn from the origin to *GM* plotted vertically at one radian (57.3°) (figure 4.14).

For smaller vessels, sailing yachts and lifeboats, it is important that *GZ* is calculated up to 180°, as these vessels are all at greater risk of capsize than larger ships, and their stability in the capsized state is often of great interest.

The area under the GZ curve to any given angle, multiplied by the gravitational weight of the ship, is the work done in heeling the ship to that angle and is known as the *dynamical stability*.

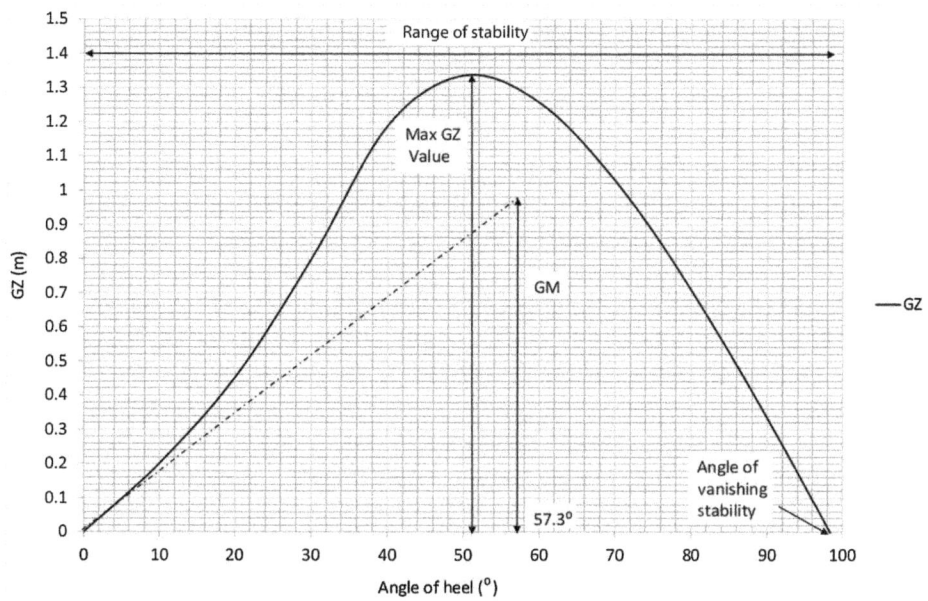

▲ **Figure 4.14** *GZ curve*

EXAMPLE: A vessel has the following righting levers at a particular draught, based on an assumed *KG* of 7.2 m:

| θ | 0° | 15° | 30° | 45° | 60° | 75° | 90° |
|---|---|---|---|---|---|---|---|
| GZ (m) | 0 | 0.43 | 0.93 | 1.21 | 1.15 | 0.85 | 0.42 |

The vessel is loaded to this draught but the actual *KG* is found to be 7.8 m and the *GM* 1.0 m.

Draw the amended statical stability curve.

$$GG_1 = 0.6 \text{ m}$$

$$G_1Z = GZ - GG_1 \sin \theta$$

(ie the vessel is *less* stable than suggested by the original values).

| Angle $\theta$ | sin $\theta$ | $GG_1 \sin \theta$ | GZ | $G_1Z$ |
|---|---|---|---|---|
| 0 | 0 | 0 | 0 | 0 |
| 15° | 0.259 | 0.15 | 0.43 | 0.28 |
| 30° | 0.500 | 0.30 | 0.93 | 0.63 |
| 45° | 0.707 | 0.42 | 1.21 | 0.79 |
| 60° | 0.866 | 0.52 | 1.15 | 0.63 |
| 75° | 0.966 | 0.58 | 0.85 | 0.27 |
| 90° | 1.000 | 0.60 | 0.42 | −0.18 |

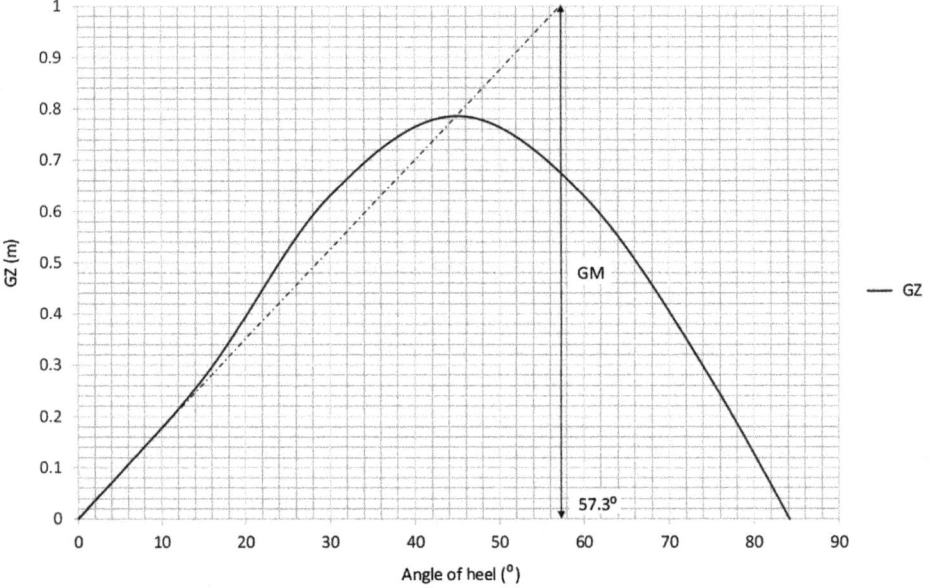

▲ **Figure 4.15** *GZ curve for above example*

The shape of the stability curve of a ship depends largely on the metacentric height and the freeboard. A tremendous change takes place in this curve when the weather deck edge becomes immersed. Thus, a ship with a large freeboard will normally have a large range of stability, while a vessel with a small freeboard will have a much smaller range. Figure 4.16 shows the effect of freeboard on two ships with the same metacentric height.

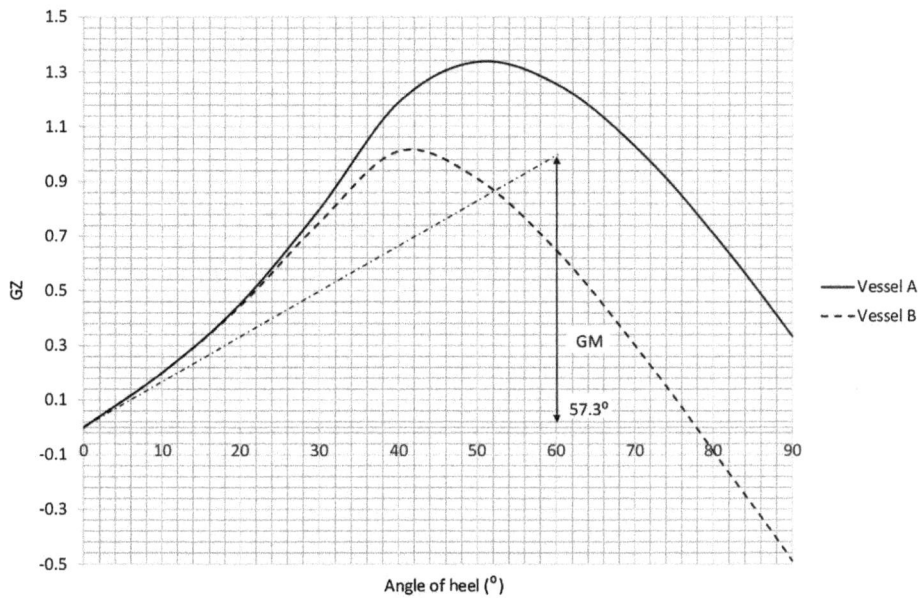

▲ **Figure 4.16** *GZ curves for ships with different freeboard*

Vessel A is a closed shelter deck ship, whereas Vessel B is a raised quarterdeck ship. It is essential for a vessel with a small freeboard, such as an oil tanker, to have a large metacentric height and thus extend the range of stability.

If a vessel is initially unstable it will not remain upright but will either heel to the *angle of loll* or capsize, depending upon the degree of instability and the shape of the stability curve, figure 4.17.

Vessel A will heel to an angle of loll of about 14° but still remains a fairly stable ship, and while this heel would be very inconvenient, the vessel would not be in a dangerous condition.

If vessel B is unstable it will capsize, since at all angles the righting lever is negative.

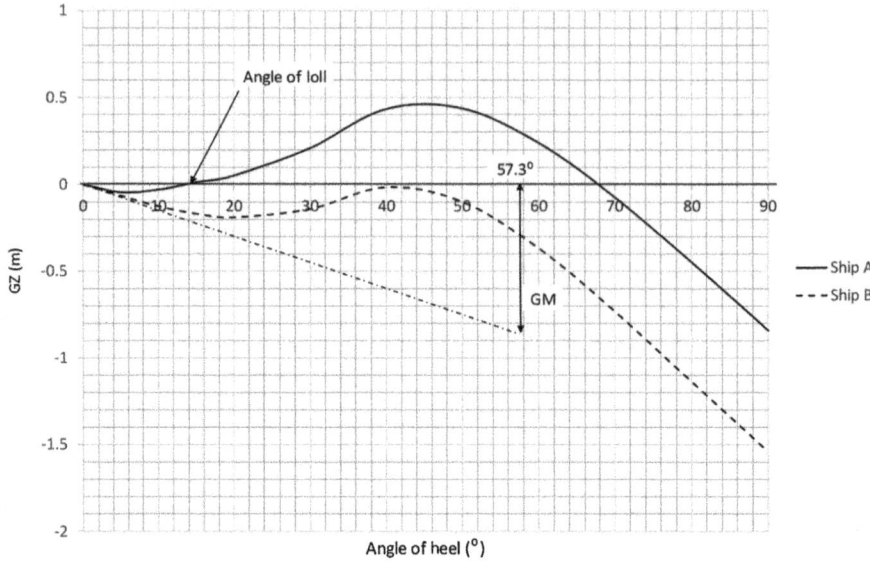

▲ **Figure 4.17** *GZ curve with angle of loll*

EXAMPLE: The righting levers of a ship of 15000 tonne displacement at angles of heel of 15°, 30°, 45° and 60° are 0.29, 0.70, 0.93 and 0.90 m respectively. Calculate the dynamical stability of the ship at 60° heel.

| Angle | GZ | SM | Product for area |
|-------|-----|----|------------------|
| 0 | 0 | 1 | 0 |
| 15° | 0.29 | 4 | 1.16 |
| 30° | 0.70 | 2 | 1.40 |
| 45° | 0.93 | 4 | 3.72 |
| 60° | 0.90 | 1 | 0.90 |
| | | | 7.18 |

*Note:* The common interval must be expressed in *radians*.

$$h = \frac{15}{57.3}$$

$$\text{Area under curve} = \frac{1}{3} - \frac{15}{57.3} - 7.18$$

$$= 0.6265$$

$$\text{Dynamical stability} = 15000 \times 9.81 \times 0.6265$$

$$= 92.19 \times 10^3 \text{ KJ}$$

$$= 92.19 \text{ MJ}$$

# Stability of a Wall-sided Ship

If a vessel is assumed to be wall-sided in the vicinity of the waterplane, the righting lever may be estimated from the expression

$$GZ = \sin\theta\left(GM + \frac{1}{2}BM\tan^2\theta\right)$$

This formula may be regarded as reasonably accurate for vessels that are deeply loaded, up to the point at which the deck edge enters the water.

If a vessel is initially unstable, it will either capsize or heel to the angle of loll. At this angle of loll θ, the vessel does not tend to return to the upright or incline to a greater angle. The righting lever is therefore zero. Hence, if the vessel is assumed to be wall-sided:

$$0 = GM + \frac{1}{2}BM\tan^2\theta$$

and since sin θ cannot be zero unless θ is zero:

$$\frac{1}{2}BM\tan^2\theta = -GM$$

$$\tan^2\theta = -2\frac{GM}{BM}$$

From which

$$\tan\theta = \pm\sqrt{-2\frac{GM}{BM}}$$

Since GM must be negative for this condition, the following must apply:

$$-2\frac{GM}{BM} > 0$$

Thus, the angle of loll may be determined for any given unstable condition.

EXAMPLE: A ship of 12000 tonne displacement has a second moment of area about the centreline of $72 \times 10^3$ m$^4$. If the metacentric height is −0.05 m, calculate the angle of loll.

$$\nabla = \frac{12000}{1.025}\,m^3$$

$$BM = \frac{I}{\nabla}$$

$$= \frac{72\times10^3 \times 1.025}{12000}\,m$$

$$= 6.150\ m$$

$$\tan\theta = \pm\sqrt{-2\frac{-0.05}{6.15}}$$

$$\tan\theta = \pm 0.1275$$

$$\theta = \pm 7.267°$$

ie the vessel may heel 7.267° either to port or to starboard.

A more practical application of this expression may be found when the vessel is listing at sea. It is necessary first to bring the ship upright and then to provide sufficient stability for the remainder of the voyage. Thus, it is essential to estimate the negative metacentric height causing the angle of loll in order to ensure that these conditions are realised.

EXAMPLE: At one point during a voyage, the above vessel is found to have an angle of loll of 13°. Calculate the initial metacentric height.

From above:

$$BM = 6.150\ m$$

$$\tan\theta = \pm\sqrt{-2\frac{GM}{BM}}$$

$$0.2309 = \pm\sqrt{-2\frac{GM}{6.15}}$$

$$GM = -\frac{0.230.9^2 \times 6.15}{2}$$

Initial metacentric height $= -0.164\ m$

# Free Surface Effect

When a tank on board a ship is not completely full of liquid and the vessel heels, the liquid moves across the tank in the same direction as the heel. There is a movement of the centre of gravity of the ship from $G$ to $G_1$, caused by the transfer of a wedge of liquid across the tank. The centre of gravity of the wedge moves from one side of the ship to the other, and at the same time moves vertically upwards, in comparison to its static position. This can be seen in figure 4.18. Accordingly, the centre of gravity of the overall ship moves away from the centreline and upwards, both of which lead to a reduction in the righting lever, and an increase in the angle of heel.

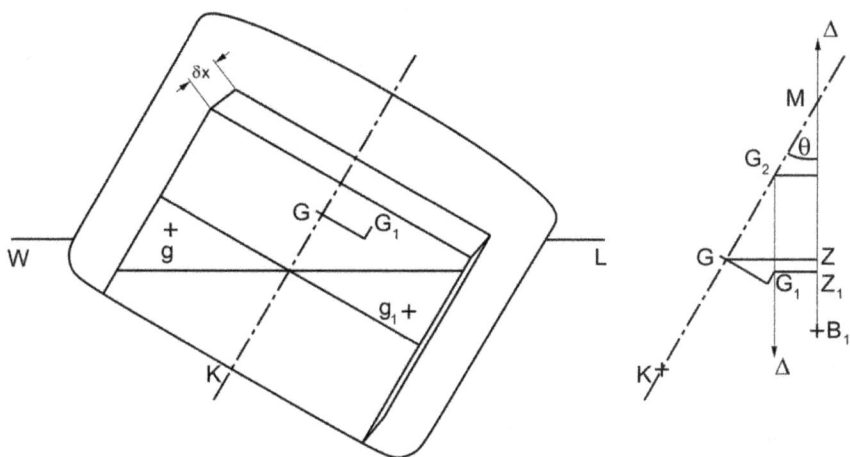

▲ **Figure 4.18** *Free surface effect*

If *m* is the mass of the wedge and $gg_1$ the distance moved by its centre, then

$$GG_1 = \frac{m \times gg_1}{\Delta}$$

But

$$m = v \times \rho_L$$

where *v* = volume of wedge

$\rho_L$ = density of liquid

$$\Delta = \nabla \times \rho_s$$

where $\nabla$ = volume of displacement

Therefore:

$$GG_1 = \frac{v \times \rho \times gg_1}{\nabla \times \rho_s}$$

Divide the tank into thin, transverse strips of length δx and let one such strip have a half width of free surface of *y*

$$\text{Volume of strip of wedge} = \frac{1}{2}y \times y \tan\theta\delta x$$

$$= \frac{1}{2}y^2 \tan\theta\delta x$$

$$\text{Mass of strip of wedge} = \rho_L \times \frac{1}{2}y^2 \tan\theta\delta x$$

$$\text{Moment of transfer of strip of wedge} = \frac{4}{3}y \times \rho_L \times \frac{1}{2}y^2 \tan\theta\delta x$$

$$= \rho_L \times \frac{2}{3}y^3 \tan\theta\delta x$$

$$= v\rho_L gg_1$$

$$= \rho_L \tan\theta\sum\frac{2}{3}y^3 \delta x$$

The second moment of area of free surface about the centreline of the tank is given by:

$$i = \sum\frac{2}{3}y^3 \delta x$$

Therefore:

$$GG_1 = \frac{\rho_L i \tan\theta}{\rho_S \nabla}$$

The righting lever has therefore been reduced from $GZ$ to $G_1Z$. But the righting lever is the perpendicular distance between the verticals through the centre of buoyancy and the centre of gravity, and this distance may be measured at any point. The vertical through $G_1$ intersects the centreline at $G_2$, and

$$G_2Z = G_1Z$$

also, $G_2Z = G_2M \sin\theta$ but $G_1Z$ does not equal $G_1M \sin\theta$.

Since the initial stability of a ship is usually measured in terms of metacentric height, it is useful to assume that the effect of a free surface of liquid is to raise the centre of gravity from $G$ to $G_2$, thus reducing the metacentric height of the vessel.

$GG_2$ is termed the *virtual reduction in metacentric height due to free surface* or, more commonly, *the free surface effect*.

Now,

$$GG_1 = GG_2 \tan\theta$$

$$GG_2 = \frac{\rho_L i \tan\theta}{\rho_S \nabla \tan\theta}$$

$$GG_2 = \frac{\rho_L i}{\rho_S \nabla}$$

$$= \frac{\rho_L i}{\Delta}$$

EXAMPLE: A ship of 5000 tonne displacement has a rectangular tank 6 m long and 10 m wide. Calculate the virtual reduction in metacentric height if this tank is partly full of oil (rd 0.8).

$$\rho_L = 1000 \times 0.8 \text{ kg/m}^3$$

$$i = \frac{1}{12} 6 \times 10^3 \text{ m}^4$$

$$\rho_S = 1025 \text{ kg/m}^3$$

$$\nabla = \frac{5000}{1.025} \text{ m}^3$$

$$GG_2 = \frac{1000 \times 0.8 \times 6 \times 10^3 \times 1.025}{1025 \times 5000 \times 12}$$

$$= 0.08 \text{ m}$$

# The Effect of Tank Divisions on Free Surface

Consider a rectangular tank of length $l$ and breadth $b$ partly full of sea water.

a) **WITH NO DIVISIONS**

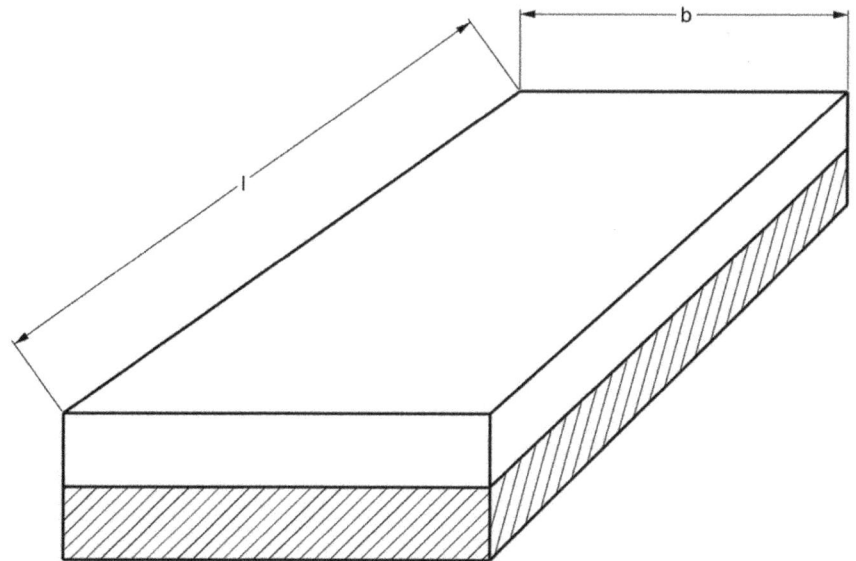

▲ Figure 4.19 *Tank with no subdivisions*

$$i = \frac{1}{12}lb^3$$

$$GG_2 = \frac{\rho_L i}{\rho_S \nabla}$$

$$= \frac{i}{\nabla}$$

since

$$\rho_L = \rho_S$$

$$i = \frac{1}{12}lb^3$$

$$GG_2 = \frac{lb^3}{12\nabla}$$

**b)** WITH A MID-LENGTH, TRANSVERSE DIVISION

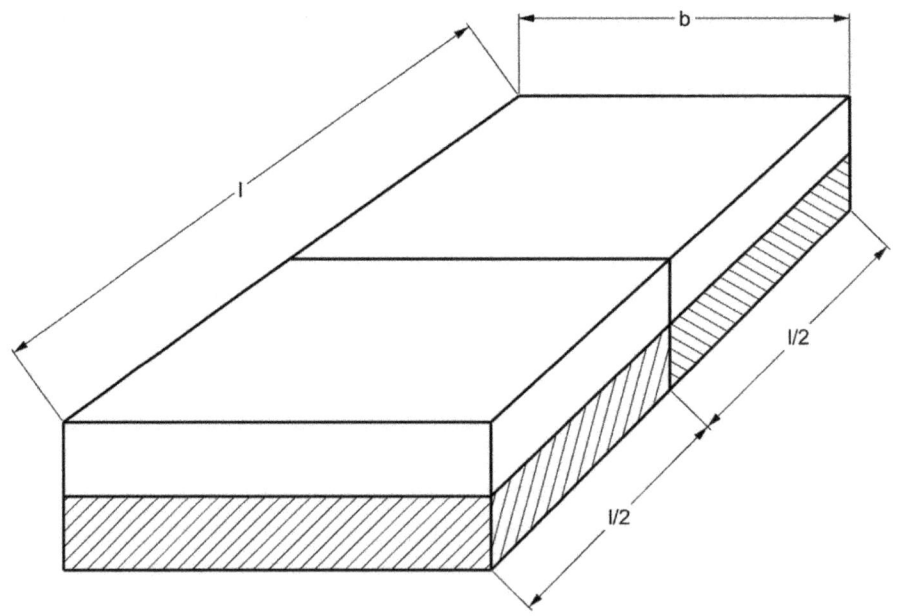

▲ **Figure 4.20** *Tank with a transverse subdivision*

For one tank:

$$i = \frac{1}{12}\frac{l}{2}b^3$$

For two tanks:

$$i = \frac{1}{12}\frac{l}{2}b^3 \times 2$$
$$= \frac{1}{12}lb^3$$

Therefore:

$$GG_2 = \frac{lb^3}{12\nabla}$$

Thus, as long as there is a free surface of liquid in both tanks, there is no reduction in free surface effect. It would, however, be possible to fill one tank completely and have a free surface effect in only one tank.

c) **WITH A LONGITUDINAL, CENTRELINE DIVISION**

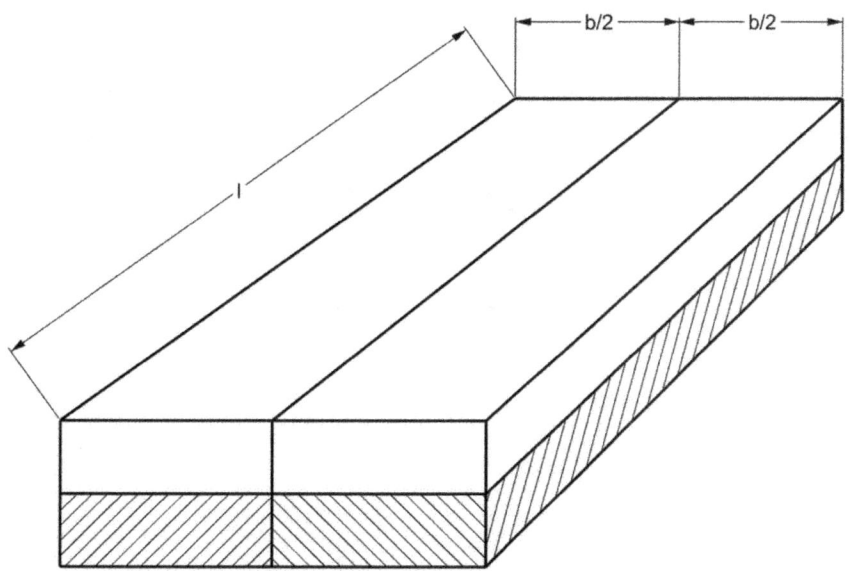

▲ **Figure 4.21** *Tank with a longitudinal subdivision*

For one tank:

$$i = \frac{1}{12}l\left(\frac{b}{2}\right)^3$$

For two tanks:

$$i = \frac{1}{12}l\left(\frac{b}{2}\right)^3 \times 2$$

$$= \frac{1}{4} \times \frac{1}{12}lb^3$$

$$GG_2 = \frac{1}{4} \times \frac{lb^3}{12\nabla}$$

Thus, the free surface effect is reduced to one-quarter of the original by introducing a longitudinal division.

d) **WITH TWO LONGITUDINAL DIVISIONS FORMING THREE EQUAL TANKS**

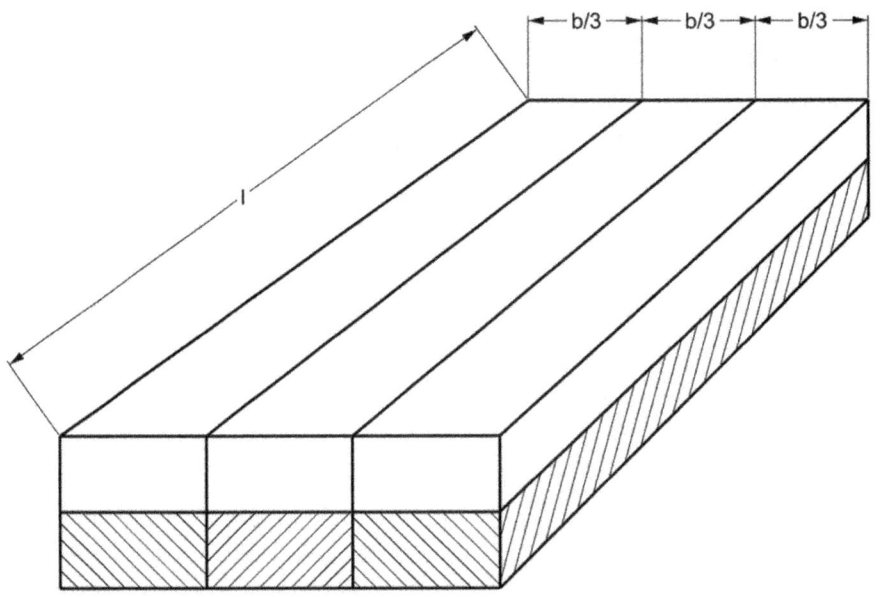

▲ **Figure 4.22** *Tank with two longitudinal subdivisions*

For one tank:

$$i = \frac{1}{12}l\left(\frac{b}{3}\right)^3$$

For three tanks:

$$i = \frac{1}{12}l\left(\frac{b}{3}\right)^3 \times 3$$

$$= \frac{1}{9} \times \frac{1}{12}lb^3$$

$$GG_2 = \frac{1}{9} \times \frac{lb^3}{12}$$

It may be seen that the free surface effect is still further reduced by the introduction of longitudinal divisions.

If a tank is subdivided by $n$ longitudinal divisions forming *equal* tanks, then:

$$GG_2 = \frac{1}{(n+1)^2}\frac{lb^3}{12\nabla}$$

# Practical Considerations

The effect of a free surface of liquid may be most dangerous in a vessel with a small metacentric height and may even cause the vessel to become unstable. In such a ship, tanks that are required to carry liquid should be pressed up tight. If the ship is initially unstable and heeling to port, then any attempt to introduce water ballast will reduce the stability. Before ballasting, therefore, an attempt should be made to lower the centre of gravity of the ship by pressing up existing tanks and lowering masses in the ship. If water is introduced into a double bottom tank on the starboard side, the vessel will flop to starboard and may possibly capsize. A small tank on the port side should therefore be filled completely before filling on the starboard side. The angle of heel will increase due to free surface and the effect of the added mass, but there will be no sudden movement of the ship.

A particularly dangerous condition may occur when a fire breaks out in the upper 'tween decks of a ship or in the accommodation of a passenger ship. If water is pumped into the space, the stability of the ship will be reduced both by the added mass of water and by the free surface effect. Any accumulation of water should be avoided. Circumstances will dictate the method used to remove the water, and will vary with the ship, cargo and position of fire, but it is likely to be via the use of a portable pump.

It is important to note that the free surface effect depends upon the displacement of the ship and the shape and dimensions of the *free surface*. It is independent of the total mass of liquid in the tank and of the position of the tank in the ship.

The ship with the greatest free surface effect is, of course, the oil tanker, since space must be left in the tanks for expansion of oil. It is common for tankers to have longitudinal bulkheads, to reduce the free surface effect and give the ships great longitudinal strength. Within the design of dry cargo vessels, it is rarely possible to have longitudinal bulkheads, due to the loading and unloading requirements. Thus, while the free surface effect in a tanker is greater than in a dry cargo ship, it is of more importance in the latter.

The importance of the free surface effect should never be underestimated. It was a contributory factor to the capsize of the *Herald of Free Enterprise* in 1987, which resulted in the loss of 193 lives.

# Inclining Experiment

While methods for calculating the centre of gravity of a ship and the overall stability have been described, these rely on an assumption of where the centre of gravity was to begin with. An experiment to establish the metacentric height, and from that the centre of gravity, is the inclining experiment.

In theory, this is a relatively simple experiment, carried out on the ship in a known and documented condition. It is preferable for that condition to be the lightship condition (ie as near empty as possible), but with all fittings and fixtures in place. (The practical difficulties associated with the experiment are highlighted in the following section.)

The experiment is commenced with the ship upright. A small mass, *m*, is moved across the ship through a distance *d*. This causes the centre of gravity to move from its original position $G$ on the centreline to $G_1$, figure 4.23.

If $\Delta$ = displacement of ship

Then

$$GG_1 = \frac{m \times d}{\Delta}$$

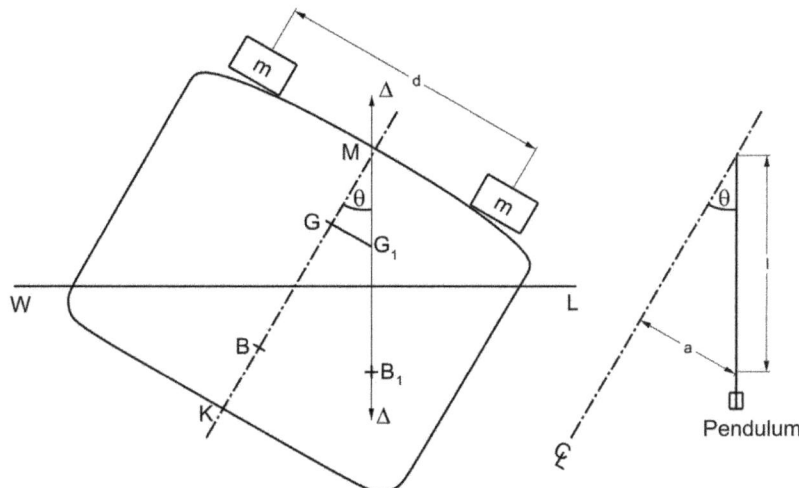

▲ **Figure 4.23** *Cross section of inclining experiment*

The ship then heels to angle $\theta$, when the centre of buoyancy moves from $B$ to $B_1$, in the same vertical line as $G_1$. But the vertical through $B_1$ intersects the centreline at $M$, the transverse metacentre.

$$GG_1 = GM\tan\theta$$

$$GM\tan\theta = \frac{m\times d}{\Delta}$$

$$GM = \frac{m\times d}{\Delta\tan\theta}$$

To determine the angle of heel, it is necessary to suspend a pendulum from, say, the underside of a hatch. The deflection $a$ of the pendulum may be measured when the mass is moved across the deck.

Thus, if $l$ = length of pendulum

$$\tan\theta = \frac{a}{l}$$

And

$$GM = \frac{m\times d\times l}{\Delta\times a}$$

The height of the transverse metacentre above the keel may be found from the metacentric diagram and hence the height of the centre of gravity of the ship may be determined.

$$KG = KM - GM$$

EXAMPLE: A mass of 6 tonne is moved transversely through a distance of 14 m on a ship of 4300 tonne displacement, when the deflection of an 11 m pendulum is found to be 120 mm. The transverse metacentre is 7.25 m above the keel.

Determine the height of the centre of gravity above the keel.

$$GM = \frac{m \times d}{\Delta \tan \theta}$$

$$= \frac{6 \times 14 \times}{4300 \times 120 \times 10^{-3}}$$

$$= 1.79 \text{ m}$$

$$KG = KM - GM$$

$$= 7.25 - 1.79$$

$$= 5.46 \text{ m}$$

# Conduct of Inclining Experiment

As described, the fundamental principles of the inclining experiment are simple. In practice, however, there are numerous places where inaccuracies can occur, making the experiment invalid. As the experiment often requires significant effort from multiple parties to conduct, it is important that the results of the experiment are valid.

The primary requirement is an accurate knowledge of the state of the ship. As inclining experiments take place following either the completion of a build or a significant refit, there can be significant amounts of yard equipment, and the vessel's own equipment may not be on board. The magnitude and position of any mass that is not included in the lightweight of the ship should be noted and it is therefore necessary to sound all tanks and inspect the whole ship. Corrections are made to the centre of gravity for any such masses.

The next phase of the experiment relates to the location and timing of the experiment. To ensure as accurate a result as possible, the ship should be in a sheltered position, eg graving dock, and the experiment should be carried out in calm weather. Often this will result in inclining experiments being conducted in early morning or in the evening, when the wind is likely to be less and the water still.

Four masses, A, B, C and D, are placed on the deck, two on each side of the ship near midships, their centres being as far as possible from the centreline. The total mass of A, B, C and D should not be greater than that required to heel the vessel by approximately

1°. Beyond 1° of heel, the assumption that the metacentre remains on the upright centreline is no longer valid, and the geometry used to establish *GM* no longer holds. The ship should be floating with minimal static heel and any remaining non-essential equipment or people are removed from the vessel. The mooring ropes are slackened and the ship-to-shore gangway removed.

In this level position, a small dinghy is used to take freeboard measurements and samples of water density around the vessel. A number of freeboard measurements should be taken along the length of the vessel, rather than just the forward and aft. Multiple readings of freeboard will allow the naval architect to account not only for the static trim of the vessel but also if there is any hogging or sagging occurring. The water density is measured at a number of points around the vessel using the marine hydrometer described in Chapter 2.

The inclining masses are then moved, one at a time, across the ship until all four are on one side, then all four on the other side, and finally two on each side. At least two pendulums are required, one forward and one aft, to measure the angle of heel. They are made as long as possible and are suspended from some convenient point, eg the underside of the hatch. The pendulum bobs are immersed in water or light oil to dampen the swing. The container of oil or water should be large enough to allow for the expected deflection of the pendulum and a wooden batten is fixed to the container so that the pendulum position can be recorded throughout the experiment. Modern electronic inclinometers can be used to measure the angle, but their use should be approved by the relevant surveyor in advance, and typically the MCA would not allow solely electronic measurement devices.

The deflections of the pendulums are recorded for each movement of mass, and thus values for tan θ against heeling moment can be plotted, figure 4.24 (Basis). The slope of this graph when divided by displacement will give the value of GM.

Some typical sources of error in the inclining experiment are as follows. Any movement of liquid will affect the results and therefore all tanks should be empty or pressed up tight, with all valves shut. Care should also be taken with large sailing vessels, where the booms or yards can move from side to side when heeling. Mooring lines that restrict the heel are to be avoided. The general result of all of these sources of error is the second line in figure 4.24 (Moving Mass), which will not be linear.

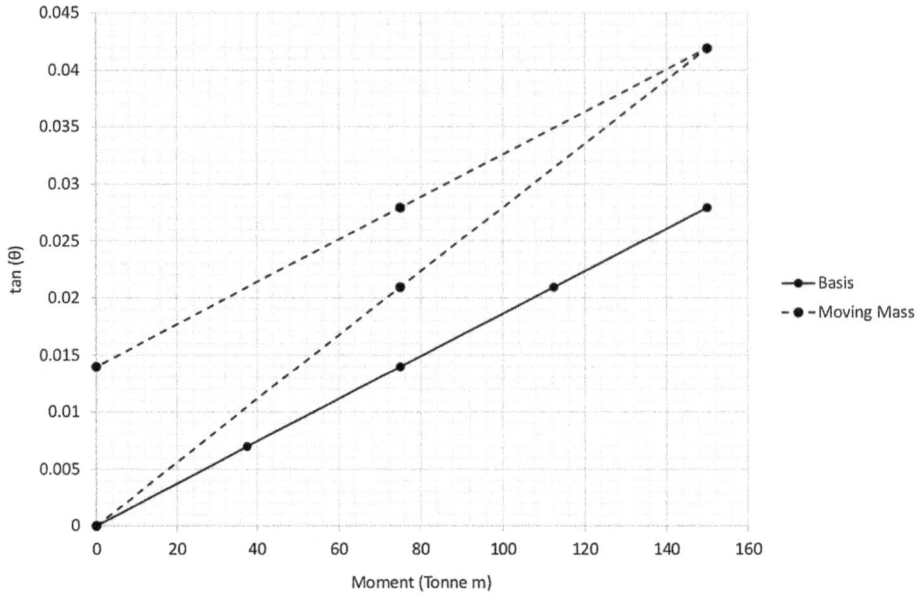

▲ **Figure 4.24** *Graph of tan (θ) versus heeling moment*

# Chapter 4 Test Examples

**Q. 4.1:** A ship displaces 12000 tonne, its centre of gravity is 6.50 m above the keel and its centre of buoyancy is 3.60 m above the keel. If the second moment of area of the waterplane about the centreline is $42.5 \times 10^3$ m$^4$, find the metacentric height.

**Q. 4.2:** A vessel of 10000 tonne displacement has a second moment of area of waterplane about the centreline of $60 \times 10^3$ m$^4$. The centre of buoyancy is 2.75 m above the keel. The following are the disposition of the masses on board the ship.
4000 tonne 6.30 m above the keel
2000 tonne 7.50 m above the keel
4000 tonne 9.15 m above the keel
Calculate the metacentric height.

**Q. 4.3:** A vessel of constant rectangular cross section has a breadth of 12 m and metacentric height of one-quarter of the draught. The vertical centre of gravity lies on the waterline. Calculate the draught.

**Q. 4.4:** A raft is made from two cylinders each 1.5 m diameter and 6 m long. The distance between the centres of the cylinders is 3 m. If the draught is 0.75 m, calculate the transverse *BM*.

**Q. 4.5:** A vessel of constant rectangular cross section is 7.2 m wide.
  **(a)** Draw the metacentric diagram using 0.5 m intervals of draught up to the 4 m waterline.
  **(b)** If the centre of gravity is 3 m above the keel, determine from the metacentric diagram the limits of draught between which the vessel will be unstable.

**Q. 4.6:** A vessel of constant triangular cross section is 9 m wide at the deck and has a depth to deck of 7.5 m. Draw the metacentric diagram using 0.5 m intervals of draught up to the 3 m waterline.

**Q. 4.7:** An inclining experiment was carried out on a ship of 8000 tonne displacement. A mass of 10 tonne was moved 14 m across the deck, causing a pendulum 8.5 m long to deflect 110 mm. The transverse metacentre was 7.15 m above the keel. Calculate the metacentric height and the height of the centre of gravity above the keel.

**Q. 4.8:** An inclining experiment was carried out on a ship of 4000 tonne displacement, when masses of 6 tonne were moved transversely through 13.5 m. The deflections of a 7.5 m pendulum were 81, 78, 85, 83, 79, 82, 84 and 80 mm respectively.
  Calculate the metacentric height.

**Q. 4.9:** A ship of 5000 tonne displacement has a rectangular double bottom tank 9 m wide and 12 m long, half full of sea water. Calculate the virtual reduction in metacentric height due to free surface.

**Q. 4.10:** A ship of 8000 tonne displacement has its centre of gravity 4.5 m above the keel and transverse metacentre 5 m above the keel when a rectangular tank 7.5 m long and 15 m wide contains sea water. A mass of 10 tonne is moved 12 m across the deck. Calculate the angle of heel:
  **a)** if there is no free surface of water
  **b)** if the water does not completely fill the tank.

**Q. 4.11:** A ship of 6000 tonne displacement has its centre of gravity 5.9 m above the keel and transverse metacentre 6.8 m above the keel. A rectangular double bottom tank 10.5 m long, 12 m wide and 1.2 m deep is now half-filled with sea water. Calculate the metacentric height.

**Q. 4.12:** An oil tanker 24 m wide displaces 25000 tonne when loaded in nine equal tanks, each 10 m long, with oil rd 0.8. Calculate the total free surface effect with:
  **a)** no longitudinal divisions

    **b)** a longitudinal centreline bulkhead

    **c)** twin longitudinal bulkheads, forming three equal tanks

    **d)** twin longitudinal bulkheads, the centre compartment having a width of 12 m.

**Q. 4.13:** A ship of 12500 tonne displacement and 15 m beam has a metacentric height of 1.1 m. A mass of 80 tonne is lifted from its position in the centre of the lower hold by one of the ship's derricks, and placed on the quay 2 m from the ship's side. The ship heels to a maximum angle of 3.5° when the mass is being moved.

    **a)** Does the *GM* alter during the operation?

    **b)** Calculate the height of the derrick head above the original centre of gravity of the mass.

**Q. 4.14:** The righting levers of a ship, for an assumed *KG* of 3.5 m, are 0, 0.25, 0.46, 0.51, 0.39, 0.10 and −0.38 m at angles of heel of 0°, 15°, 30°, 45°, 60°, 75° and 90° respectively.

When the ship is loaded to the same displacement, the centre of gravity is 3 m above the keel and the metacentric height 1.25 m. Draw the amended curve of statical stability.

**Q. 4.15:** The righting moments of a ship at angles of heel of 0°, 15°, 30°, 45° and 60° are 0, 1690, 5430, 9360 and 9140 kN m respectively. Calculate the dynamical stability at 60°.

**Q. 4.16:** A ship of 18000 tonne displacement has *KB* 5.25 m, *KG* 9.24 m and second moment of area about the centreline of $82 \times 10^3$ m⁴.

Using the wall-sided formula, calculate the righting levers at intervals of 5° heel up to 20° and sketch the stability curve up to this angle.

**Q. 4.17:** A ship of 7200 tonne displacement has *KG* 5.2 m, *KB* 3.12 m and *KM* 5.35 m. 300 tonne of fuel at *Kg* 0.6 m are now used. Ignoring free surface effect and assuming the *KM* remains constant, calculate the angle to which the vessel will heel.

# 5

# TRIM

## Definitions

**TRIM** is the difference between the draughts forward and aft. Thus, if a ship floats at draughts of 6 m forward and 7 m aft, it is said to trim 1 m by the *stern*. If the draught forward is greater than the draught aft, the vessel is said to trim by the *head*.

**CENTRE OF FLOTATION (LCF)** is the centroid of the waterplane and is the axis about which a ship changes trim when a mass is added, removed or moved longitudinally.

If a small mass $m$ is added to a ship at the centre of flotation, there is an increase in mean draught but no change in trim, since the centre of gravity of the added mass is at the same position as the centre of the added layer of buoyancy. A large mass (eg one exceeding, say, $1/20$ of the displacement) will cause a considerable increase in draught and hence a change in waterplane area and centre of flotation.

**MEAN DRAUGHT** is the draught at which the vessel would lie in level keel conditions. Since the vessel changes trim about the LCF, the draught at this point remains constant for any given displacement, whether the vessel is at level keel or trimmed. Hence the mean draught may be taken as the draught at the LCF.

The mean of the end draughts may be compared with the actual draught amidships to determine whether the vessel is hogging or sagging, but is of little relevance in hydrostatic calculations.

# Effect of Adding Small Masses

It is useful to assume that when a small mass is added to the ship, it is first placed at the centre of flotation and then moved forward or aft to its final position. Thus the effect of an added mass on the draughts may be divided into:

a) a bodily increase in draught

b) a change in trim due to the movement of the mass from the centre of flotation to its final position.

The bodily increase in draught may be found by dividing the mass by the TPC.

The change in trim due to any longitudinal movement of mass may be found by considering its effect on the centre of gravity of the ship.

Consider a ship of displacement $\Delta$ and length $L$, lying at waterline WL and having a mass $m$ on the deck (figure 5.1). The centre of gravity $G$ and the centre of buoyancy $B$ lie in the same vertical line.

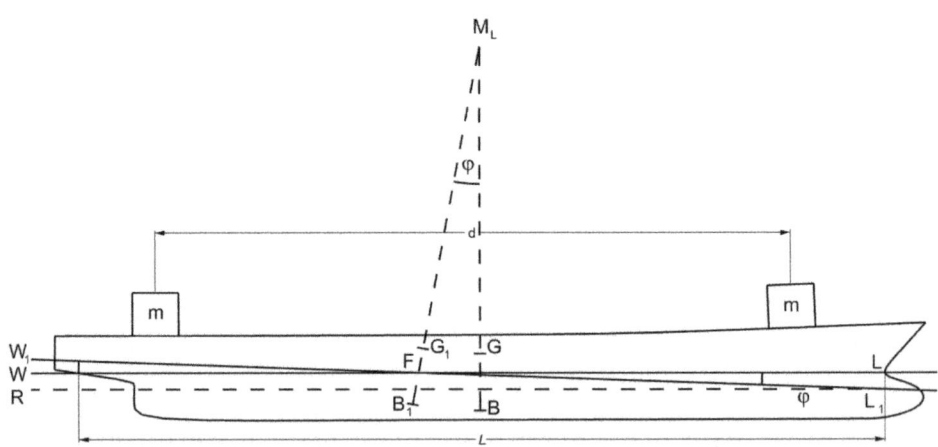

▲ **Figure 5.1** *Longitudinal movement of mass*

If the mass is moved a distance $d$ aft, the centre of gravity moves aft from $G$ to $G_1$, and:

$$GG_1 = \frac{m \times d}{\Delta}$$

The ship then changes trim through the centre of flotation $F$ until it lies at waterline $W_1L_1$. This change in trim causes the centre of buoyancy to move aft from $B$ to $B_1$, in the same vertical line as $G_1$. The vertical through $B_1$ intersects the original vertical through $B$ at $M_L$, the *longitudinal metacentre*. $GM_L$ is known as the *longitudinal metacentric height*:

$$GM_L = KB + BM_L - KG$$

$$BM_L = \frac{I_F}{\nabla}$$

Where $I_F$ = second moment of area of the waterplane about a transverse axis through the centre of flotation $F$. If the vessel trims through an angle $\varphi$, then

$$GG_1 = GM_L \tan\varphi$$

And

$$GM_L \tan\varphi = \frac{m \times d}{\Delta}$$

$$\tan\varphi = \frac{m \times d}{\Delta \times GM_L}$$

Draw $RL_1$ parallel to WL.

$$\text{Change in trim} = W_1W + LL_1$$
$$= W_1R$$
$$= \frac{t}{100}$$

Where $t$ = change in trim in cm over length $L$ m.

But

$$\tan\varphi = \frac{t}{100L}$$

$$\frac{t}{100L} = \frac{m \times d}{\Delta \times GM_L}$$

$$t = \frac{m \times d \times 100L}{\Delta \times GM_L} \text{ cm}$$

The change in trim may therefore be calculated from this expression. $m \times d$ is known as the *trimming moment*.

It is useful to know the moment that will cause a change in trim of 1 cm.

$$m \times d = \frac{t \times \Delta \times GM_L}{100L} \text{ tonne m}$$

Let $t = 1$ cm

Then moment to change trim 1 cm

$$MCT\ 1\ cm = \frac{\Delta \times GM_L}{100L}\ \text{tonne m}$$

$$t = \frac{\text{trimming moment}}{MCT\ 1\ cm}\ cm$$

$$\text{Change in trim} = \frac{m \times d}{MCT\ 1\ cm}\ \text{cm by the stern}$$

It is now possible to determine the effect of this change in trim on the end draughts. Since the vessel changes trim by the stern, the forward draught will be reduced while the after draught will be increased.

By similar triangles:

$$\frac{t}{L} = \frac{LL_1}{FL} = \frac{W_1W}{WF}$$

$t$, $LL_1$ and $W_1W$ may be expressed in cm while $L$, FL and WF are expressed in m.

$$\text{Change in draught forward}\quad LL_1 = -\frac{t}{L} \times FL\ \text{cm}$$

$$\text{Change in draught aft}\quad W_1W = +\frac{t}{L} \times WF\ \text{cm}$$

EXAMPLE: A ship of 5000 tonne displacement, 96 m long, floats at draughts of 5.6 m forward and 6.3 m aft. The TPC is 11.5, $GM_L$ 105 m and centre of flotation 2.4 m aft of midships.

Calculate:

**a)** the MCT 1 cm

**b)** the new end draughts when 88 tonne are added 31 m forward of midships.

a)

$$MCT\ 1\ cm = \frac{\Delta \times GM_L}{100L}$$
$$= \frac{5000 \times 105}{100 \times 96}$$
$$= 54.69\ \text{tonne m}$$

b)

$$\text{Bodily sinkage} = \frac{88}{11.5}$$
$$= 7.65\ \text{cm}$$
$$d = 31 + 2.4$$
$$= 33.4\ \text{m from } F$$

$$\text{Trimming moment} = 88 \times 33.4 \text{ tonne m}$$

$$\text{Change in trim} = \frac{88 \times 33.4}{54.69}$$

$$= 53.74 \text{ cm by the head}$$

$$\text{Distance from } F \text{ to fore end} = \frac{96}{2} + 2.4$$

$$= 50.4 \text{ m}$$

$$\text{Distance from } F \text{ to aft end} = \frac{96}{2} - 2.4$$

$$= 45.6 \text{ m}$$

$$\text{Change in trim forward} = +\frac{53.74}{96} \times 50.4$$

$$= +28.21 \text{ cm}$$

$$\text{Change in trim aft} = -\frac{53.74}{96} \times 45.6$$

$$= -25.53 \text{ cm}$$

$$\text{New draught forward} = 5.60 + 0.076 + 0.282$$

$$= 5.958 \text{ m}$$

$$\text{New draught aft} = 6.30 + 0.076 - 0.255$$

$$= 6.121 \text{ m}$$

If a number of items are added to the ship at different positions along its length, the total mass and net trimming moment may be used to determine the final draughts.

EXAMPLE: A ship 150 m long has draughts of 7.7 m forward and 8.25 m aft, MCT 1 cm 250 tonne m, TPC 26 and LCF 1.8 m forward of midships. Calculate the new draughts after the following masses have been added:

50 tonne, 70 m aft of midships

170 tonne, 36 m aft of midships

100 tonne, 5 m aft of midships

130 tonne, 4 m forward of midships

40 tonne, 63 m forward of midships

| Mass (tonne) | Distance from $F$ (m) | Moment forward (tonne m) | Moment aft (tonne m) |
|---|---|---|---|
| 50 | 71.8A | — | 3590 |
| 170 | 37.8A | — | 6426 |
| 100 | 6.8A | — | 680 |

| Mass (tonne) | Distance from $F$ (m) | Moment forward (tonne m) | Moment aft (tonne m) |
|---|---|---|---|
| 130 | 2.2F | 286 | — |
| 40 | 61.2F | 2448 | — |
| **Total**  490 | | 2734 | 10696 |

$$\text{Excess moment aft} = 10696 - 2734$$
$$= 7962 \text{ tonne m}$$
$$\text{Change in trim} = \frac{7962}{250}$$
$$= 31.85 \text{ cm by the stern}$$
$$\text{Change in trim forward} = -\frac{31.85}{150}\left(\frac{150}{2} - 1.8\right)$$
$$= -15.54 \text{ cm}$$
$$\text{Change in trim aft} = +\frac{31.85}{150}\left(\frac{150}{2} + 1.8\right)$$
$$= +16.31 \text{ cm}$$
$$\text{Bodily sinkage} = \frac{490}{26}$$
$$= 18.85 \text{ cm}$$
$$\text{New draught forward} = 7.70 + 0.1885 - 0.155$$
$$= 7.734 \text{ m}$$
$$\text{New draught aft} = 8.25 + 0.1885 + 0.155$$
$$= 8.594 \text{ m}$$

# Determination of Draughts After the Addition of Large Masses

When a large mass is added to a ship, the resultant increase in draught is sufficient to cause changes in all the hydrostatic details. It then becomes necessary to calculate the final draughts from first principles. Such a problem exists every time a ship loads or discharges the major part of its deadweight.

The underlying principle is that after loading or discharging, the vessel is in equilibrium and hence the final centre of gravity is in the same vertical line as the final centre of buoyancy.

For any given condition of loading, it is possible to calculate the displacement Δ and the longitudinal position of the centre of gravity G relative to midships.

From the hydrostatic curves or data, the mean draught may be obtained at this displacement, and hence the value of MCT 1 cm and the distance of the LCB and LCF from midships. These values are calculated for the level keel condition and it is unlikely that the LCB will be in the same vertical line as G. Thus a trimming moment acts on the ship. This trimming moment is the displacement multiplied by the longitudinal distance between B and G, known as the *trimming lever*.

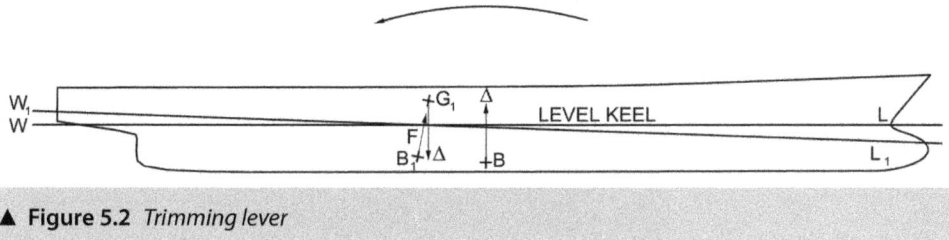

▲ **Figure 5.2** *Trimming lever*

The trimming moment, divided by the MCT 1 cm, gives the change in trim from the level keel condition, ie the total trim of the vessel. The vessel changes trim about the LCF and hence it is possible to calculate the end draughts. When the vessel has changed trim in this manner, the new centre of buoyancy $B_1$ lies in the same vertical line as $G_1$.

EXAMPLE: A ship 125 m long has a light displacement of 4000 tonne with LCG 1.60 m aft of midships. The following items are now added:

Cargo 8500 tonne LCG 3.9 m forward of midships

Fuel 1200 tonne LCG 3.1 m aft of midships

Water 200 tonne LCG 7.6 m aft of midships

Stores 100 tonne LCG 30.5 m forward of midships.

At 14000 tonne displacement, the mean draught is 7.8 m, MCT 1 cm 160 tonne m, LCB 2 m forward of midships and LCF 1.5 m aft of midships.

Calculate the final draughts.

| Item | Mass (t) | Lcg (m) | Moment forward | Moment aft |
|------|----------|---------|----------------|------------|
| Cargo | 8500 | 3.9F | 33150 | — |
| Fuel | 1200 | 3.1A | — | 3720 |

| Item | Mass (t) | Lcg (m) | Moment forward | Moment aft |
|------|---------|---------|----------------|-----------|
| Water | 200 | 7.6A | — | 1520 |
| Stores | 100 | 30.5F | 3050 | — |
| Lightweight | 4000 | 1.6A | — | 6400 |
| Displacement | 14000 | | 36200 | 11640 |

$$\text{Excess moment forward} = 36200 - 11640$$
$$= 24560 \text{ tonne m}$$
$$\text{LCG from midships} = \frac{24560}{14000}$$
$$= 1.754 \text{ m forward}$$
$$\text{LCB from midships} = 2.000 \text{ m forward}$$
$$\text{trimming lever} = 1.754 - 2.000$$
$$= 0.246 \text{ m aft}$$
$$\text{trimming moment} = 14000 \times 0.246 \text{ tonne m}$$
$$\text{trim} = \frac{14000 \times 0.246}{160}$$
$$\text{Change in draught forward} = -\frac{21.5}{125}\left(\frac{125}{2} + 1.5\right)$$
$$= -11.0 \text{ cm}$$
$$\text{Change in draught aft} = +\frac{21.5}{125}\left(\frac{125}{2} - 1.5\right)$$
$$= +10.5 \text{ cm}$$
$$\text{Draught forward} = 7.80 - 0.110$$
$$= 7.690 \text{ m}$$
$$\text{Draught aft} = 7.80 + 0.105$$
$$= 7.905 \text{ m}$$

# Change in Mean Draught Due to Change in Density

The displacement of a ship floating freely at rest is equal to the mass of the volume of water that it displaces. For any given displacement, the volume of water displaced must depend upon the density of the water. When a ship moves from sea water into river water without change in displacement, there is a slight increase in draught.

Consider a ship of displacement $\Delta$ tonne, waterplane area $A_w$ m², which moves from sea water of $\rho_S$ t/m³ into river water of $\rho_R$ t/m³ without change in displacement.

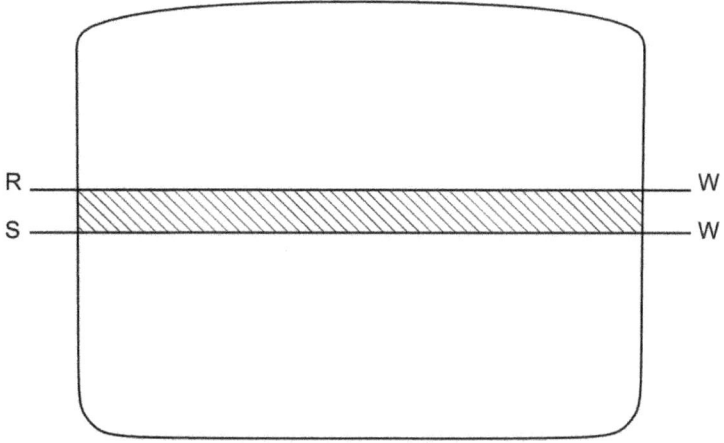

▲ **Figure 5.3** *Change in draught due to change in density*

Volume of displacement in sea water:

$$\nabla_S = \frac{\Delta}{\rho_S}\,m^3$$

Volume of displacement in river water:

$$\nabla_R = \frac{\Delta}{\rho_R}\,m^3$$

Change in volume of displacement:

$$\begin{aligned}
v &= \nabla_R - \nabla_S \\
&= \frac{\Delta}{\rho_R} - \frac{\Delta}{\rho_S} \\
&= \Delta\left(\frac{1}{\rho_R} - \frac{1}{\rho_S}\right)
\end{aligned}$$

This change in volume causes an increase in draught. Since the increase is small, the waterplane area may be assumed to remain constant and the increase in mean draught may therefore be found by dividing the change in volume by the waterplane area.

$$\text{Increase in draught} = \frac{\Delta}{A_W}\left(\frac{1}{\rho_R} - \frac{1}{\rho_S}\right)\text{m}$$

$$= \frac{100\Delta}{A_W}\left(\frac{1}{\rho_R} - \frac{1}{\rho_S}\right)\text{cm}$$

Tonnes per cm immersion is given by:

$$\text{TPC} = \frac{A_W}{100} \times \rho_S$$

Therefore:

$$A_W = 100 \times \frac{\text{TPC}}{\rho_S}\text{m}^2$$

Substituting for $A_W$ in the formula for increase in draught:

$$\text{Increase in draught} = \frac{100\Delta\rho_S}{100\text{TPC}}\left(\frac{\rho_S - \rho_R}{\rho_S \times \rho_R}\right)$$

$$= \frac{\Delta}{\text{TPS}}\left(\frac{\rho_S - \rho_R}{\rho_R}\right)$$

A particular case occurs when a ship moves from sea water of 1.025 t/m³ into fresh water of 1.000 t/m³, the TPC being given in the sea water.

$$\text{Increase in draught} = \frac{\Delta}{\text{TPC}}\left(\frac{1.025 - 1.000}{1.000}\right)$$

$$\text{Increase in draught} = \frac{\Delta}{40\text{TPC}}$$

This is known as the *fresh water allowance,* used when computing the freeboard of a ship, and is the difference between the S line and the F line on the freeboard markings.

EXAMPLE: A ship of 10000 tonne displacement has a waterplane area of 1300 m². The ship loads in water of 1.010 t/m³ and moves into water of 1.026 t/m³. Find the change in mean draught.

Since the vessel moves into water of a greater density, there will be a reduction in mean draught.

$$\text{Reduction in draught} = \frac{100\Delta}{A_W}\left(\frac{\rho_s - \rho_R}{\rho_s \times \rho_R}\right) \text{ cm}$$

$$= \frac{100 \times 10000}{1300}\left(\frac{1.026 - 1.010}{1.026 \times 1.010}\right) \text{ cm}$$

$$= 11.88 \text{ cm}$$

When a vessel moves from water of one density to water of a different density, there may be a change in displacement due to the consumption of fuel and stores, causing an additional change in mean draught. If the vessel moves from sea water into river water, it is possible in certain circumstances for the increase in draught due to change in density to be equal to the reduction in draught due to the removed mass. In such a case, there will be no change in mean draught.

EXAMPLE: 215 tonne of oil fuel and stores are used in a ship while passing from sea water of 1.026 t/m³ into river water of 1.002 t/m³. If the mean draught remains unchanged, calculate the displacement in the river water.

Let $\Delta$ = displacement in river water

Then $\Delta + 215$ = displacement in sea water

Since the draught remains unaltered, the volume of displacement in the river water must be equal to the volume of displacement in the sea water.

$$\nabla_R = \frac{\Delta}{\rho_R}$$

$$= \frac{\Delta}{1.002} \text{ m}^3$$

$$\nabla_S = \frac{\Delta + 215}{\rho_S}$$

$$= \frac{\Delta + 215}{1.026} \text{ m}^3$$

Hence:

$$\nabla_R = \nabla_S$$

$$\frac{\Delta}{1.002} = \frac{\Delta + 215}{1.026}$$

$$1.026\Delta = 1.002\Delta + 1.002 \times 215$$

$$\Delta = \frac{1.002 \times 215}{0.024}$$

$$= 8976 \text{ tonne}$$

# Change in Trim Due to Change in Density

If a box barge passes from sea water into river water, or vice versa, without change in displacement, there will be a change in draught. As the barge is symmetric, and the centre of buoyancy is in line with the centre of flotation, there is no trimming moment, and the barge stays level. This does not happen in the case of a ship, as due to its shape, the centre of flotation is not in line with the centre of buoyancy, and there will be a change in trim in addition to the change in mean draught. This change in trim is always small.

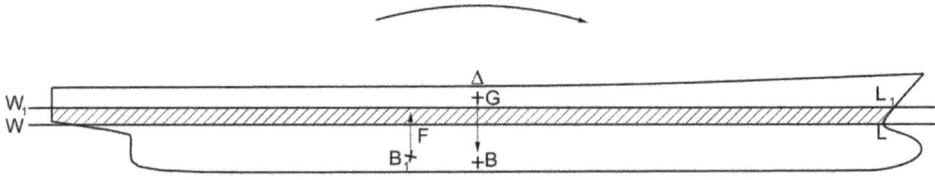

▲ **Figure 5.4** *Change in trim due to change in density*

Consider a ship of displacement Δ lying at waterline WL in sea water of density $\rho_S$ t/m³. The centre of gravity G and the centre of buoyancy B are in the same vertical line.

If the vessel now moves into river water of $\rho_R$ t/m³, there is a bodily increase in draught and the vessel lies at waterline $W_1L_1$. The volume of displacement has been increased by a layer of volume $v$, whose centre of gravity is at the centre of flotation F. This causes the centre of buoyancy to move from B to $B_1$, the centre of gravity remaining at G.

Volume of displacement in sea water:

$$\nabla_{SW} = \frac{\Delta}{\rho_{SW}} \ m^3$$

Volume of displacement in river water:

$$\nabla_{RW} = \frac{\Delta}{\rho_{RW}} \ m^3$$

Change in volume of displacement:

$$v = \nabla_{RW} - \nabla_{SW}$$

$$= \Delta\left(\frac{1}{\rho_{RW}} - \frac{1}{\rho_{SW}}\right)$$

$$= \Delta\left(\frac{\rho_{SW} - \rho_{RW}}{\rho_{RW}\,\rho_{SW}}\right) m^3$$

Shift in centre of buoyancy:

$$BB_1 = \frac{v \times FB}{\nabla_{RW}}$$

$$= \Delta\left(\frac{\rho_{SW} - \rho_{RW}}{\rho_{RW}\,\rho_{SW}}\right) FB \times \frac{\rho_{RW}}{\Delta}$$

$$= FB\left(\frac{\rho_{SW} - \rho_{RW}}{\rho_{SW}}\right) m$$

Since $B_1$ is no longer in line with $G$:

Moment causing trim by head $= \Delta \times BB_1$

$$\text{Change in trim} = \frac{\Delta \times BB_1}{MCT\ 1\ cm}\ cm$$

$$= \frac{\Delta FB}{MCT\ 1\ cm}\left(\frac{\rho_{SW} - \rho_{RW}}{\rho_{RW}}\right)\ cm\ \text{by the head}$$

*Note:* If the ship moves from the river water into sea water, it will change trim by the stern, and:

$$= \frac{\Delta FB}{MCT\ 1\ cm}\left(\frac{\rho_{SW} - \rho_{RW}}{\rho_{RW}}\right)\ cm\ \text{by the stern}$$

EXAMPLE: A ship 120 m long and 9100 tonne displacement floats at a level keel draught of 6.5 m in fresh water of 1.000 t/m³. MCT 1 cm 130 tonne m, TPC in sea water 16.5, LCB 2.3 m forward of midships. LCF 0.6 m aft of midships.

Calculate the new draughts if the vessel moves into sea water of 1.024 t/m³ without change in displacement.

$$\text{Reduction in mean draught} = \frac{\Delta}{TPC}\left(\frac{\rho_{SW} - \rho_{RW}}{\rho_{RW}}\right)$$

$$= \frac{9100}{16.5}\left(\frac{1.025 - 1.000}{1.000}\right)$$

$$= 13.24\ cm$$

$$\text{Change in trim} = \frac{\Delta FB}{\text{MCT 1 cm}}\left(\frac{\rho_{sw} - \rho_{RW}}{\rho_{RW}}\right)$$

$$= \frac{9100 \times (2.30 + 0.60)}{130}\left(\frac{1.024 - 1.000}{1.000}\right)$$

$$= 4.87 \text{ cm}$$

$$\text{Change forward} = -\frac{4.87}{120}\left(\frac{120}{2} + 0.6\right)$$

$$= -2.46 \text{ cm}$$

$$\text{Change aft} = -\frac{4.87}{120}\left(\frac{120}{2} - 0.6\right)$$

$$= +2.41 \text{ cm}$$

$$\text{New draught forward} = 6.50 - 0.132 - 0.025$$

$$= 6.343 \text{ m}$$

$$\text{New draught aft} = 6.50 - 0.132 - 0.024$$

$$= 6.392 \text{ m}$$

# Change in Mean Draught Due to Bilging

**Buoyancy** is the upthrust exerted by the water on the ship and depends upon the volume of water displaced by the ship up to the waterline.

**Reserve buoyancy** is the potential buoyancy of a ship and depends upon the intact, watertight volume above the waterline. When a mass is added to a ship, or buoyancy is lost due to bilging, the reserve buoyancy is converted into buoyancy by increasing the draught. If the loss in buoyancy exceeds the reserve buoyancy, the vessel will sink.

**Permeability** $\mu$ is the volume of a compartment into which water may flow if the compartment is laid open to the sea, expressed as a ratio or percentage of the total volume of the compartment. Thus, if a compartment is completely empty, the permeability is 100%. The permeability of a machinery space is about 85% and accommodation about 95%. The permeability of a cargo hold varies considerably with the type of cargo, but an average value may be taken as 60%.

The effects of bilging a mid-length compartment may be shown most simply by considering a box barge of length $L$, breadth $B$ and draught $d$, having a mid-length compartment of length $l$, permeability $\mu$.

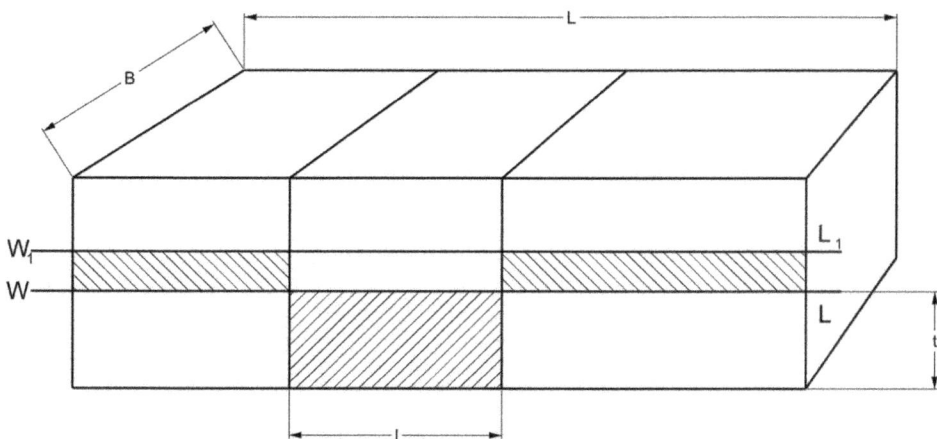

▲ **Figure 5.5** *Bilging of a compartment*

If this compartment is bilged, buoyancy is lost and must be replaced by increasing the draught. The volume of buoyancy lost is the volume of the compartment up to waterline WL, less the volume of water excluded by the cargo in the compartment.

$$\text{Volume of lost buoyancy} = \mu l B t$$

This is replaced by the increase in draught multiplied by the area of the intact part of the waterplane, ie the area of waterplane on each side of the bilged compartment plus the area of cargo that projects through the waterplane in the bilged compartment.

$$\begin{aligned}
\text{Area of intact waterplane} &= (L - l)B + lB(1 - \mu) \\
&= LB - lB + lB - \mu l B \\
&= (L - \mu l)B
\end{aligned}$$

$$\begin{aligned}
\text{Increase in draught} &= \frac{\text{volume of lost buoyancy}}{\text{area of intact waterplane}} \\
&= \frac{\mu l B t}{(L - \mu l)B} \\
&= \frac{\mu l t}{L - \mu l}
\end{aligned}$$

$\mu l$ may be regarded as the *effective length* of the bilged compartment.

EXAMPLE: A box barge 30 m long and 8 m beam floats at a level keel draught of 3 m and has a mid-length compartment 6 m long. Calculate the new draught if this compartment is bilged:

**a)** with $\mu = 100\%$

**b)** with $\mu = 75\%$.

a)

$$\text{Volume of lost buoyancy} = 6 \times 8 \times 3 \text{ m}^3$$
$$\text{Area of intact waterplane} = (30 - 6) \times 8 \text{ m}^2$$
$$\text{Increase in draught} = \frac{6 \times 8 \times 3}{24 \times 8}$$
$$= 0.75 \text{ m}$$
$$\text{New draught} = 3 + 0.75$$
$$= 3.75 \text{ m}$$

b)

$$\text{Volume of lost buoyancy} = 0.75 \times 6 \times 8 \times 3 \text{ m}^3$$
$$\text{Area of intact waterplane} = (30 - 0.75 \times 6) \times 8 \text{ m}^2$$
$$\text{Increase in draught} = \frac{0.75 \times 6 \times 8 \times 3}{25.5 \times 8}$$
$$= 0.529 \text{ m}$$
$$\text{New draught} = 3 + 0.529$$
$$= 3.529 \text{ m}$$

# Change in Draughts Due to Bilging an End Compartment

If a bilged compartment does not lie at the mid-length, then there is a change in trim in addition to the change in mean draught.

Consider a box barge of length $L$, breadth $B$ and draught $t$, having an empty compartment of length $l$ at the extreme fore end.

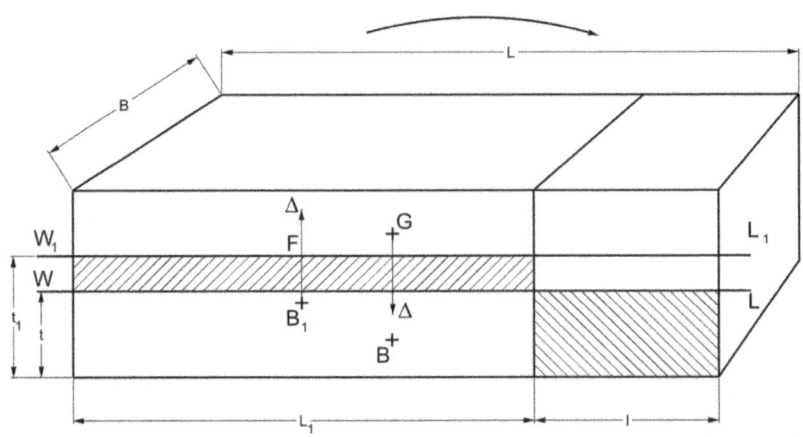

▲ **Figure 5.6** *Bilging of an end compartment*

Before bilging, the vessel lies at waterline WL, the centre of gravity $G$ and the centre of buoyancy $B$ lying in the same vertical line.

After bilging the end compartment, the vessel lies initially at waterline $W_1L_1$. The new mean draught $t_1$ may be calculated as shown previously, assuming that the compartment is amidships.

The volume of lost buoyancy has been replaced by a layer whose centre is at the middle of the length $L_1$. This causes the centre of buoyancy to move aft from $B$ to $B_1$, a distance of ½ $l$. Thus, a moment of $\Delta \times BB_1$ acts on the ship, causing a *considerable* change in trim by the head. The vessel changes trim about the centre of flotation $F$, which is the centroid of the *intact* waterplane, ie the midpoint of $L_1$.

$$\text{Trimming moment} = \Delta \times BB_1$$

$$\text{Change in trim} = \frac{\Delta \times BB_1}{\text{MCT 1 cm}} \text{ cm by the head}$$

$$\text{MCT 1 cm} = \frac{\Delta \times GM_L}{100L} \text{ tonne m}$$

$GM_L$ must be calculated for the *intact* waterplane

$$KB_1 = \frac{t}{2}$$

$$B_1M_L = \frac{L_1^3 B}{12\nabla}$$

Where:

$$\Delta = LBt$$
$$= L_1Bt_1$$
$$GM_L = KB_1 + B_1M_L - KG$$

$$\text{Change in trim} = \frac{\Delta \times \frac{1}{2}l}{\Delta \times GM_L} \times 100L$$

$$= \frac{50Ll}{GM_L} \text{ cm by the head}$$

EXAMPLE: A box barge 120 m long and 8 m beam floats at an even keel draught of 3 m and has an empty compartment 6 m long at the extreme fore end. The centre of gravity is 2.8 m above the keel. Calculate the final draughts if this compartment is bilged.

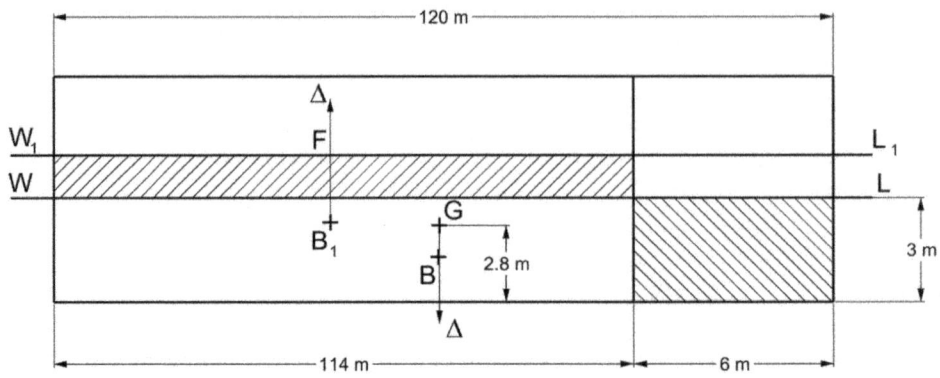

$$\text{Increase in mean draught} = \frac{6 \times 8 \times 3}{(120-6) \times 8}$$

$$= 0.158 \text{ m}$$

$$\text{New draught } t_1 = 3.158 \text{ m}$$

$$KB_1 = \frac{t_1}{2}$$

$$= 1.579 \text{ m}$$

$$B_1 M_L = \frac{114^3 \times 8}{12 \times 120 \times 8 \times 3}$$

$$= 342.94 \text{ m}$$

$$GM_L = 1.58 + 342.94 - 2.80$$

$$= 341.72 \text{ m}$$

$$\text{Change in trim} = \frac{50 \times 120 \times 6}{341.72}$$

$$= 105.3 \text{ cm by the head}$$

$$\text{Change forward} = +\frac{105.3}{120} \times \left(\frac{120}{2} + 3\right)$$

$$= +55.3 \text{ cm}$$

$$\text{Change aft} = -\frac{105.3}{120} \times \left(\frac{120}{2} - 3\right)$$

$$= -50.0 \text{ cm}$$

$$\text{New draught forward} = 3.158 + 0.553$$

$$= 3.711 \text{ m}$$

$$\text{New draught aft} = 3.158 - 0.500$$

$$= 2.658 \text{ m}$$

# Chapter 5 Test Examples

**Q. 5.1:** A ship 125 m long displaces 12000 tonne. When a mass of 100 tonne is moved 75 m from forward to aft there is a change in trim of 65 cm by the stern. Calculate:

**a)** MCT 1 cm

**b)** the longitudinal metacentric height

**c)** the distance moved by the centre of gravity of the ship.

**Q. 5.2:** A ship 120 m long floats at draughts of 5.5 m forward and 5.8 m aft; MCT 1 cm 80 tonne m, TPC 13, LCF 2.5 m forward of midships. Calculate the new draughts when a mass of 110 tonne is added 24 m aft of midships.

**Q. 5.3:** A ship 130 m long displaces 14000 tonne when floating at draughts of 7.5 m forward and 8.1 m aft. $GM_L$ 125 m, TPC 18, LCF 3 m aft of midships.

Calculate the final draughts when a mass of 180 tonne lying 40 m aft of midships is removed from the ship.

**Q. 5.4:** The draughts of a ship 90 m long are 5.8 m forward and 6.4 m aft. MCT 1 cm 50 tonne m; TPC 11 and LCF 2 m aft of midships. Determine the point at which a mass of 180 tonne should be placed so that the after draught remains unaltered, and calculate the final draught forward.

**Q. 5.5:** A ship 150 m long floats at draughts of 8.2 m forward and 8.9 m aft. MCT 1 cm 260 tonne m, TPC 28 and LCF 1.5 m aft of midships. It is necessary to bring the vessel to an even keel and a double bottom tank 60 m forward of midships is available. Calculate the mass of water required and the final draught.

**Q. 5.6:** A ship whose length is 110 m has MCT 1 cm 55 tonne m; TPC 9; LCF 1.5 m forward of midships and floats at draughts of 4.2 m forward and 4.45 m aft. Calculate the new draughts after the following masses have been added:

20 tonne 40 m aft of midships

50 tonne 23 m aft of midships

30 tonne 2 m aft of midships

70 tonne 6 m forward of midships

15 tonne 30 m forward of midships.

**Q. 5.7:** The draughts of a ship 170 m long are 6.85 m forward and 7.5 m aft. MCT 1 cm 300 tonne m; TPC 28; LCF 3.5 m forward of midships.

Calculate the new draughts after the following changes in loading have taken place:

160 tonne added 63 m aft of midships

200 tonne added 27 m forward of midships

120 tonne removed 75 m aft of midships

70 tonne removed 16 m aft of midships.

**Q. 5.8:** A ship 80 m long has a light displacement of 1050 tonne and LCG 4.64 m aft of midships.

The following items are then added:

Cargo 2150 tonne, LCG 4.71 m forward of midships

Fuel 80 tonne, LCG 32.55 m aft of midships

Water 15 tonne, LCG 32.90 m aft of midships

Stores 5 tonne, LCG 33.6 m forward of midships.

The following hydrostatic particulars are available:

| Draught m | Displacement tonne | MCT 1 cm tonne m | LCB from midships m | LCF from midships |
|-----------|--------------------|-------------------|---------------------|-------------------|
| 5.00 | 3533 | 43.10 | 1.00F | 1.27A |
| 4.50 | 3172 | 41.26 | 1.24F | 0.84A |

Calculate the final draughts of the loaded vessel.

**Q. 5.9:** A ship of 15000 tonne displacement has a waterplane area of 1950 m². It is loaded in river water of 1.005 t/m³ and proceeds to sea, where the density is 1.022 t/m³.

Calculate the change in mean draught.

**Q. 5.10:** A ship of 7000 tonne displacement has a waterplane area of 1500 m². In passing from sea water into river water of 1005 kg/m³, there is an increase in draught of 10 cm. Find the density of the sea water.

**Q. 5.11:** The $\frac{1}{2}$ ordinates of the waterplane of a ship of 8200 tonne displacement,

90 m long, are 0, 2.61, 3.68, 4.74, 5.84, 7.00, 7.30, 6.47. 5.35, 4.26, 3.16, 1.88 and 0 m respectively. It floats in sea water of 1.024 t/m³. Calculate:

a) TPC

b) mass necessary to increase the mean draught by 12 cm

c) change in mean draught when moving into water of 1.005 t/m³.

**Q. 5.12:** A ship consumes 360 tonne of fuel, stores and water when moving from sea water of 1.025 t/m³ into fresh water of 1.000 t/m³, and on arrival it is found that the draught has remained constant.

Calculate the displacement in the sea water.

**Q. 5.13:** A ship 90 m long displaces 5200 tonne and floats at draughts of 4.95 m forward and 5.35 m aft when in sea water of 1023 kg/m³. The waterplane area is 1100 m², $GM_L$ 95 m, LCB 0.6 m forward of midships and LCF 2.2 m aft of midships.

Calculate the new draughts when the vessel moves into fresh water of 1002 kg/m³.

**Q. 5.14:** A ship of 22000 tonne displacement is 160 m long, MCT 1 cm 280 tonne m, waterplane area 3060 m², centre of buoyancy 1 m aft of midships and centre of flotation 4 m aft of midships. It floats in water of 1.007 t/m³ at draughts of 8.15 m forward and 8.75 m aft.

Calculate the new draughts if the vessel moves into sea water of 1.026 t/m³.

**Q. 5.15:** A box barge 60 m long and 10 m wide floats at an even keel draught of 4 m. It has a compartment amidships 12 m long.

Calculate the new draught if this compartment is laid open to the sea when:

**a)** $\mu$ is 100%

**b)** $\mu$ is 85%

**c)** $\mu$ is 60%.

**Q. 5.16:** A box barge 50 m long and 8 m wide floats at a draught of 3 m and has a mid-length compartment 9 m long containing coal (rd 1.28), which stows at 1.22 m³/t.

Calculate the new draught if this compartment is bilged.

**Q. 5.17:** A vessel of constant rectangular cross section is 60 m long, 12 m beam and floats at a draught of 4.5 m. It has a mid-length compartment 9 m long, which extends right across the ship and up to the deck, but is subdivided by a horizontal watertight flat 3 m above the keel.

Find the new draught if this compartment is bilged:

**a)** below the flat

**b)** above the flat.

**Q. 5.18:** A box barge 25 m long and 4 m wide floats in fresh water at a draught of 1.2 m and has an empty mid-length compartment 5 m long. The bottom of the barge is lined with teak (rd 0.805) 120 mm thick. After grounding, all the teak is torn off and the centre compartment is laid open to the sea. Calculate the final draught.

**Q. 5.19:** A box barge 100 m long, 12 m beam and 4 m draught has a compartment at the extreme fore end 8 m long, subdivided by a horizontal watertight flat 2 m above the keel. The centre of gravity is 3 m above the keel.

Calculate the end draughts if the compartment is bilged:

**a)** at the flat, water flowing into both compartments

**b)** below the flat

**c)** above the flat.

# 6

# FLUID LOADING

In previous editions of this book, the term 'hydrostatics' was used to cover much of the content of this chapter. As this is often used to describe the parameters and calculations associated with flotation, detailed in Chapter 2, this chapter has been titled Fluid Loading, and will solely consider the load on either a ship or structural component due to static fluid pressure. The field of hydrodynamics considers the pressure loading on a ship due to motion and is not required within the Engineering syllabus of merchant seamen.

## Pressure Exerted by a Liquid

Liquid pressure is the load per unit area exerted by the liquid and may be expressed in multiples of $N/m^2$.

eg

$$10^3 N/m^2 = 1 kN/m^2$$
$$10^5 N/m^2 = 1 bar$$

This pressure acts equally in all directions and perpendicular to the surface of any immersed plane.

Consider a trough containing liquid of density $\rho$ kg/m$^3$

Let $A$ = cross-sectional area of a cylinder of this liquid in m$^2$

and $h$ = height of cylinder in m, figure 6.1.

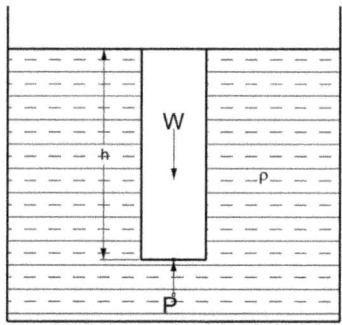

▲ **Figure 6.1** *Pressure exerted by a liquid*

The cylinder is in equilibrium under the action of two vertical forces:

    a)   the gravitational force $W$ acting vertically down.

    b)   the upthrust $P$ exerted by the liquid on the cylinder.

Thus

$$P = W$$

But

$$P = pA$$

Where

$$p = \text{liquid pressure at a depth } h \text{ m}$$
$$W = \rho gAh$$
$$pA = \rho gAh$$
$$P = \rho gh$$

Thus it may be seen that the liquid pressure depends upon the density $\rho$ and the vertical distance $h$ from the point considered to the surface of the liquid. Distance $h$ is known as the *head*.

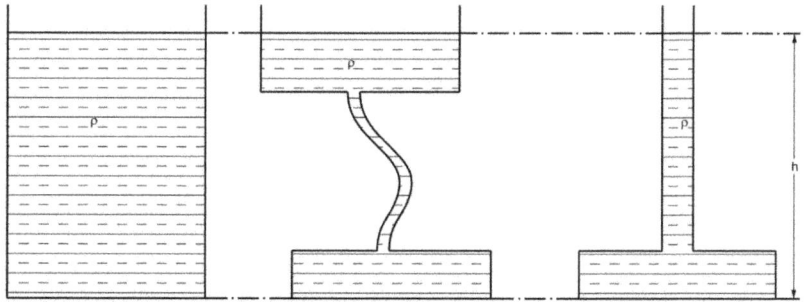

▲ **Figure 6.2** *Pressure within containers*

The pressure at the base of each of the containers shown in figure 6.2 is $\rho gh$, although it may be seen that the total mass of the liquid is different in each case. Container (ii) could represent a supply tank and header tank used in most domestic hot water systems. The pressure at the supply tank depends upon the height of the header tank.

Container (iii) could represent a double bottom tank having a vertical overflow pipe. The pressure inside the tank depends upon the height to which the liquid rises in the pipe.

The total load exerted by a liquid on a *horizontal* plane is the product of the pressure and the area of the plane.

$$P = pA$$

EXAMPLE: A rectangular double bottom tank is 20 m long, 12 m wide and 1.5 m deep, and is full of sea water having a density of 1.025 tonne/m³.

Calculate the pressure in kN/m³ and the load in MN on the top and bottom of the tank if the water is:

**a)** at the top of the tank

**b)** 10 m up the sounding pipe above the tank top.

a)

$$\text{Pressure on top} = \rho gh$$
$$= 1.025 \times 9.81 \times 0$$
$$\text{Load on top} = 0 \text{ N}$$
$$\text{Pressure on bottom} = 1.025 \times 9.81 \times 1.5$$
$$= 15.09 \text{ kN/m}^2$$
$$\text{Load on bottom} = 15.09 \times 20 \times 12$$
$$= 3622 \text{ kN}$$
$$= 3.622 \text{ MN}$$

b)

$$\text{Pressure on top} = 1.025 \times 9.81 \times 10$$
$$= 100.6 \text{ kN/m}^2$$
$$\text{Load on top} = 100.6 \times 20 \times 12$$
$$= 24144 \text{ kN}$$
$$= 24.144 \text{ MN}$$
$$\text{Pressure on bottom} = 1.025 \times 9.81 \times 11.5$$
$$= 115.6 \text{ kN/m}^2$$
$$\text{Load on bottom} = 115.6 \times 20 \times 12$$
$$= 27744 \text{ kN}$$
$$= 27.744 \text{ MN}$$

This example shows clearly the effect of a head of liquid. It should be noted that a very small volume of liquid in a vertical pipe may cause a considerable increase in load.

# Load on an Immersed Plane

The pressure on any horizontal plane is constant, but if the plane is inclined to the horizontal there is a variation in pressure over the plane due to the difference in head. The total load on such a plane may be determined as follows.

Consider an irregular plane of area $A$, totally immersed in a liquid of density $\rho$ and lying at an angle $\theta$ to the surface of the liquid as shown in figure 6.3.

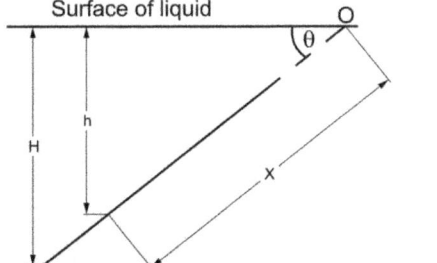
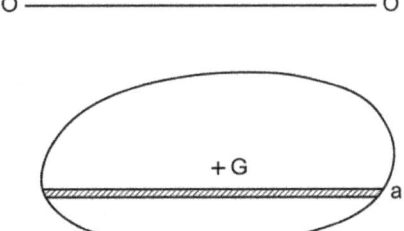

▲ **Figure 6.3** *Pressure on an immersed plane*

Divide the plane into thin strips parallel to the surface of the liquid. Let one such strip, distance $h$ below the surface of the liquid, have an area $a$. Since the strip is thin, the pressure at the centre of the strip can be assumed to act over the whole strip; the small variations can be ignored.

$$\text{Load on strip} = \rho gah$$
$$\text{Load on plane} = \rho g\left(a_1h_1 + a_2h_2 + a_3h_3 + \ldots\right)$$
$$= \rho g\sum_{ah}$$

But $\Sigma ah$ is the first moment of area of the plane about the surface of the liquid.

If $H$ is the distance of the centroid of the plane from the liquid surface, then:

$$\sum_{ah} = AH$$
$$\text{Load on plane} = \rho gAH$$

EXAMPLE: A rectangular bulkhead is 10 m wide and 8 m deep.

It is loaded on one side only with oil of relative density 0.8.

Calculate the load on the bulkhead if the oil is:

**a)** just at the top of the bulkhead
**b)** 3 m up the sounding pipe.

a)

$$\text{Load on bulkhead} = \rho g A H$$
$$= 0.8 \times 1.0 \times 9.81 \times 10 \times 8 \times \frac{8}{2}$$
$$= 2511\,\text{kN}$$

b)

$$\text{Load on bulkhead} = 0.8 \times 1.0 \times 9.81 \times 10 \times 8 \times \left(\frac{8}{2} + 3\right)$$
$$= 4395\,\text{kN}$$

# Centre of Pressure

The centre of pressure on an immersed plane is the point at which the whole liquid load may be regarded as acting.

Consider again figure 6.3.

Let the strip be distance $x$ from the axis 0-0.

$$h = x\sin\theta$$
$$\text{Load on strip} = \rho g a h$$
$$= \rho g a x \sin\theta$$
$$\text{Load on plane} = \rho g \sin\theta(a_1 x_1 + a_2 x_2 + a_3 x_3 + \ldots)$$
$$= \rho g \sin\theta \sum ax$$

Taking moments about axis 0-0:

$$\text{Moment of load on strip} = x \times \rho a x \sin\theta$$
$$= \rho g a x^2 \sin\theta$$

$$\text{Moment of load on plane} = \rho g \sin \theta \left( a_1 x_1^2 + a_2 x_2^2 + a_3 x_3^2 + \ldots \right)$$

$$\text{Centre of pressure from O} - \text{O} = \frac{\text{moment}}{\text{load}}$$

$$= \frac{\rho g \sin \theta \sum_{ax^2}}{\rho g \sin \theta \sum_{ax}}$$

$$= \frac{\sum_{ax^2}}{\sum_{ax}}$$

But $\sum_{ax}$ is the first moment of area of the plane about O-O and $\sum ax^2$ is the second moment of area of the plane about O-O.

If the plane is vertical, then O-O represents the surface of the liquid, and thus:

$$\text{Centre of pressure from surface} = \frac{\text{second moment of area about surface}}{\text{first moment of area about surface}}$$

The second moment of area may be calculated using the *theorem of parallel axes*. If $I_{NA}$ is the second moment about an axis through the centroid (the neutral axis), then the second moment about an axis O-O parallel to the neutral axis and distance $H$ from it is given by:

$$I_{OO} = I_{NA} + AH^2$$

where $A$ is the area of the plane.

Thus:

$$\text{Centre of pressure from O} - \text{O} = \frac{I_{OO}}{AH}$$

$$= \frac{I_{NA} + AH^2}{AH}$$

$$= \frac{I_{NA}}{AH} + H$$

$$I_{NA} \text{ for a rectangle is } \frac{BD^3}{12}$$

$$I_{NA} \text{ for a triangle is } \frac{BD^3}{36}$$

$$I_{NA} \text{ for a circle is } \frac{\pi D^4}{64}$$

The following examples show how these principles may be applied.

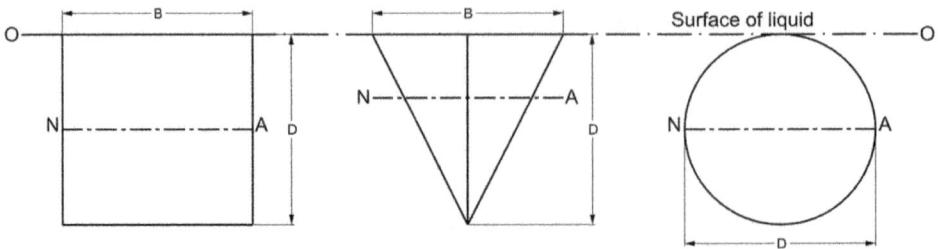

▲ **Figure 6.4** *Different geometric planes*

a) **RECTANGULAR PLANE WITH EDGE IN SURFACE**

$$\text{Centre of pressure from O} - \text{O} = \frac{I_{NA}}{AH} + H$$

$$= \frac{\dfrac{BD^3}{12}}{BD \times \dfrac{D}{2}} + \frac{D}{2}$$

$$= \frac{D}{6} + \frac{D}{2}$$

$$= \frac{2D}{3}$$

b) **TRIANGULAR PLANE WITH EDGE IN SURFACE**

$$\text{Centre of pressure from O} - \text{O} = \frac{I_{NA}}{AH} + H$$

$$= \frac{\dfrac{BD^3}{36}}{\dfrac{BD}{2} \times \dfrac{D}{3}} + \frac{D}{3}$$

$$= \frac{D}{6} + \frac{D}{3}$$

$$= \frac{D}{2}$$

c) **CIRCULAR PLANE WITH EDGE IN SURFACE**

$$\text{Centre of pressure from O} - \text{O} = \frac{I_{NA}}{AH} + H$$

$$= \frac{\dfrac{\pi D^4}{64}}{\dfrac{\pi D^2}{4} \times \dfrac{D}{2}} + \frac{D}{2}$$

$$= \frac{D}{8} + \frac{D}{2}$$

$$= \frac{5D}{8}$$

If the top edge of the plane is below the surface of the liquid, these figures change considerably.

EXAMPLE: A peak bulkhead is in the form of a triangle, apex down, 6 m wide at the top and 9 m deep. The tank is filled with sea water. Calculate the load on the bulkhead and the position of the centre of pressure relative to the top of the bulkhead if the water is:

**a)** at the top of the bulkhead

**b)** 4 m up the sounding pipe.

a)

$$\text{Load on bulkhead} = \rho g A H$$

$$= 1.025 \times 9.81 \times \frac{6 \times 9}{2} \times \frac{9}{3}$$

$$= 814.5 \text{ kN}$$

$$\text{Centre of pressure from top} = \frac{D}{2}$$

$$= \frac{9}{2}$$

$$= 4.5 \text{ m}$$

b)

$$\text{Load on bulkhead} = 1.025 \times 9.81 \times \frac{6 \times 9}{2} \times \left( \frac{9}{3} + 4 \right)$$

$$= 1901 \text{ kN}$$

$$\text{Centre of pressure from surface} = \frac{I_{NA}}{AH} + H$$

$$= \frac{\frac{1}{36} \times 6 \times 9^3}{\frac{1}{2} \times 6 \times 9 \times 7} + 7$$

$$= 0.624 + 7$$

$$= 7.642 \text{ m}$$

$$\text{Centre of pressure from top} = 7.642 - 4$$

$$= 3.642 \text{ m}$$

# Fluid Loading on a Bulkhead

An application of first and second moments of area is the calculation of the load exerted by a liquid on a bulkhead and the position of the centre of pressure. Let the widths of a bulkhead at intervals of $h$, commencing from the top, be $y_0, y_1, y_2, \ldots y_6$ (figure 6.5).

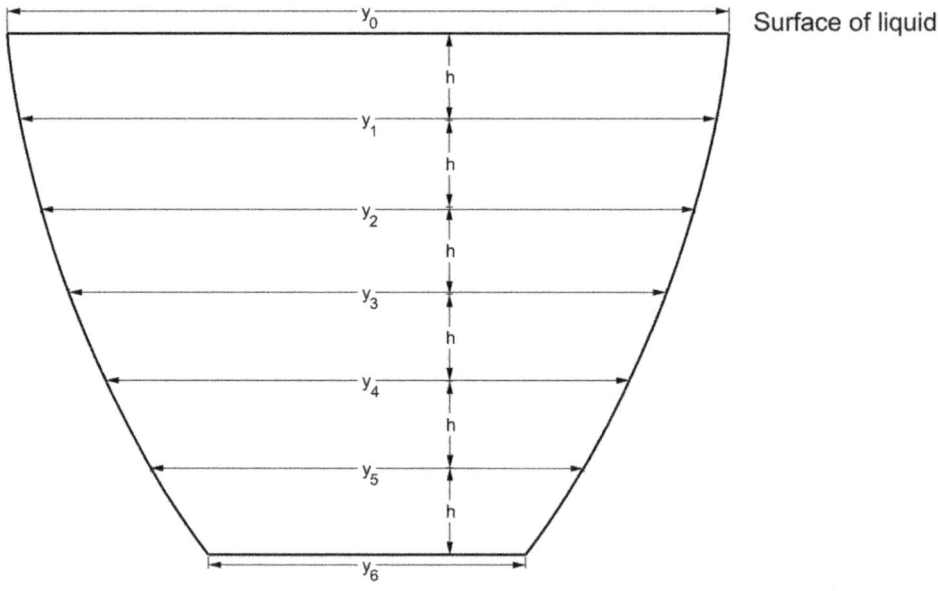

▲ **Figure 6.5** *Bulkhead within ship*

Assume the bulkhead to be flooded to the top edge with liquid of density $\rho$ on one side only.

| Width | SM | Product for area | Lever | Product for 1st moment | Lever | Product for 2nd moment |
|-------|-----|------------------|-------|------------------------|-------|------------------------|
| $y_0$ | 1 | $1y_0$ | 0 | — | 0 | — |
| $y_1$ | 4 | $4y_1$ | 1 | $4y_1$ | 1 | $4y_1$ |
| $y_2$ | 2 | $2y_2$ | 2 | $4y_2$ | 2 | $8y_2$ |
| $y_3$ | 4 | $4y_3$ | 3 | $12y_3$ | 3 | $36y_3$ |
| $y_4$ | 2 | $2y_4$ | 4 | $8y_4$ | 4 | $32y_4$ |
| $y_5$ | 4 | $4y_5$ | 5 | $20y_5$ | 5 | $100y_5$ |

| Width | SM | Product for area | Lever | Product for 1st moment | Lever | Product for 2nd moment |
|-------|-----|------------------|-------|------------------------|-------|------------------------|
| $y_6$ | 1 | $1y_6$ | 6 | $6y_6$ | 6 | $36y_6$ |
| | | $\Sigma_a$ | | $\Sigma_m$ | | $\Sigma_i$ |

$$\text{Area of bulkhead} = \frac{h}{3}\Sigma_a$$

$$\text{First moment of area of bulkhead about surface of liquid} = \frac{h^2}{3}\Sigma_m$$

It was shown previously that:

$$\text{Load on bulkhead} = \rho g \times \text{first moment of area}$$

$$= \rho g \times \frac{h^2}{3}\Sigma_m$$

$$\text{Second moment of area of bulkhead about surface of liquid} = \frac{h^3}{3}\Sigma_i$$

$$\text{Centre of pressure from surface of liquid} = \frac{\text{Second moment of area}}{\text{First moment of area}}$$

$$= 0.642 + 7$$

(*Note:* It is not necessary to calculate the area unless requested to do so).

EXAMPLE: A fore peak bulkhead is 4.8 m deep and 5.5 m wide at the deck. At regular intervals of 1.2 m below the deck, the horizontal widths are 5.0, 4.0, 2.5 and 0.5 m respectively. The bulkhead is flooded to the top edge with sea water on one side only. Calculate:

a) area of bulkhead

b) load on bulkhead

c) position of centre of pressure.

| Depth | Width | SM | Product for area | Lever | Product for 1st moment | Lever | Product for 2nd moment |
|-------|-------|-----|------------------|-------|------------------------|-------|------------------------|
| 4.8 | 5.5 | 1 | 5.5 | 0 | — | 0 | — |
| 3.6 | 5.0 | 4 | 20.0 | 1 | 20.0 | 1 | 20.0 |
| 2.4 | 4.0 | 2 | 8.0 | 2 | 16.0 | 2 | 32.0 |

| Depth | Width | SM | Product for area | Lever | Product for 1st moment | Lever | Product for 2nd moment |
|-------|-------|----|----|-------|----|-------|----|
| 1.2 | 2.5 | 4 | 10.0 | 3 | 30.0 | 3 | 90.0 |
| 0 | 0.5 | 1 | 0.5 | 4 | 2.0 | 4 | 8.0 |
| | | | 44.0 | | 68.0 | | 150.0 |

Common interval $= 1.2$ m

a)

$$\text{Area of bulkhead} = \frac{1.2}{3} \times 44$$
$$= 17.6 \text{ m}^2$$

b)

$$\text{Load on bulkhead} = 1.025 \times 9.81 \times \frac{1.2^2}{3} \times 68.0$$
$$= 328.2 \text{ kN}$$

c)

$$\text{Centre of pressure from surface} = 1.2 \times \frac{150}{68}$$
$$= 2.647 \text{ m}$$

# Load Diagram

If the pressure at any point in an immersed plane is multiplied by the width of the plane at this point, the load per unit depth of plane is obtained. If this is repeated at a number of points, the resultant values may be plotted to form the load diagram for the plane.

The *area* of this load diagram represents the *load* on the plane, while its *centroid* represents the position of the *centre of pressure*. For a rectangular plane with its edge in the surface, the load diagram is in the form of a triangle. For a rectangular plane with its edge below the surface, the load diagram is in the form of a trapezoid. The load diagrams for triangular planes are parabolic.

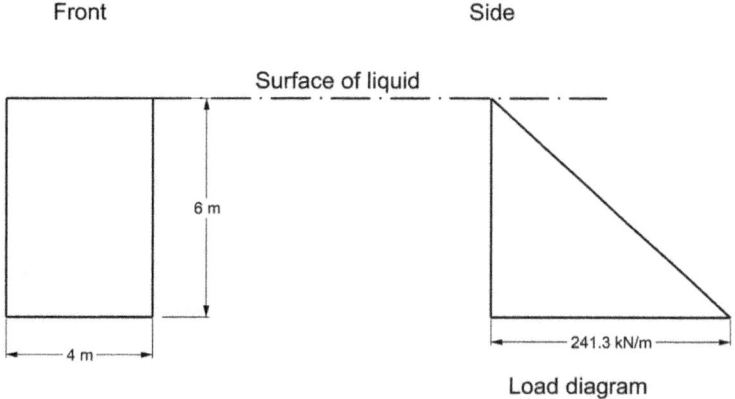

Front    Side

Surface of liquid

6 m

4 m

241.3 kN/m

Load diagram

▲ **Figure 6.6** *Load diagram for bulkhead*

Consider a rectangular bulkhead 4 m wide and 6 m deep loaded to its top edge with sea water.

$$\text{Load/m at top of bulkhead} = \rho gh \times \text{width}$$
$$= 1.025 \times 9.81 \times 0 \times 4$$
$$= 0$$
$$\text{Load/m at bottom of bulkhead} = 1.025 \times 9.81 \times 6 \times 4$$
$$\text{Load on bulkhead} = \text{area of load diagram}$$
$$\text{Load on bulkhead} = \text{area of load diagram}$$
$$= \frac{1}{2} \times 6 \times 241.3$$
$$= 723.9 \text{ kN}$$
$$\text{Centre of pressure from surface} = \text{centroid of load diagram}$$
$$= \frac{2}{3} \times 6$$
$$= 4 \text{ m from top}$$
$$\text{Load on bulkhead} = \rho gAH$$
$$= 1.025 \times 9.81 \times 4 \times 6 \times \frac{1}{2} \times 6$$
$$= 724.0 \text{ kN}$$
$$\text{Centre of pressure} = \frac{2}{3} D$$
$$= 4 \text{ m from top}$$

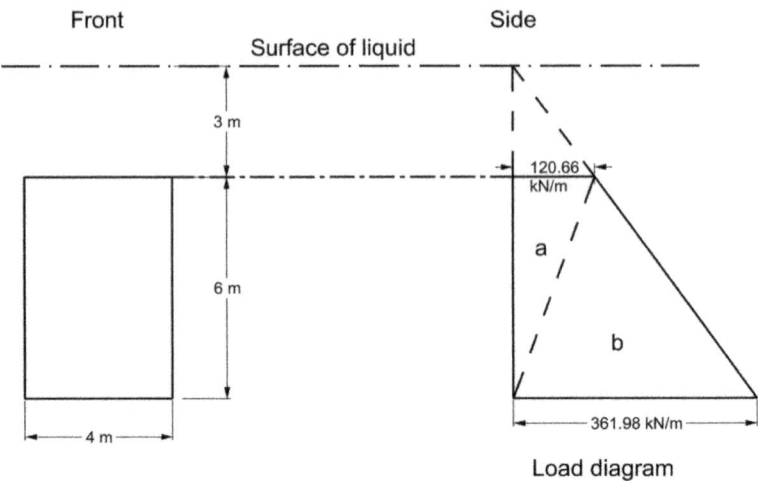

**▲ Figure 6.7** *Load diagram for submerged bulkhead*

If, in the above example, there is a 3 m head of water above the bulkhead, then:

$$\text{Load/m at top of bulkhead} = 1.025 \times 9.81 \times 3 \times 4$$
$$= 120.66 \text{ kN}$$
$$\text{Load/m at bottom of bulkhead} = 1.025 \times 9.81 \times 9 \times 4$$
$$= 361.99 \text{ kN}$$

The load diagram may be divided into two triangles $a$ and $b$:

$$\text{Area } a = \frac{1}{2} \times 6 \times 120.66$$
$$= 361.98 \text{ kN}$$
$$\text{Area } b = \frac{1}{2} \times 6 \times 361.98$$
$$= 1085.94 \text{ kN}$$
$$\text{Total load} = 361.98 + 1085.94$$
$$= 1447.92 \text{ kN}$$

Taking moments about the top of bulkhead:

$$\text{Centre of pressure} = \frac{361.98 \times \frac{1}{3} \times 6 + 1085.94 \times \frac{2}{3} \times 6}{361.98 + 1085.94}$$
$$= \frac{723.96 + 4343.76}{1447.92}$$
$$= 3.5 \text{ m from top of bulkhead}$$

These results may again be checked by calculation.

$$\text{Load on bulkhead} = 1.025 \times 9.81 \times 4 \times 6 \times \left(\frac{1}{2} \times 6 + 3\right)$$

$$= 1448 \text{ kN}$$

$$\text{Centre of pressure from surface} = \frac{I_{NA}}{AH} + H$$

$$= \frac{\frac{1}{12} \times 4 \times 6^3}{4 \times 6 \times \left(\frac{1}{2} \times 6 + 3\right)} + \left(\frac{1}{2} \times 6 + 3\right)$$

$$= 0.5 + 6$$

$$= 6.5 \text{ m}$$

$$\text{Centre of pressure} = 6.5 - 3$$

$$= 3.5 \text{ m from top of bulkhead}$$

# Shearing Force on Bulkhead Stiffeners

A bulkhead stiffener supports a rectangle of plating equal to the length of the stiffener times the spacing of the stiffeners. If the bulkhead has liquid on one side to the top edge, the stiffener supports a load that increases uniformly from zero at the top to a maximum at the bottom (figure 6.8).

Let $l$ = length of stiffener in m

$s$ = spacing of stiffeners in m

$\rho$ = density of liquid in kg/m³

$P$ = load on stiffener

$W$ = load / m at bottom of stiffener

Then:

$$P = \rho g l s \times \frac{l}{2}$$

$$= \frac{1}{2} \rho g l^2 s$$

$$W = \rho g l s$$

$$P = W \times \frac{l}{2}$$

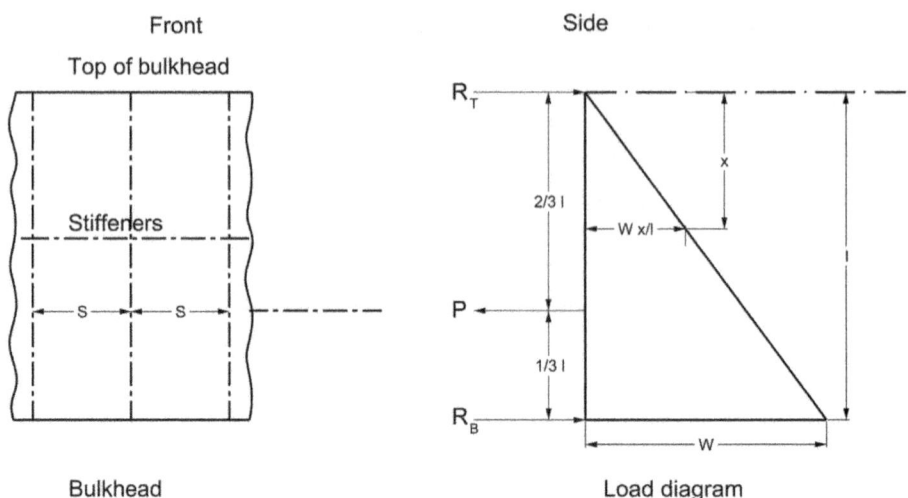

▲ **Figure 6.8** *Load diagram for bulkhead with stiffeners*

The load $P$ acts at the centre of pressure, which is $\frac{2}{3}l$ from the top. Reactions are set up by the end connections at the top ($R_T$) and at the bottom ($R_B$).

Taking moments about the top,

$$R_B \times l = P \times \frac{2}{3}l$$

$$R_B = \frac{2}{3}P$$

and

$$R_T = \frac{1}{3}P$$

The shearing force at a distance $x$ from the top will be the reaction at the top, less the area of the load diagram from this point to the top, ie

$$SF_x = R_T - \frac{Wx}{l} \times \frac{x}{2}$$

$$= \frac{1}{3}P - \frac{Wx^2}{2l}$$

$$= \frac{Wl}{6} - \frac{Wx^2}{2l}$$

Let $x = 0$

$$SF \text{ at top} = \frac{Wl}{6}$$

$$= \frac{1}{3}P$$

Let $x = l$

$$SF \text{ at bottom} = \frac{Wl}{6} - \frac{Wl^2}{2l}$$

$$= \frac{Wl}{6} - \frac{Wl}{2}$$

$$= -\frac{Wl}{3}$$

$$= \frac{2}{3}P$$

Since the shearing force is positive at the top and negative at the bottom, there must be some intermediate point at which the shearing force is zero. This is also the position of the maximum bending moment.

Let SF = 0

$$0 = \frac{Wl}{6} - \frac{Wx^2}{2l}$$

$$\frac{Wx^2}{2l} = \frac{Wl}{6}$$

$$x^2 = \frac{l^2}{3}$$

$$\text{Position of zero shear } = \frac{l}{\sqrt{3}}$$

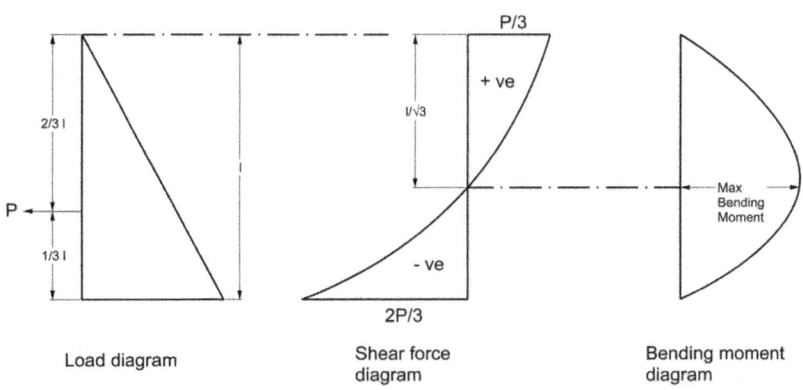

| Load diagram | Shear force diagram | Bending moment diagram |

▲ **Figure 6.9**  *Load, shear force and bending moment diagrams*

EXAMPLE: A bulkhead 9 m deep is supported by vertical stiffeners 750 mm apart. The bulkhead is flooded to the top edge with sea water on one side only. Calculate:

**a)**   shearing force at top

**b)**   shearing force at bottom

**c)**   position of zero shear.

$$\text{Load on stiffener } P = \rho g A H$$

$$= 1.025 \times 9.81 \times 9 \times 0.75 \times \frac{9}{2}$$

$$= 305.4 \text{ kN}$$

a)

$$\text{Shearing force at top} = \frac{1}{3}P$$

$$= \frac{305.4}{3}$$

$$= 101.8 \text{ kN}$$

b)

$$\text{Shearing force at bottom} = \frac{2}{3}P$$

$$= \frac{2}{3} \times 305.4$$

$$= 203.6 \text{ kN}$$

c)

$$\text{Position of zero shear} = \frac{l}{\sqrt{3}}$$

$$= \frac{9}{\sqrt{3}}$$

$$= 5.197 \text{ m from the top}$$

# Shearing Force and Bending Moment

Consider a loaded ship lying in still water. The upthrust over any unit length of the ship depends upon the immersed cross-sectional area of the ship at that point. If the values of upthrust at different positions along the length of the ship are plotted on a base representing the ship's length, a *buoyancy curve* is formed (figure 6.10). This curve increases from zero at each end to a maximum value in way of the parallel midship portion. The area of this curve represents the total upthrust exerted by the water on the ship.

The total weight of a ship consists of a number of independent weights concentrated over short lengths of the ship, such as cargo, machinery, accommodation, cargo handling gear, poop and forecastle and a number of items that form continuous material over the length of the ship, such as decks, shell and tank top. A *curve of weights* is shown in the diagram.

The difference between the weight and buoyancy at any point is the *load* at that point. In some cases, the load is an excess of weight over buoyancy and in other cases an excess of buoyancy over weight. A load diagram is formed by plotting these differences. Because of this unequal loading, however, shearing forces and bending moments are set up in the ship.

The *shearing force* at any point is represented by the *area* of the load diagram on one side of that point. A *shearing force diagram* may be formed by plotting these areas on a base of the ship's length.

The *bending moment* at any point is represented by the *area* of the shearing force diagram on one side of that point. A *bending moment diagram* may be formed by plotting such areas on a base of the ship's length.

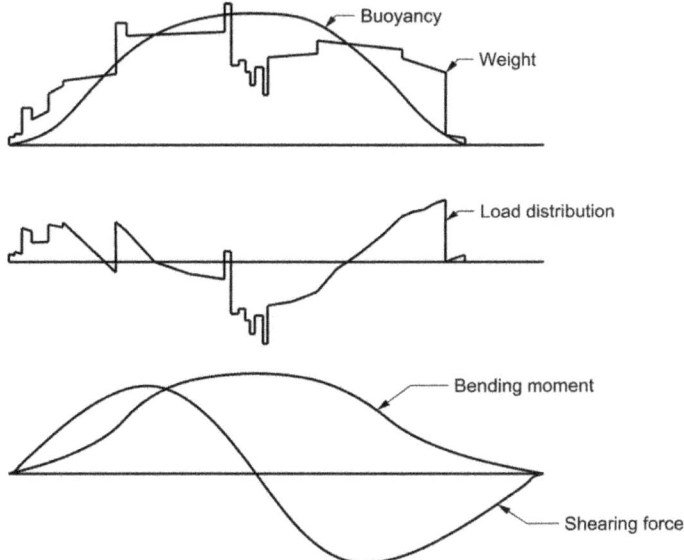

▲ **Figure 6.10** *Load distribution within ship*

The maximum bending moment occurs where the shearing force is zero and this is usually near amidships.

EXAMPLE: A box barge 200 m long is divided into five equal compartments. The weight is uniformly distributed along the vessel's length.

500 tonne of cargo are added to each of the end compartments. Sketch the shearing force and bending moment diagrams and state their maximum values.

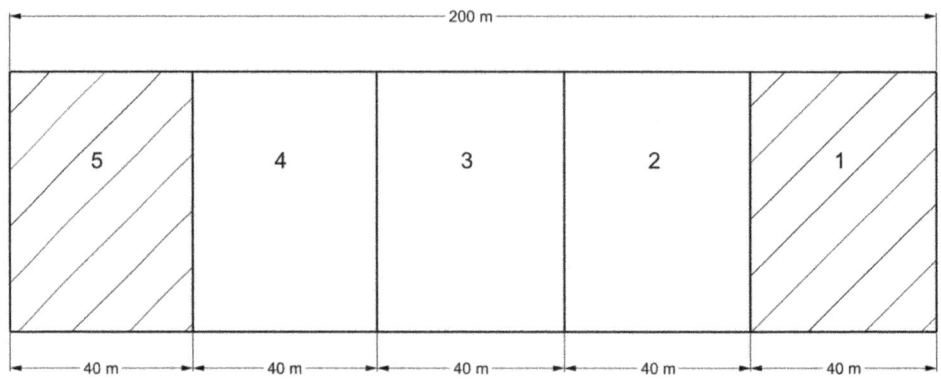

▲ **Figure 6.11** *Loading within box barge*

Before adding the cargo, the buoyancy and weight were equally distributed and produced no shearing force or bending moment. It is therefore only necessary to consider the added cargo and the additional buoyancy required.

$$\text{Additional buoyancy/m} = \frac{1000\,g}{200}$$
$$= 5\,g \text{ kN}$$
$$\text{Additional weight/m} = \frac{500\,g}{40}$$
$$= 12.5\,g \text{ kN}$$
$$\text{Load/m} = 12.5\,g - 5\,g$$
$$= 7.5\,g \text{ kN excess weight}$$

Compartments 2, 3 and 4

$$\text{Load/m} = 5\,g \text{ kN excess buoyancy}$$
$$\text{Shearing force at A} = 0$$

Shearing force at B $= -7.5\,g \times 40$

$= -300\,g$

$= 2943$ kN

Shearing force at C $= -300\,g + 5\,g \times 40$

$= -100\,g$

$= 981$ kN

Shearing force at D $= -300\,g + 5\,g \times 60$

$= 0$

Shearing force at E $= +100\,g$

Shearing force at F $= +300\,g$

Shearing force at G $= 0$

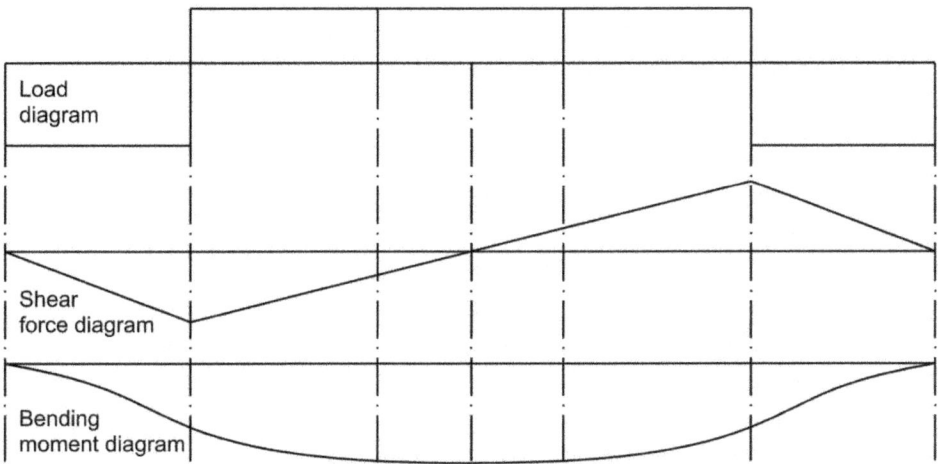

▲ **Figure 6.12** *Shear force and bending moment diagram*

Bending moment at A and G $= 0$

Bending moment at B and F $= -300\,g \times \dfrac{40}{2}$

$= -6000\,g$

$= 58.86$ MNm

Bending moment at C and E $= -15000\,g + 100\,g \times \dfrac{20}{2}$

$= -14000\,g$

$= 137.34$ MNm

Bending moment at D $= -300\,g \times \dfrac{100}{2}$

$= -15000\,g$

$= 147.15$ MNm

# Chapter 6 Test Examples

**Q. 6.1:** A forward deep tank 12 m long extends from a longitudinal bulkhead to the ship's side. The widths of the tank surface measured from the longitudinal bulkhead at regular intervals are 10, 9, 7, 4 and 1 m. Calculate the second moment of area of the tank surface about a longitudinal axis passing through its centroid.

**Q. 6.2:** A rectangular double bottom tank 12 m long and 10 m wide is full of sea water. Calculate the head of water above the tank top if the load due to water pressure on the tank top is 9.6 MN.

**Q. 6.3:** A ballast tank is 15 m long, 12 m wide and 1.4 m deep and is filled with fresh water. Calculate the load on the top and short side, if:
**a)** the tank is just completely full
**b)** there is a head of 7 m of water above the tank top.

**Q. 6.4:** A vertical bulkhead 9 m wide and 8 m deep has sea water on one side only to a depth of 6 m. Calculate the pressure in $kN/m^2$ at the bottom of the bulkhead and the load on the bulkhead.

**Q. 6.5:** A bulkhead is in the form of a trapezoid 9 m wide at the deck, 5 m wide at the bottom and 8 m deep. Find the load on the bulkhead if it has oil (rd 0.85) on one side only:
**a)** to a depth of 6 m
**b)** with 4 m head to the top edge.

**Q. 6.6:** The end bulkhead of an oil fuel bunker is in the form of a rectangle 10 m wide and 12 m high. Find the total load and the position of the centre of pressure relative to the top of the bulkhead if the tank is filled with oil (rd 0.9):
**a)** to the top edge
**b)** with 3 m head to the top edge.

**Q. 6.7:** A dock gate 6 m wide and 5 m deep has fresh water on one side to a depth of 3 m and sea water on the other side to a depth of 4 m. Calculate:
**a)** the resultant load
**b)** the position of the centre of pressure.

**Q. 6.8:** A triangular bulkhead is 5 m wide at the top and 7 m deep. It is loaded to a depth D with sea water, when it is found that the load on the bulkhead is 190 kN. Find the depth D and the distance from the top of the bulkhead to the centre of pressure.

**Q. 6.9:** A triangular bulkhead is 7 m wide at the top and has a vertical depth of 8 m. Calculate the load on the bulkhead and the position of the centre of pressure if the bulkhead is flooded with sea water on only one side:
a) to the top edge
b) with 4 m head to the top edge.

**Q. 6.10:** A bulkhead is supported by vertical stiffeners. The distance between the stiffeners is one-ninth of the depth of the bulkhead. When the bulkhead is flooded to the top with sea water on one side only, the maximum shearing force in the stiffeners is 200 kN. Calculate:
a) the height of the bulkhead
b) the shearing force at the top of the stiffeners
c) the position of zero shear.

**Q. 6.11:** The widths of a deep tank bulkhead at equal intervals of 1.2 m, commencing at the top, are 8.0, 7.5, 6.5, 5.7, 4.7, 3.8 and 3.0 m. Calculate the load on the bulkhead and the position of the centre of pressure, if the bulkhead is flooded to the top edge with sea water on one side only.

# 7

# RESISTANCE

When a ship moves through the water at any speed, a force or resistance is exerted by the water on the ship. The ship must therefore exert an equal thrust to overcome the resistance to travel at that speed. If, for example, the resistance of the water on the ship at 17 knots is 800 kN, and the ship provides a thrust of 800 kN, then the vessel will travel at 17 knots.

The total resistance or tow-rope resistance $R_t$ of a ship may be divided into two main sections:

a) frictional resistance $R_f$
b) residuary resistances $R_r$

Hence:

$$R_t = R_f + R_r$$

## Frictional Resistance $R_f$

As the ship moves through the water, the hull experiences a resistive force due to the friction between the hull surface and the water molecules. The water molecules directly adjacent to the hull are brought into motion and travel at the same speed as the hull. So too are the neighbouring molecules, but at a progressively slower speed the further away from the hull one moves, until far away from the hull, where the water molecules are unaffected by the hull's movement. The region of water molecules that are brought into motion is known as the *boundary layer*, and as one moves aft from the bow, the thickness of this boundary layer gets larger and larger. When the boundary layer forms at the bow, and is still thin, the water molecules all move in an ordered manner, known as *laminar flow*. As the boundary layer thickness increases, it becomes far harder for the molecules to move in this ordered manner, and their movement and velocity fluctuates, known as *turbulent flow*.

The frictional resistance the hull experiences is due to the energy lost in bringing these water molecules into motion. The more turbulent flow that occurs over a hull, the greater the energy lost, and so the hull experiences a larger frictional resistance. The amount of frictional resistance the hull experiences depends upon:

i.    the speed of the ship

ii.   the wetted surface area

iii.  the length of the ship

iv.   the roughness of the hull

v.    the density of the water

vi.   the viscosity of the water (viscosity is a measure of how 'sticky' the water is).

An early investigator into ship resistance was William Froude (1810–1879), who developed a formula for the frictional resistance of a ship based on experiments conducted measuring the resistance of a series of flat plates:

$$R_f = fSV^n \, (\text{N})$$

where $f$ is a coefficient that depends upon the length of the ship $L$, the roughness of the hull and the density of the water

$S$ is the wetted surface area in $\text{m}^2$

$V$ is the ship speed in knots

$n$ is an index of about 1.825

The value of $f$ for a mild steel hull in sea water is given by:

$$f = 0.417 + \frac{0.773}{L + 2.862}$$

Thus $f$ is reduced as the length of the ship is increased.

In a slow or medium-speed ship, the frictional resistance forms the major part of the total resistance, and may be as much as 75% of $R_t$. The importance of surface roughness may be seen when a ship is badly fouled with marine growth or heavily corroded, when the speed of the ship may be considerably reduced.

When conducting powering calculations using empirical formulae such as Froude's, it is important to be aware of the units required for variables within the formula. The use of knots is only suitable if the empirical equation specifically states it. Otherwise, it is advisable to keep calculations in SI units.

$$1 \text{ knot} = 1.852 \text{ km/h} = 0.5144 \text{ m/s}$$

EXAMPLE: A ship whose wetted surface area is 5150 m² travels at 15 knots. Calculate the frictional resistance and the power required to overcome this resistance.

$$f = 0.422, n = 1.825$$

$$R_f = fSV^n$$
$$= 0.422 \times 5150 \times 15^{1.825}$$
$$= 303700 \text{ N}$$

Power to overcome frictional resistance $= R_f \times V$
$$= 303700 \times 15 \times 0.5144 \text{ W}$$
$$= 2344 \text{ kW}$$

EXAMPLE: A plate drawn through fresh water at 3 m/s has a frictional resistance of 12 N/m². Estimate the power required to overcome the frictional resistance of a ship at 12 knots if the wetted surface area is 3300 m² and the index of speed is 1.9.

$$= 6.173 \text{ m/s}$$
$$= 6.175 \text{ m/s}$$
$$R_f = 12 \times 3300$$
$$= 39600 \text{ N in FW}$$

$$R_f = 39600 \times 1.025 \times \left(\frac{6.173}{3}\right)^{1.9}$$

$$= 159894 \text{ N}$$
Power $= 159894 \times 6.173$
$$= 987024 \text{ W}$$
$$= 987.0 \text{ kW}$$
$$= 988.0 \text{ kW}$$

The work of William Froude formed the foundation of ship model testing and since then, many have developed further principles and formulae to assist in predicting full-scale ship resistance. The underlying principle that the overall resistance has a frictional component and a residuary component remains the same, but the ship type and experience of those conducting the test will determine exactly how these components are established. Significant amounts of experimental research have been done to more accurately estimate the frictional resistance of ships, and it is common to use the International Towing Tank Convention (ITTC) formula for frictional resistance, which is:

$$C_f = \frac{0.075}{(\log_{10} R_n - 2)^2}$$

where $C_f$ is the non-dimensional coefficient of frictional resistance. $R_n$ is known as the *Reynolds number* and is given by the following formula:

$$R_n = \frac{VL}{v}$$

where v is the kinematic viscosity of the water. The Reynolds number is dimensionless, and is a way of comparing flows of different sizes to ensure they are equivalent. For a model-scale ship and a full-scale ship that are both operating at the same Reynolds number, the proportions of the overall boundary layer that is either laminar or turbulent is the same.

$C_f$ can be given dimensions using $\frac{1}{2}\rho S V^2$.

$$R_f = \frac{1}{2}\rho S V^2 C_f$$

For first level approximations and for simplicity, in that it does not require the use of logarithmic functions, the examples given within this text will use Froude's formula for skin friction, as that remains more typical of exam questions asked. The ITTC formula is included here for general knowledge.

## Residuary Resistance $R_r$

The residuary resistance of a ship is used to describe every type of resistance that is not due to friction. It will typically be divided into:

(i) Resistance caused by the formation of waves as the ship passes through the water. In slow or medium-speed ships, the wave-making resistance is small compared with the frictional resistance. As speed increases, so too does the proportion of wave-making resistance, and at high speeds it may be 50% or 60% of the total resistance.

The wave-making resistance of a ship may be reduced by the incorporation of a bulbous bow into the design. The bulb produces its own wave system, which interferes with the wave produced by the stem of the ship, resulting in a reduced height of bow wave and consequent reduction in the energy required to produce the wave. As a bulbous bow requires the hull to have more wetted surface area, there is an increase in frictional resistance, but this is typically less than the reduction in wave resistance.

(ii) Viscous pressure resistance or form drag occurs due to the pressure changes in the water flowing around the hull. As the ship passes through the water, its shape, or form, causes a pressure variation in the surrounding fluid, which results in a resistive force on the ship. The more streamlined a hull is, the smaller this component of resistance will be. If the water changes direction abruptly, such as round a box barge, the resistance may be considerable, but in modern, well-designed ships it should be small.

Some higher-speed ships may well have a transom that ends abruptly and is immersed at rest. At lower speeds, this can be the source of a large form drag, as it represents an abrupt change in direction for the surrounding flow. As the speed increases, however, the water will begin to leave the transom in a smooth manner and the transom will no longer be immersed, which results in an improved high-speed performance.

(iii) Eddy resistance is similar to form drag, but is concerned with the smaller-scale changes in form, such as at the trailing edge of a rudder or around shaft and brackets. This resistance will be small in a ship where careful attention is paid to detail and underwater fittings are streamlined. It can, however, be the source of vibration on a ship.

The relation between the frictional resistance and the residuary resistances is shown in figure 7.1.

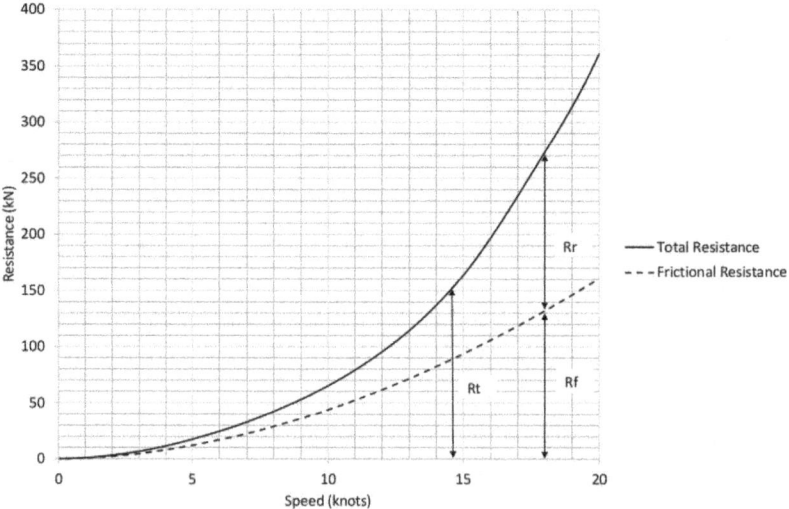

▲ **Figure 7.1** *Ship resistance with speed*

Residuary resistances follow Froude's Law of Comparison:

> *The residuary resistances of similar ships are in the ratio of the cube of their linear dimensions if their speeds are in the ratio of the square root of their linear dimensions.*

Thus:

$$\frac{R_{r_1}}{R_{r_2}} = \left(\frac{L_1}{L_2}\right)^3 \quad if \quad \frac{V_1}{V_2} = \sqrt{\frac{L_1}{L_2}}$$

or

$$\frac{R_{r_1}}{R_{r_2}} = \frac{\Delta_1}{\Delta_2} \quad if \frac{V_1}{V_2} = \sqrt[6]{\frac{\Delta_1}{\Delta_2}} = \left(\frac{\Delta_1}{\Delta_2}\right)^{\frac{1}{6}}$$

$V_1$ and $V_2$ are termed *corresponding speeds*. Thus at corresponding speeds:

$$\frac{V_1}{\sqrt{L_1}} = \frac{V_2}{\sqrt{L_2}}$$

$V/\sqrt{L}$ is known as the *speed-length ratio*.

It may therefore be seen that at corresponding speeds the wave-making characteristics of similar ships are the same. At high speeds, the speed-length ratio is high and the wave-making resistance is large. To give the same wave-making characteristics, the corresponding speed of a much smaller, similar ship will be greatly reduced and may not be what is popularly regarded to be a high speed. A ship is therefore considered slow or fast in relation to its speed-length ratio.

If $V/\sqrt{L}$ is below 1.0, the ship is said to be slow ($V$ in knots: $L$ in m)

If $V/\sqrt{L}$ is above 1.5, the ship is said to be fast.

Thus, a speed of 15 knots would be regarded as slow for a ship 225 m long, but fast for a ship 100 m long. This is something that we feel as passengers on board as well. A ferry travelling at 20 knots does not 'feel' as fast as a small 5 m speed boat travelling at the same speed.

EXAMPLE: The residuary resistance of a model 7 m long is 20 N when towed at 3½ knots.

Calculate the power required to overcome the residuary resistance of a similar ship 140 m long at its corresponding speed.

$$V_2 = V_1 \sqrt{\frac{L_2}{L_1}}$$

$$= 3.5 \sqrt{\frac{140}{7}}$$

$$= 15.652 \text{ knots}$$

$$R_{r_2} = R_{r_1} \left(\frac{L_2}{L_1}\right)^3$$

$$= 20 \left(\frac{140}{7}\right)^3$$

$$= 160000 \text{ N}$$

$$\text{Power} = R_r \times V$$

$$= 160000 \times 15.652 \times 0.5144$$

$$= 1288 \text{ kW}$$

Establishing the residuary resistance for a ship is not a trivial task. Typically, it is based on the results of model experiments and it may well be a contractual build requirement that model tests are conducted. A scale model of the ship is built and towed through a range of speeds, corresponding to the same speed-length ratios that are of interest for the full-scale ship. The experiments are conducted in long, straight tanks of water, known as *towing tanks* or *model basins*, which have moving carriages above them. The model is attached to the carriage via a dynamometer, an accurate force-measurement device. The dynamometer measures the total resistance, and by calculating the frictional resistance of the model at each speed-length ratio, the residuary resistance of the model is established.

Unfortunately, as we test models in effectively the same fluid, water, as the full-scale ship will operate in, it is not possible for us to run the model at a speed where we have the same speed-length ratio and Reynolds number. Since the frictional resistance can be calculated, we choose to run at the equivalent speed-length ratio. It is important to account for the fact that, as we are not at the same Reynolds number, the proportions of our hull that will have laminar flow or turbulent flow in the boundary layer will not be the same as the full-scale ship. This is typically done by adding some additional roughness to the surface of the model, such as sandpaper, to force the boundary layer to become turbulent at this point.

Once the model residuary resistance has been calculated, it can then be scaled for the full-size ship at the equivalent speed-length ratio. This will then be added to the ship frictional resistance, calculated using the same empirical formulae, to give a total ship resistance.

By knowing the total resistance of the ship, it is possible to determine the power required to overcome this resistance. This is known as the *effective power* (ep) of the ship.

The model is tested without appendages such as rudder and bilge keels. An allowance must therefore be made for these appendages and also the general disturbance of the water at sea compared with tank conditions. This allowance is known as the *ship correlation factor* (SCF). The ship correlation factor may also include an allowance to account for fouling, or weather and the additional resistance associated with travelling through waves.

The power obtained directly from the model tests is known as the *effective power (naked)* ($ep_n$). The true effective power is the $ep_n$ multiplied by the ship correlation factor.

EXAMPLE: A 6 m model of a ship has a wetted surface area of 8 m². When towed at a speed of 3 knots in fresh water, the total resistance is found to be 38 N.

If the ship is 130 m long, calculate the effective power at the corresponding speed.

Take $n = 1.825$ and calculate $f$ from the formula. SCF 1.15

Model:

$$R_t = 38 \text{ N in fresh water}$$
$$= 38 \times 1.025$$
$$= 38.95 \text{ N in sea water}$$
$$f = 0.417 + \frac{0.773}{L + 2.862}$$
$$f = 0.417 + \frac{0.773}{8.862}$$
$$= 0.504$$
$$R_f = 0.504 \times 8 \times 3^{1.825}$$
$$= 29.94 \text{ N}$$
$$R_r = R_t - R_f$$
$$= 38.95 - 29.94$$
$$= 9.01 \text{ N}$$

Ship:

$$R_r \propto L^3$$
$$R_r = 9.01 \times \left(\frac{130}{6}\right)^3$$
$$= 91600 \text{ N}$$

$$S \propto L^2$$

$$S = 8 \times \left(\frac{130}{6}\right)^2$$

$$= 3755 \text{ m}^2$$

$$V \propto \sqrt{L}$$

$$V = 3\sqrt{\frac{130}{6}}$$

$$= 13.96 \text{ knots}$$

$$f = 0.417 + \frac{0.773}{132.862}$$

$$= 0.4228$$

$$R_f = 0.4228 \times 3755 \times 13.96^{1.825}$$

$$= 195000 \text{ N}$$

$$R_t = 195000 + 91600$$

$$R_t = 286600 \text{ N}$$

$$ep_n = 286600 \times 13.96 \times \frac{1852}{3600}$$

$$= 2059 \text{ kW}$$

$$\text{Effective power } ep = 2059 \times 1.15$$

$$= 2368 \text{ kW}$$

# Admiralty Coefficient

It is sometimes necessary to obtain an approximation of the power of a ship without resorting to model experiments, and several methods are available. One system that has been in use for several years is the Admiralty Coefficient method. This is based on the assumption that for small variations in speed, the total resistance may be expressed in the form:

$$R_t \propto \rho S V^n$$

It was seen earlier that

$$S \propto \Delta^{\frac{2}{3}}$$

Hence with constant density

$$\text{Admiralty Coefficient } C = \frac{\Delta^{\frac{2}{3}} V^{n+1}}{sp}$$

But,

$$\text{power} \propto R_t \times V$$

Therefore:

$$\text{power} \propto \Delta^{\frac{2}{3}} V^{n+1}$$

$$\text{power} = \frac{\Delta^{\frac{2}{3}} V^{n+1}}{\text{a coefficient}}$$

The coefficient is known as the *Admiralty Coefficient*. Originally, this method was used to determine the power supplied by the engine. Since types of machinery vary considerably, it is now considered that the relation between displacement, speed and shaft power (sp) is of more practical value.

Most merchant ships may be classed as slow- or medium-speed, and for such vessels the index *n* may be taken as 2; for fast ships, *n* can be taken as 3. Thus:

$$\text{Admiralty Coefficient } C = \frac{\Delta^{\frac{2}{3}} V^{n+1}}{\text{shaft power}}$$

where $\Delta$ = displacement in tonne

$V$ = ship speed in knots

sp = shaft power in kW

The Admiralty Coefficient may be regarded as constant for similar ships at their corresponding speeds. Values of C vary between about 350 and 600 for different ships, the higher values indicating more efficient ships. For small changes in speed, the value of C may be regarded as constant for any ship at constant displacement. Calculations using the Admiralty Coefficient are not sufficiently accurate that the installation of an engine within a ship would be made based on them, but it can be used at a preliminary design stage and for assessing small changes in operation.

At corresponding speeds:

$$V \propto \Delta^{\frac{1}{6}}$$

Therefore:

$$V^3 \propto \Delta^{\frac{3}{6}}$$

$$sp \propto \Delta^{\frac{2}{3}} V^3$$

$$\propto \Delta^{\frac{2}{3}} \times \Delta^{\frac{3}{6}}$$

$$\propto \Delta^{\frac{7}{6}}$$

$$\frac{sp_1}{sp_2} = \left(\frac{\Delta_1}{\Delta_2}\right)^{\frac{7}{6}}$$

Thus, if the shaft power of one ship is known, the shaft power for a similar ship may be obtained at the corresponding speed.

EXAMPLE: A ship of 14000 tonne displacement has an Admiralty Coefficient of 450. Calculate the shaft power required at 16 knots.

$$sp = \frac{\Delta^{\frac{2}{3}} V^3}{C}$$

$$= \frac{14000^{\frac{2}{3}} 16^3}{450}$$

$$= 5286 \text{ kW}$$

EXAMPLE: A ship of 15000 tonne displacement requires 3500 kW at a particular speed.

Calculate the shaft power required by a similar ship of 18000 tonne displacement at its corresponding speed.

$$sp \propto \Delta^{\frac{7}{6}}$$

$$sp = 3500 \times \left(\frac{18000}{15000}\right)^{\frac{7}{6}}$$

$$= 4330 \text{ kW}$$

The index of speed $n$ for high-speed ships may be considerably more than 2 and thus the shaft power may vary with speed to some index greater than 3 (eg 4). This higher index, however, is only applicable within the high-speed range.

$$\frac{sp_1}{sp_2} = \left(\frac{V_1}{V_2}\right)^4$$

where both $V_1$ and $V_2$ are within the high-speed range.

EXAMPLE: A ship travelling at 20 knots requires 12000 kW shaft power. Calculate the shaft power at 22 knots if, within this speed range, the index of speed is 4.

$$\frac{sp_1}{sp_2} = \left(\frac{V_1}{V_2}\right)^4$$

$$sp_2 = 12000 \times \left(\frac{22}{20}\right)^4$$

$$= 17570 \text{ kW}$$

# Fuel Coefficient and Fuel Consumption

The fuel consumption of a ship depends upon the power developed, indeed the overall efficiency of power plant is often measured in terms of the *specific fuel consumption*, which is the consumption per unit of power, expressed in kg/h. Efficient large ship engines may have a specific fuel consumption as low as 0.16 kg/kWh, while a gas turbine is typically 0.27 kg/kWh. The specific fuel consumption of a ship at different speeds follows the form shown in figure 7.2.

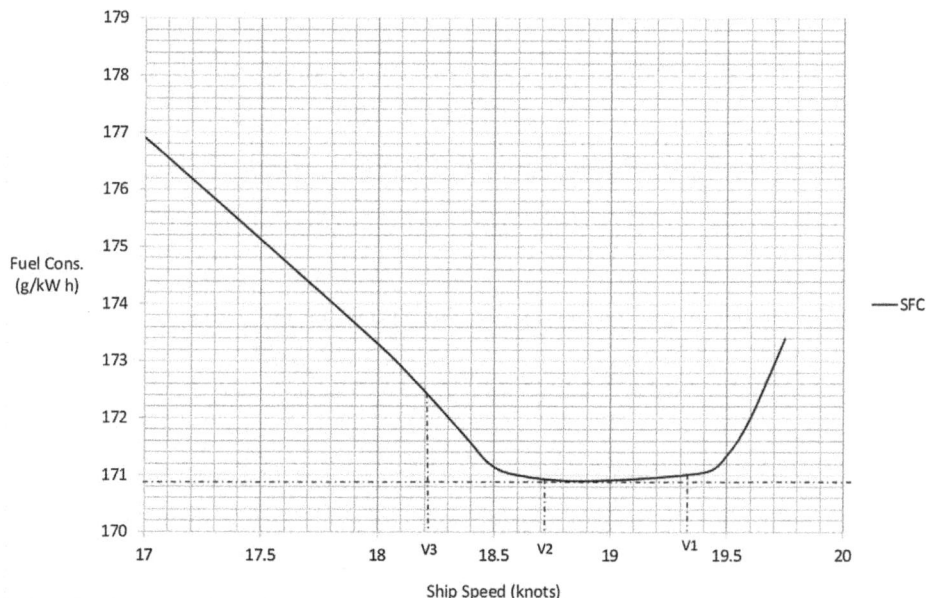

▲ **Figure 7.2** *Fuel consumption with ship speed*

Between $V_1$ and $V_2$ the specific consumption may be regarded as constant for practical purposes, and if the ship speed varies only between these limits, then:

Fuel consumption/unit time is proportional to the power developed and since:

$$sp \propto \Delta^{\frac{2}{3}} V^3$$

$$\text{Fuel consumption/unit time} \propto \Delta^{\frac{2}{3}} V^3$$

$$\text{Fuel consumption/unit time} = \frac{\Delta^{\frac{2}{3}} V^3}{\text{fuel coefficient}}$$

Values of fuel coefficient vary between about 40000 and 130000, the higher values indicating more efficient ships.

EXAMPLE: The fuel coefficient of a ship of 14000 tonne displacement is 75000. Calculate the fuel consumption per day if the vessel travels at 12.5 knots.

$$\text{Fuel consumption/day} = \frac{14000^{\frac{2}{3}} \times 12.5^3}{75000}$$
$$= 15.12 \text{ tonne}$$

If the displacement and fuel coefficient remain constant, ie between $V_1$ and $V_2$ (figure 7.2):

$$\text{Fuel consumption/unit time} \propto \text{speed}^3$$

Hence:

$$\frac{\text{cons}_1}{\text{cons}_2} = \left(\frac{V_1}{V_2}\right)^3$$

EXAMPLE: A ship uses 20 tonne of fuel per day at 13 knots. Calculate the daily consumption at 11 knots.

$$\text{New daily consumption} = 20 \times \left(\frac{11}{13}\right)^3$$
$$= 12.11 \text{ tonne}$$

The total fuel consumption for any voyage may be found by multiplying the daily consumption by the number of days required to complete the voyage.

If $D$ is the distance travelled at $V$ knots, then:

$$\text{Number of days} \propto \frac{D}{V}$$

But

$$\text{daily consumption} \propto V^3$$

Therefore:

$$\text{total voyage consumption} \propto V^3 \times \frac{D}{V}$$

$$\frac{\text{voy. cons}_1}{\text{voy. cons}_2} = \left(\frac{V_1}{V_2}\right)^2 \times \frac{D_1}{D_2}$$

Hence, for any given distance travelled, the voyage consumption varies as the speed *squared*.

EXAMPLE: A vessel uses 125 tonne of fuel on a voyage when travelling at 16 knots. Calculate the mass of fuel saved if, on the return voyage, the speed is reduced to 15 knots, the displacement of the ship remaining constant.

$$\text{New voyage consumption} = 125 \times \left(\frac{15}{16}\right)^3$$
$$= 110 \text{ tonne}$$

Therefore:

$$\text{Saving in fuel} = 125 - 110$$
$$= 15 \text{ tonne}$$

A general expression for voyage consumption is:

$$\frac{\text{new voy. cons.}}{\text{old voy. cons.}} = \left(\frac{\text{new displ.}}{\text{old displ.}}\right)^{\frac{2}{3}} \times \left(\frac{\text{new speed}}{\text{old speed}}\right)^2 \times \frac{\text{new dist.}}{\text{old dist.}}$$

All of the above calculations are based on the assumption that the ship speed lies between $V_1$ and $V_2$ (figure 7.2). If the speed is reduced to $V_3$, however, the specific consumption may be increased by $x$%. In this case, the daily consumption and voyage consumption are also increased by $x$%.

EXAMPLE: A ship has a daily fuel consumption of 30 tonne at 15 knots. The speed is reduced to 12 knots and at this speed the consumption per unit power is 8% more than at 15 knots. Calculate the new consumption per day.

$$\text{New daily consumption} = 1.08 \times 30 \times \left(\frac{12}{15}\right)^3$$
$$= 16.6 \text{ tonne}$$

It should be noted that if a formula for fuel consumption is given in any question, the formula must be used for the *complete question*.

# Chapter 7 Test Examples

**Q. 7.1:** A ship has a wetted surface area of 3200 m². Calculate the power required to overcome frictional resistance at 17 knots if $n = 1.825$ and $f = 0.424$.

**Q. 7.2:** A plate towed edgeways in sea water has a resistance of 13 N/m² at 3 m/s. A ship travels at 15 knots and has a wetted surface area of 3800 m². If the frictional resistance varies as speed$^{1.97}$, calculate the power required to overcome frictional resistance.

**Q. 7.3:** The frictional resistance per square metre of a ship is 12 N at 180 m/min. The ship has a wetted surface area of 4000 m² and travels at 14 knots. Frictional resistance varies as speed$^{1.9}$. If frictional resistance is 70% of the total resistance, calculate the effective power.

**Q. 7.4:** A ship is 125 m long, 16 m beam and floats at a draught of 7.8 m. Its block coefficient is 0.72. Calculate the power required to overcome frictional resistance at 17.5 knots if $n = 1.825$ and $f = 0.423$. Use Taylor's formula for wetted surface, with $c = 2.55$.

**Q. 7.5:** The residuary resistance of a one-twentieth scale model of a ship in sea water is 36 N when towed at 3 knots. Calculate the residuary resistance of the ship at its corresponding speed and the power required to overcome residuary resistance at this speed.

**Q. 7.6:** A ship of 14000 tonne displacement has a residuary resistance of 113 kN at 16 knots. Calculate the corresponding speed of a similar ship of 24000 tonne displacement and the residuary resistance at this speed.

**Q. 7.7:** The frictional resistance of a ship in fresh water at 3 m/s is 11 N/m². The ship has a wetted surface area of 2500 m² and the frictional resistance is 72% of the total resistance and varies as speed$^{1.92}$. If the effective power is 1100 kW, calculate the speed of the ship.

**Q. 7.8:** A 6 m model of a ship has a wetted surface area of 7 m², and when towed in fresh water at 3 knots, has a total resistance of 35 N. Calculate the effective power of the ship, 120 m long, at its corresponding speed.
$n = 1.825$: $f$ from formula: SCF = 1.15.

**Q. 7.9:** A ship of 12000 tonne displacement has an Admiralty Coefficient of 550. Calculate the shaft power at 16 knots.

**Q. 7.10:** A ship requires a shaft power of 2800 kW at 14 knots, and the Admiralty Coefficient is 520. Calculate:

a) the displacement

b) the shaft power if the speed is reduced by 15%.

**Q. 7.11:** A ship of 8000 tonne displacement has an Admiralty Coefficient of 470. Calculate its speed if the shaft power provided is 2100 kW.

**Q. 7.12:** A ship 150 m long and 19 m beam floats at a draught of 8 m and has a block coefficient of 0.68.

a) If the Admiralty Coefficient is 600, calculate the shaft power required at 18 knots.

b) If the speed is now increased to 21 knots, and within this speed range resistance varies as speed³, find the new shaft power.

**Q. 7.13:** A ship of 15000 tonne displacement has a fuel coefficient of 62500. Calculate the fuel consumption per day at 14.5 knots.

**Q. 7.14:** A ship of 9000 tonne displacement has a fuel coefficient of 53500. Calculate the speed at which it must travel to use 25 tonne of fuel per day.

**Q. 7.15:** A ship travels 2000 nautical miles at 16 knots and returns with the same displacement at 14 knots. Find the saving in fuel on the return voyage if the consumption per day at 16 knots is 28 tonne.

**Q. 7.16:** The daily fuel consumption of a ship at 15 knots is 40 tonne. 1100 nautical miles from port, it is found that the bunkers are reduced to 115 tonne. If the ship reaches port with 20 tonne of fuel on board, calculate the reduced speed and the time taken in hours to complete the voyage.

**Q. 7.17:** A ship uses 23 tonne of fuel per day at 14 knots. Calculate the speed if the consumption per day is:

a) increased by 15%

b) reduced by 12%

c) reduced to 18 tonne.

**Q. 7.18:** The normal speed of a ship is 14 knots and the fuel consumption per hour is given by $0.12 + 0.001\ V^3$ tonne, with $V$ in knots, Calculate:

a) the total fuel consumption over a voyage of 1700 nautical miles

b) the speed at which the vessel must travel to save 10 tonne of fuel per day.

**Q. 7.19:** A ship's speed is increased by 20% above normal for 8 hours, reduced by 10% below normal for 10 hours and for the remaining 6 hours of the day the speed is normal. Calculate the percentage variation in fuel consumption in that day from normal.

**Q. 7.20:** A ship's speed was 18 knots. A reduction of 3.5 knots gave a saving in fuel consumption of 22 tonne per day. Calculate the consumption per day at 18 knots.

**Q. 7.21:** The daily fuel consumption of a ship at 17 knots is 42 tonne. Calculate the speed of the ship if the consumption is reduced to 28 tonne per day, and the specific consumption at the reduced speed is 18% more than at 17 knots.

# PROPELLERS

Propellers provide the thrust to move the vast majority of ships and marine craft. While for high-speed applications some vessels may use water jets, over a wide speed range standard propellers are the most efficient and effective means of propulsion.

The propeller's function is to convert the torque provided by the propeller shaft into a linear thrust. The propeller will be formed from a number of blades (three to seven), being attached to a central hub or boss. The blades are shaped on a helix and will have a wing-like cross section. The angle and shape of the blade causes the water passing the back face of the blades (the face that can be seen when viewed from astern) to be at a greater pressure than water passing the front face. The pressure difference resolves into the thrust, which propels the vessel. The naval architect will look to select a propeller for the typical service speed of the ship. The aim is to have a propeller that provides the required thrust for the minimum amount of input torque, thereby being as efficient as possible, and burning less fuel.

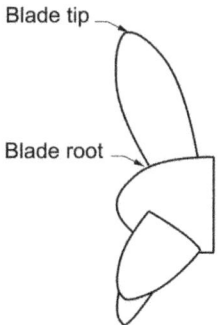
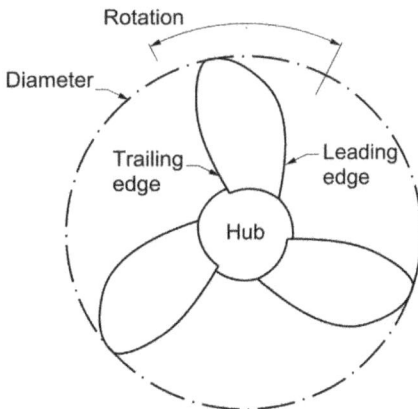

▲ **Figure 8.1** *Parts of a propeller*

# Features of Propellers

**DIAMETER** $D$ The diameter of the propeller is the diameter of the circle or disc cut out by the blade tips. Increasing a propeller's diameter will generally increase its efficiency, but this will be limited by the draught and form of the aft part of the ship.

**PITCH** $P$ is defined as the distance the propeller would move forwards in one revolution if it were working in an unyielding fluid. This is similar to how a screw thread progresses into wood as it is turned. In reality, water is not an unyielding fluid, and so the propeller would not progress as far as the pitch.

**PITCH RATIO** $p$, or face pitch ratio, is the face pitch divided by the diameter. Thus:

$$p = \frac{P}{D}$$

**THEORETICAL SPEED** $V_T$ is the distance the propeller would advance in unit time if working in an unyielding fluid. Thus, if the propeller turns at $N$ rev/min,

$$V_T = P \times N \text{ m/min}$$
$$= \frac{P \times N \times 60}{1852} \text{ knots}$$

**APPARENT SLIP:** As described previously, water is not an unyielding fluid, and so the ship speed $V$ will normally be less than the theoretical speed. The difference between the two speeds is known as the *apparent slip* and is usually expressed as a ratio or percentage of the theoretical speed.

$$\text{Apparent slip speed} = V_T - V \text{ knots}$$
$$\text{Apparent slip} = \frac{V_T - V}{V_T} \times 100\%$$

If the ship speed is measured relative to the surrounding water, ie by means of a log line or paddle wheel speedometer, the theoretical speed will invariably exceed the ship speed, giving a *positive* apparent slip. If, however, the ship speed is measured relative to the land, then any movement of water will affect the apparent slip, and should the vessel be travelling in a following current the ship speed may exceed the theoretical speed, resulting in a *negative* apparent slip.

**WAKE:** In its passage through the water, the ship sets in motion particles of water in its neighbourhood, as described in Chapter 7. This moving water is known as the *wake* and is important in propeller calculations since the propeller works in wake water. The speed of the ship relative to the wake is termed the speed of advance.

**SPEED OF ADVANCE** $V_a$. The wake speed is often expressed as a fraction of the ship speed.

$$\text{Wake fraction } w_t = \frac{V - V_a}{V}$$

The wake fraction may be obtained approximately from the expression

$$w_t = 0.5C_b - 0.05$$

where $C_b$ is the block coefficient. The subscript $t$ is because the wake fraction is also known as the *Taylor wake fraction*.

**REAL SLIP** or **TRUE SLIP** is the difference between the theoretical speed and the speed of advance, expressed as a ratio or percentage of the theoretical speed.

$$\text{Real slip speed} = V_T - V_a \text{ knots}$$
$$\text{Real slip} = \frac{V_T - V_a}{V_T} \times 100\%$$

The real slip is always positive and is independent of current.

EXAMPLE: A propeller of 4.5 m pitch turns at 120 rev/min and drives the ship at 15.5 knots. If the wake fraction is 0.30, calculate the apparent slip and the real slip.

$$\text{Theoretical speed } V_T = \frac{4.5 \times 120 \times 60}{1852}$$
$$= 17.49 \text{ knots}$$
$$= 11.38\%$$
$$w_t V = V - V_a$$
$$w_t = \frac{V - V_a}{V}$$

Therefore:

$$V_a = V(1 - w_t)$$
$$= 15.5 \times (1 - 0.3)$$
$$= 15.5 \times 0.7$$
$$= 10.85 \text{ knots}$$
$$\text{Real slip} = \frac{17.49 - 10.85}{17.49} \times 100\%$$
$$= 37.96\%$$

The relation between the different speeds may be shown clearly by the following line diagram.

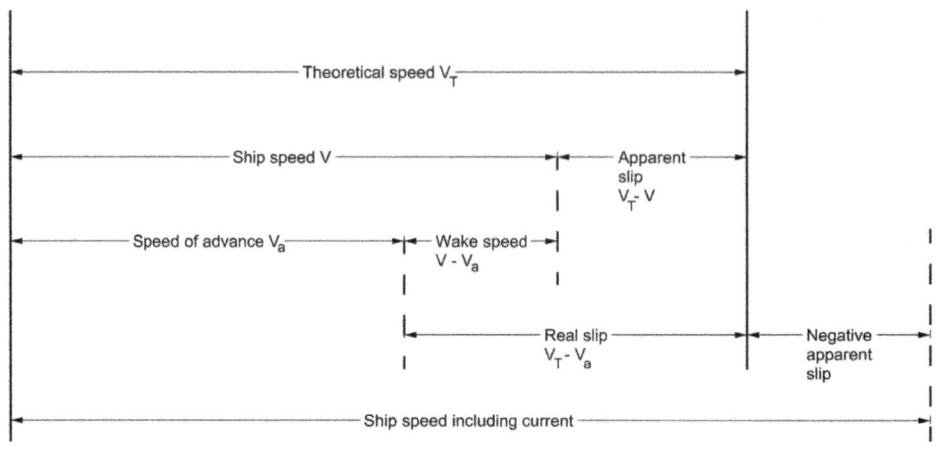

▲ **Figure 8.2** *Propeller slip*

**PROJECTED AREA** $A_p$ is the sum of the blade areas projected on to a plane that is perpendicular to the axis of the screw.

**DEVELOPED AREA** is the actual area of the driving faces

a)   clear of the boss $A_D$
b)   including the boss area $A_B$.

**BLADE AREA RATIO** BAR is the developed area excluding boss divided by the area of the circle cut out by the blade tips

$$BAR = \frac{A_D}{\frac{\pi}{4}D^2}$$

**DISC AREA RATIO** DAR is the developed area including boss divided by the area of the circle cut out by the blade tips

$$DAR = \frac{A_B}{\frac{\pi}{4}D^2}$$

# Thrust

The thrust exerted by a propeller may be calculated approximately by regarding the propeller as a disc that increases the momentum of the water coming into it. Water is received into the propeller disc at the speed of advance and projected aft at the theoretical speed.

Consider a time interval of 1 second.

Let

$$A = \text{effective disc area in m}^2$$
$$= \text{disc area} - \text{boss area}$$
$$\rho = \text{density of water in kg/m}^3$$
$$P = \text{pitch of propeller in m}$$
$$n = \text{revs/s}$$
$$V_a = \text{speed of advance in m/s}$$

Mass of water passing through disc in 1 second.

$$M = \rho APn \text{ kg}$$
$$\text{Change in velocity} = Pn - V_a \text{ m/s}$$

Since this change in velocity occurs within 1 second, the acceleration, $a$, is:

$$a = Pn - V_a \text{ m/s}^2$$

But real slip, $s$, is:

$$s = \frac{Pn - V_a}{Pn}$$
$$pn - V_a = sPn$$
$$a = sPn$$

Since,

$$\text{Force} = \text{mass} \times \text{acceleration}$$
$$\text{Thrust } T = M \times a$$
$$= \rho APn \times sPn$$
$$= \rho AP^2 n^2 s \text{ N}$$

It is interesting to note that increased slip leads to increased thrust and that the propeller will not exert a thrust with zero slip. The power produced by the propeller is known as the *thrust power* (tp).

$$\text{tp} = \text{Thrust (N)} \times \text{speed of advance (m/s) W}$$
$$= T \times V_a \text{ W}$$

Hence:

$$\frac{tp_1}{tp_2} = \frac{T_1 V_{a_1}}{T_2 V_{a_2}}$$

If the power remains constant, but the external conditions vary, then

$$T_1 V_{a_1} = T_2 V_{a_2}$$

and since the speed of advance depends upon rev/min,

$$T_1 N_1 = T_2 N_2$$

Now, the thrust is absorbed by the thrust collars and hence the thrust varies directly as the pressure $t$ on the thrust collars.

Therefore:

$$t_1 N_1 = t_2 N_2$$

This indicates that if, with constant power, the ship meets a head wind, the speed will reduce but the pressure on the thrust collars will increase.

EXAMPLE: The tp of a ship is 2000 kW and the pressure on the thrust 20 bar at 120 rev/min.
Calculate the pressure on the thrust when the tp is 1800 kW at 95 rev/min.

$$\frac{20 \times 120}{2000} = \frac{t_2 \times 95}{1800}$$

$$t_2 = \frac{20 \times 120 \times 1800}{2000 \times 95}$$

$$= 22.74 \text{ bar}$$

# Relation Between Powers

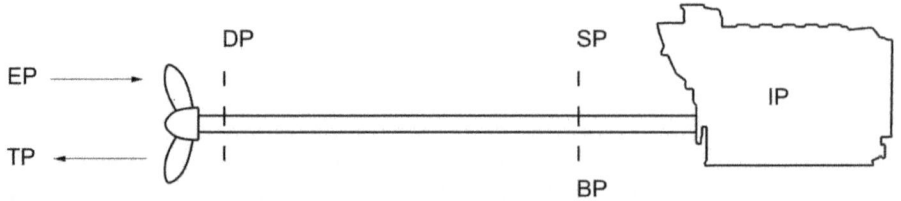

▲ **Figure 8.3** *Power diagram*

The power produced by the engine is the *indicated power* (ip). The mechanical efficiency of the engine is usually between about 80% and 90% and therefore only this percentage of the ip is transmitted to the shaft, giving the *shaft power* (sp) or *brake power* (bp).

$$\text{sp or bp} = \text{ip} \times \text{mechanical efficiency}$$

Shaft losses vary between about 3% and 5% and therefore the power delivered to the propeller, the delivered power dp, is almost 95% of the sp.

$$\text{dp} = \text{sp} \times \text{transmission efficiency}$$

The delivered power may be calculated from the torque on the shaft:

$$\text{dp} = \text{torque} \times 2\pi n$$

The propeller has an efficiency of 60% to 70% and hence the thrust power tp is given by:

$$\text{tp} = \text{dp} \times \text{propeller efficiency}$$

The action of the propeller in accelerating the water creates a suction on the after end of the ship. The thrust exerted by the propeller must exceed the total resistance by this amount. The relation between thrust and resistance may be expressed in the form:

$$R_t = T(1-t)$$

where *t* is the thrust deduction factor.

The thrust power will therefore differ from the effective power. The ratio of ep to tp is known as the *hull efficiency*, which is a little more than unity for single-screw ships and about unity for twin-screw ships.

$$\text{ep} = \text{tp} \times \text{hull efficiency}$$

In an attempt to estimate the power required by the machinery from the calculation of ep, a *quasi-propulsive coefficient* (QPC) is introduced. This is the ratio of ep to dp and obviates the use of hull efficiency and propeller efficiency. The prefix 'quasi' is used to show that the mechanical efficiency of the machinery and the transmission losses have not been taken into account.

$$\text{ep} = \text{dp} \times \text{QPC}$$

The true propulsive coefficient is the relation between the ep and the ip, although in many cases sp is used in place of ip,

ie

$$\text{ep} = \text{ip} \times \text{propulsive coefficient}$$

or

$$\text{ep} = \text{sp} \times \text{propulsive coefficient}$$

EXAMPLE: The total resistance of a ship at 13 knots is 180 kN, the QPC is 0.70, shaft losses 5% and the mechanical efficiency of the machinery 87%.

Calculate the indicated power.

$$ep = R_t \times V$$
$$= 180 \times 10^3 \times 13 \times 0.5144$$
$$= 1204 \text{ kW}$$
$$dp = \frac{ep}{QPC}$$
$$= \frac{1204}{0.7}$$
$$= 1720 \text{ kW}$$
$$sp = \frac{dp}{\text{transmission efficiency}}$$
$$= \frac{1720}{0.95}$$
$$= 1810 \text{ kW}$$
$$ip = \frac{sp}{\text{mechanical efficiency}}$$
$$= \frac{1810}{0.87}$$
$$\text{indicated power} = 2080 \text{ kW}$$
$$\tan \theta = \frac{P}{2\pi R}$$
$$\text{Pitch} = \tan \theta \times 2\pi R$$

# Measurement of Pitch

If the propeller is assumed to have no forward motion, then a point on the blade, distance $R$ from the centre of the boss, will move a distance of $2\pi R$ in one revolution. As stated previously, the pitch, $P$, is the distance the propeller will advance in one revolution in an unyielding fluid. The pitch angle $\theta$ may be defined as:

$$\tan \theta = \frac{P}{2\pi R}$$
$$\text{Pitch} = \tan \theta \times 2\pi R$$

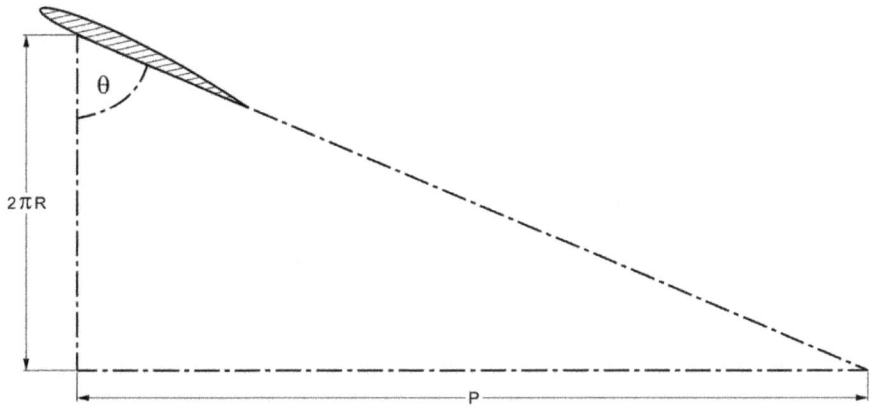

▲ **Figure 8.4** *Propeller pitch*

The pitch of a propeller may be measured without removing the propeller from the ship, by means of a simple instrument known as a *pitchometer*. One form of this instrument consists of a protractor with an adjustable arm. The face of the boss is used as a datum, and a spirit level is set horizontal when the pitchometer is set on the datum. The instrument is then set on the propeller blade at the required distance from the boss and the arm containing the level moved until it is horizontal. A reading of pitch angle or pitch may then be read from the protractor at the required radius (figure 8.5).

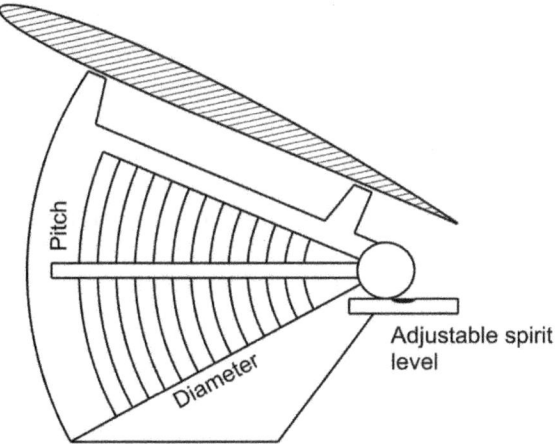

▲ **Figure 8.5** *Propeller pitch measurement gauge*

An alternative method is to turn the propeller until one blade is horizontal. A weighted cord is draped over the blade at any given radius, as shown in figure 8.6.

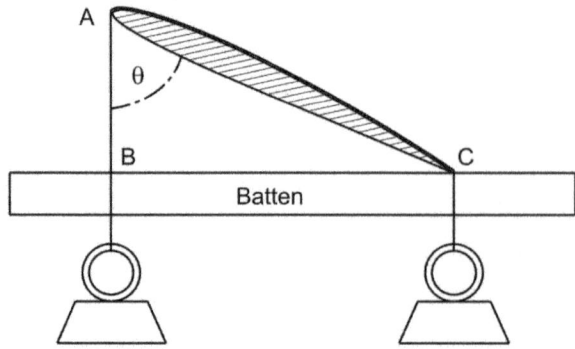

▲ **Figure 8.6** *Propeller pitch measurement*

A batten is placed horizontally at the lower edge of the blade with the aid of a spirit level. The distances AB and BC are then measured. $\theta$ is the pitch angle, and

$$\tan\theta = \frac{BC}{AB}$$

Therefore:

$$\text{Pitch} = \frac{BC}{AB} \times 2\pi R$$

# Cavitation

The thrust of a propeller varies approximately as the square of the revolutions. Thus, as the speed of rotation is increased there is a considerable increase in thrust. The distribution of pressure due to thrust over the blade section is approximately as shown in figure 8.7.

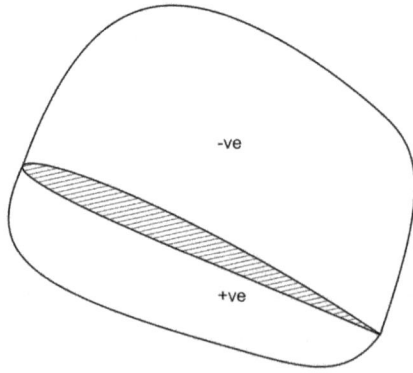

▲ **Figure 8.7** *Propeller pitch measurement*

The net pressure at any point on the back of the blade is the algebraic sum of the atmospheric pressure, water pressure and negative pressure or suction, which generates the thrust. When this suction is high at any point, the net pressure may fall below the vapour pressure of the water at water temperature, causing a cavity or bubble to form on the blade. This cavity is filled with water vapour and with air, which disassociates from the sea water. As the blade turns, the bubble moves across the blade to a point where the net pressure is higher, causing the cavity to collapse. The forming and collapsing of these cavities is known as *cavitation*.

When the cavity collapses, the water pounds the blade material, and since the breakdown occurs at the same position each time, causes severe erosion of the blades and may produce holes in the blade material several millimetres deep. Cavitation also causes reduction in thrust and efficiency, vibration and noise. It may be reduced or avoided by reducing the revolutions and by increasing the blade area for constant thrust, thus reducing the negative pressure.

The propeller designer must balance increasing the blade area against reductions in efficiency, as a larger blade area often requires more torque to turn it. The number of blades is also a design parameter that propeller designers will vary. For a given amount of thrust, the greater the number of blades, the less thrust each propeller blade must provide. This reduction in thrust per blade means a reduced pressure difference across the blades, which thereby reduces the likelihood of cavitation or vibration. As with increasing blade area, there is often an increase in the torque required to turn a propeller with more blades.

Since cavitation is affected by pressure and temperature, it is more likely to occur in propellers operating near the surface than in those deeply submerged, and will occur more readily in the tropics than in cold regions.

# Chapter 8 Test Examples

**Q. 8.1:** A ship travels at 14 knots when the propeller, 5 m pitch, turns at 105 rev/min. If the wake fraction is 0.35, calculate the apparent and real slip.

**Q. 8.2:** A propeller of 5.5 m diameter has a pitch ratio of 0.8. When turning at 120 rev/min, the wake fraction is found to be 0.32 and the real slip 35%. Calculate the ship speed, speed of advance and apparent slip.

**Q. 8.3:** A ship of 12400 tonne displacement is 120 m long, 17.5 m beam and floats at a draught of 7.5 m. The propeller has a face pitch ratio of 0.75 and, when turning at 100 rev/min, produces a ship speed of 12 knots with a real slip of 30%. Calculate the apparent slip, pitch and diameter of the propeller. The wake fraction $w$ may be found from the expression:

$w = 0.5C_b - 0.05$

**Q. 8.4:** When a propeller of 4.8 m pitch turns at 110 rev/min, the apparent slip is found to be $-s$% and the real slip $+ 1.5s$%. If the wake speed is 25% of the ship speed, calculate the ship speed, the apparent slip and the real slip.

**Q. 8.5:** A propeller 4.6 m diameter has a pitch of 4.3 m and boss diameter of 0.75 m. The real slip is 28% at 95 rev/min.

Calculate:

    **a)** the speed of advance

    **b)** thrust

    **c)** thrust power.

**Q. 8.6:** The pressure exerted on the thrust is 17.5 bar at 115 rev/min. Calculate the thrust pressure at 90 rev/min.

**Q. 8.7:** The power required to drive a ship at a given speed was 3400 kW and the pressure on the thrust 19.5 bar. Calculate the new thrust pressure if the speed is reduced by 12% and the corresponding power is 2900 kW.

**Q. 8.8:** A ship of 15000 tonne displacement has an Admiralty Coefficient, based on shaft power, of 420. The mechanical efficiency of the machinery is 83%, shaft losses 6%, propeller efficiency 65% and QPC 0.71. At a particular speed the thrust power is 2550 kW.

Calculate:

    **a)** indicated power

    **b)** effective power

    **c)** ship speed.

**Q. 8.9:** A propeller of 4 m pitch has an efficiency of 67%. When turning at 125 rev/min, the real slip is 36% and the delivered power 2800 kW. Calculate the thrust of the propeller.

**Q. 8.10:** The pitch angle, measured at a distance of 2 m from the centre of the boss, was found to be 21.5°. Calculate the pitch of the propeller.

**Q. 8.11:** The pitch of a propeller is measured by means of a batten and cord. The horizontal ordinate is found to be 40 cm while the vertical ordinate is 1.15 m at a distance of 2.6 m from the centre of the boss. Calculate the pitch of the propeller and the blade width at that point.

# 9

# RUDDER THEORY

The vast majority of vessels have a rudder as the primary means for manoeuvring. The rudder does this by generating a force transverse to the direction of travel, which, by being located away from the ship's centre of gravity, creates a turning moment that will cause the vessel to change direction.

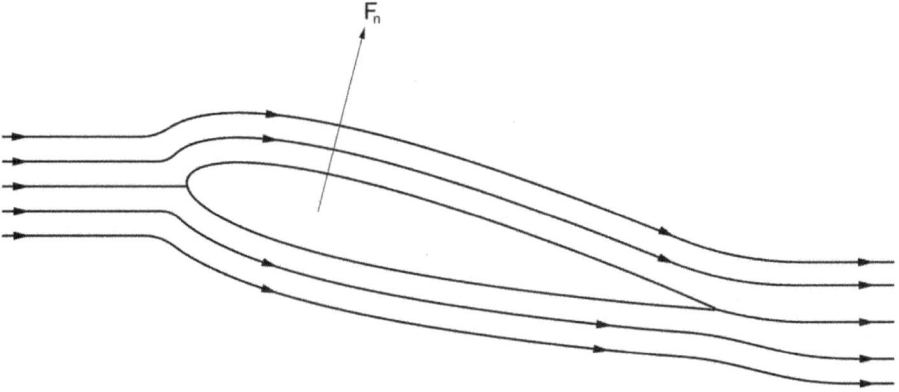

▲ **Figure 9.1** *Cross section through rudder with streamlines and force acting*

The rudder behaves in a similar way to a wing in that, when angled to the flow, the water travelling over the side presented to the flow moves more slowly around the rudder than the side that has been turned away from the incoming flow (figure 9.1). This creates a difference in the fluid dynamic pressure acting over the rudder, which results in a force acting on the rudder. By assuming this force is normal to the rudder, it can be described by the following equation:

$$F_n = kAV^2 \sin\alpha$$

where $k$ = a coefficient that depends upon the shape of the rudder and the density of the water. When the ship speed is expressed in m/s, average values of $k$ for sea water vary between about 570 and 610.

$A$ = rudder area

$V$ = ship speed.

$a$ = rudder angle

The area of rudder is specified by Classification Societies, and is usually given by a formula incorporating various parameters relating to ship size, type and rudder type. The largest influence in defining the rudder area is the area of the ship's middle-line plane (ie length of ship × draught). Values of one-sixtieth for fast ships and one-seventieth for slow ships are typical:

$$\text{area of rudder} = \frac{L \times d}{60} \text{ for fast ships}$$
$$= \frac{L \times d}{70} \text{ for slow ships}$$

The force $F_n$ acts at the centre of effort of the rudder. The position of the centre of effort varies with the shape of the rudder and the rudder angle. For rectangular rudders, the centre of effort is between 20% and 38% of the width of the rudder from the leading edge. The rudder stock is usually positioned forward of this point, so that the effect of the normal force is to tend to push the rudder back to its centreline position. Such movement is resisted by the rudder stock and the steering gear. By positioning the stock in front of the steering gear, the rudder should return to the centreline in the event of failure of the steering gear. From the equations above, it is possible to calculate the turning moment or torque on the rudder stock.

If the centre of effort is $b$m from the centre of the rudder stock, then at any angle $a$

$$\text{Torque on stock } T = F_n \times b$$
$$= kAV^2 b \sin\alpha$$

From the basic torsion equation, the diameter of the stock may be found for any given allowable stress.

$$\frac{T}{J} = \frac{q}{r}$$

where $q$ = allowable stress in N/m$^2$

$r$ = radius of stock in m

$J$ = second moment of area about a polar axis in m$^4$

$$J = \frac{\pi d^4}{32}$$

$$= \frac{\pi r^4}{2}$$

For any rudder, at constant ship speed, values of torque may be plotted on a base of rudder angle. The area under this curve up to any angle is the work done in turning the rudder to this angle, and may be found by the use of Simpson's Rule. Care must be taken to express the common interval in radians, not degrees.

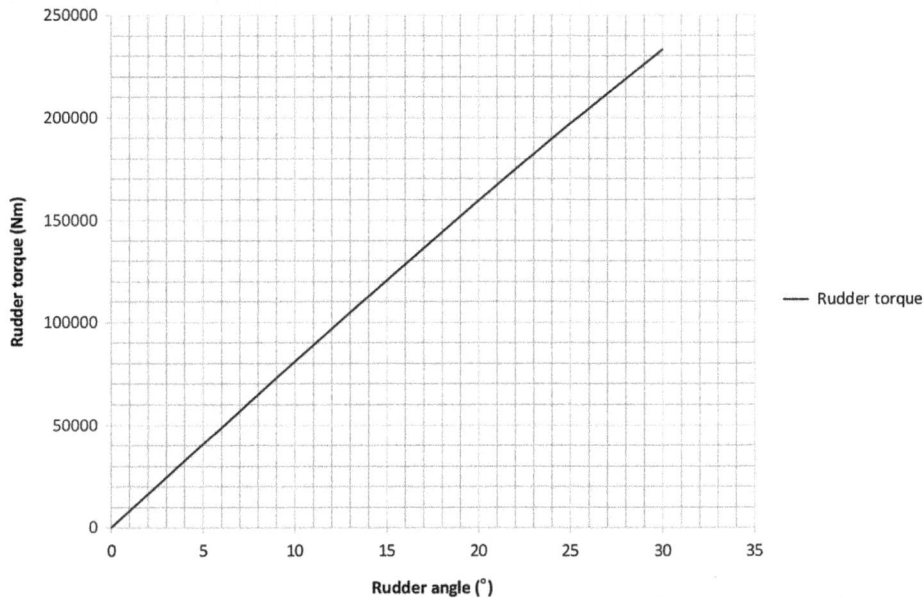

▲ **Figure 9.2** *Torque versus rudder angle*

If the centre of the rudder stock is between 20% and 38% of the width of the rudder from the leading edge, then at a given angle the centre of stock will coincide with the centre of effort and thus there will be no torque. The rudder is then said to be *balanced*. At any other rudder angle, the centres of stock and effort will not coincide and there will be a torque of reduced magnitude.

While it may be seen that the diameter of stock and power of the steering gear may be reduced if a balanced rudder is fitted, this may lead to complications with other elements of the steering gear. As the rudder is located in the wake of the propeller, the force on it is always unsteady. With a balanced rudder, the variation in torque occurs about a zero mean, which requires rapid changing in the direction in which the steering control system applies a resistive torque. The control system will have to react less, to a varying torque of larger magnitude, but which does not change direction.

It is usual to limit the rudder angle to 35° on each side of the centreline, since, if this angle is exceeded, the diameter of the turning circle is increased.

EXAMPLE: A rudder has an area of 15 m² with its centre of effort 0.9 m from the centre of stock. The maximum rudder angle is 35° and it is designed for a service speed of 15 knots. Calculate the diameter of the rudder stock if the maximum allowable stress in the stock is 55 MN/m² and the normal rudder force is given by:

$$F_n = 580AV^2 \sin \alpha$$
$$\text{Ship speed } V = 15 \times 0.5144$$
$$= 7.716 \text{ m/s}$$
$$F_n = 580 \times 15 \times 7.716^2 \times 0.5736$$
$$= 297107 \text{ N}$$
$$= \text{Torque } T = F_n b$$
$$= 297107 \times 0.9$$
$$= 267396 \text{ Nm}$$
$$\frac{T}{J} = \frac{q}{r}$$
$$J = \frac{Tr}{q}$$
$$\frac{\pi r^4}{2} = \frac{Tr}{q}$$
$$r^3 = \frac{2T}{\pi q}$$
$$= \frac{2 \times 267396}{3.142 \times 55 \times 10^6}$$
$$= 0.0031 \text{ m}^3$$
$$r = 0.146 \text{ m}$$

# Angle of Heel Due to Force on Rudder

When the rudder is turned from its central position, a transverse component of the normal rudder force acts on the rudder.

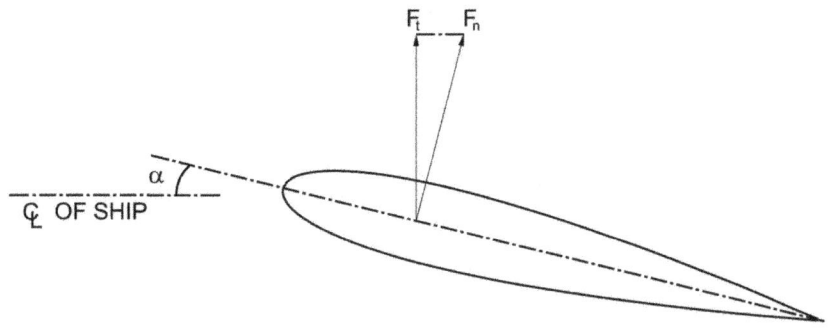

▲ **Figure 9.3** *Transverse component of rudder force*

$$F_t = \text{transverse rudder force}$$
$$F_t = F_n \cos \alpha$$
$$= kAV^2 \sin \alpha \cos \alpha$$

This transverse force acts at the centre of the rudder N, and tends to push the ship sideways. A resistance R is exerted by the water on the ship, and acts at the *centre of lateral resistance L*, which is the centroid of the projected, immersed plane of the ship (sometimes taken as the centre of buoyancy). This resistance is increased as the ship moves, until it reaches its maximum value when it is equal to the transverse force. At this point, a moment acts on the ship, causing it to heel to an angle θ when the heeling moment is equal to the righting moment.

$$\text{Heeling moment} = F_t \times NL \cos \theta$$
$$\text{Righting moment} = \Delta_g \times GZ$$

if θ is small:

$$= \Delta g \times GM \sin \theta$$

For equilibrium:

$$\text{Righting moment} = \text{heeling moment}$$
$$\Delta g \times GM \sin \theta = F_t \times NL \cos \theta$$
$$\frac{\sin \theta}{\cos \theta} = \frac{F_t \times NL}{\Delta g \times GM}$$
$$\tan \theta = \frac{F_t \times NL}{\Delta g \times GM}$$

From this, the angle of heel may be obtained.

The angle of heel due to the force on the rudder is small unless the speed is excessive or the metacentric height small. In most merchant ships, this angle is hardly noticeable.

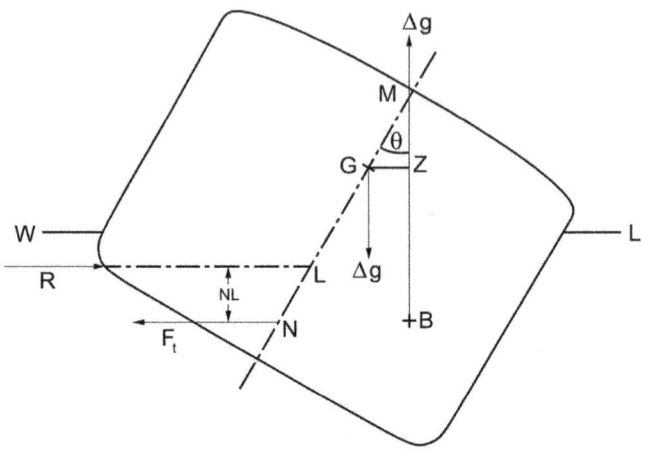

▲ **Figure 9.4** *Heeling effect of rudder angle in cross section*

EXAMPLE: A ship of 8000 tonne displacement has a rudder of area 18 m². The centre of lateral resistance is 4 m above the keel, while the centroid of the rudder is 2.35 m above the keel. The maximum rudder angle is 35°. Calculate the angle of heel due to the force on the rudder if the latter is put hard over to port when travelling at 21 knots with a metacentric height of 0.4 m.

$$F_n = 580AV^2 \sin\alpha$$
$$\text{Ship speed } V = 21\times0.5144$$
$$= 10.80 \text{ m/s}$$
$$\text{Transverse force } F_t = 580AV^2 \sin\alpha\cos\alpha$$
$$= 580\times18\times10.80^2 \times 0.5736\times0.8192$$
$$= 572200 \text{ N}$$
$$\tan\theta = \frac{F_t \times NL}{\Delta g \times GM}$$
$$= \frac{527200\times(4.0-2.35)}{8000\times10^3 \times9.81\times0.4}$$
$$= 0.03007$$
$$\text{Angle of heel } \theta = 1.722° \text{ to port}$$

# Angle of Heel When Turning

As the ship commences to turn, a centrifugal force acts in addition to the rudder force. The effect of this force is to create a moment opposing the rudder force, ie tending to heel the ship in the opposite direction.

It is convenient to ignore the rudder force and consider only the centrifugal force. This force acts at the centre of gravity of the ship and may be calculated from the formula:

$$\text{Centrifugal force } CF = \frac{\Delta V^2}{\beta}$$

where $V$ is the ship speed in m/s
$\beta$ is the radius of the turning circle in m

Once again, a resistance $R$ is exerted by the water on the ship due to the transverse movement, and has its maximum value when it is equal to the centrifugal force. This resistance is known as the *centripetal force*. A moment then acts on the ship, causing it to heel.

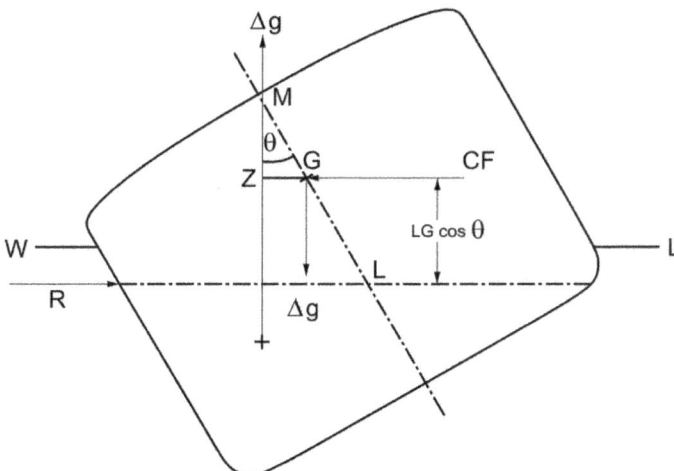

▲ **Figure 9.5** *Heeling effect due to centrifugal force*

The ship will be in equilibrium when the heeling moment is equal to the righting moment.

$$\text{Heeling moment} = CF \times LG\cos\theta$$

$$= \frac{\Delta V^2}{\beta} \times LG\cos\theta$$

$$\text{Righting moment} = \Delta g \times GZ$$

if $\theta$ is small:

$$= \Delta g \times GM\sin\theta$$

$$\Delta g \times GM\sin\theta = \frac{\Delta V^2}{\beta} \times LG\cos\theta$$

$$\frac{\sin\theta}{\cos\theta} = \frac{\Delta V^2}{\beta} \times \frac{LG}{\Delta g \times GM}$$

$$\tan\theta = \frac{V^2 \times LG}{g \times \beta \times GM}$$

From this expression, the angle of heel may be calculated.

It is usual in calculations of heel when turning to ignore the heel due to the rudder force and consider it to be a small factor of safety, ie the actual angle of heel will be less than that calculated. If, when the ship is turning in a circle to port, the rudder is put hard over to starboard, the heel due to the rudder force is added to the previous heel due to centrifugal force, causing an increase in angle of heel. This may prove dangerous, especially in a small, high-speed vessel.

EXAMPLE: A ship with a metacentric height of 0.4 m has a speed of 21 knots. The centre of gravity is 6.2 m above the keel, while the centre of lateral resistance is 4 m above the keel. The rudder is put hard over to port and the vessel turns in a circle 1100 m radius. Calculate the angle to which the ship will heel.

$$\text{Ship speed } V = 21 \times 0.5144$$

$$= 10.80 \text{ m/s}$$

$$\tan\theta = \frac{V^2 \times LG}{g \times \beta \times GM}$$

$$= \frac{10.80^2 \times (6.2 - 4.0)}{9.81 \times 1100 \times 0.4}$$

$$= 0.1189$$

$$\text{Angle of heel } \theta = 6.78°$$

Using the details from the previous example, it may be seen that if $\theta_1$ is the final angle of heel to starboard:

$$\tan\theta_1 = 0.1189 - 0.03007$$
$$= 0.08883$$

Final angle of heel $\theta_1 = 5.076°$

# Chapter 9 Test Examples

*Note:* In the following questions, the normal rudder force should be taken as 580 $AV^2 \sin \alpha$.

**Q. 9.1:** A ship whose maximum speed is 18 knots has a rudder of area 25 m². The distance from the centre of stock to the centre of effort of the rudder is 1.2 m and the maximum rudder angle 35°. If the maximum allowable stress in the stock is 85 MN/m², calculate the diameter of the stock.

**Q. 9.2:** The service speed of a ship is 14 knots and the rudder, 13 m² in area, has its centre of effort 1.1 m from the centre of stock. Calculate the torque on the stock at 10° intervals of rudder angle up to 40° and estimate the work done in turning the rudder from the centreline to 40°.

**Q. 9.3:** A ship 150 m long and 8.5 m draught has a rudder whose area is one-sixtieth of the middle-line plane and diameter of stock 320 mm. Calculate the maximum speed at which the vessel may travel if the maximum allowable stress is 70 MN/m², the centre of stock is 0.9 m from the centre of effort and the maximum rudder angle is 35°.

**Q. 9.4:** A ship displaces 5000 tonne and has a rudder of area 12 m². The distance between the centre of lateral resistance and the centre of the rudder is 1.6 m and the metacentric height 0.24 m. Calculate the initial angle of heel if the rudder is put over to 35° when travelling at 16 knots.

**Q. 9.5:** A vessel travelling at 17 knots turns with a radius of 450 m when the rudder is put hard over. The centre of gravity is 7 m above the keel, the transverse metacentre 7.45 m above the keel and the centre of buoyancy 4 m above the keel. If the centripetal force is assumed to act at the centre of buoyancy, calculate the angle of heel when turning. The rudder force may be ignored.

# 10

# ALTERNATIVE FUELS AND POWERING

As concern about and understanding of climate change has grown, society has looked to take a long term move away from being reliant on fossil fuels. In support of the United Nations Sustainable Development Goals, the International Maritime Organization (IMO) has set the target to have net-zero Greenhouse Gas (GHG) emissions from international shipping by or close to 2050. This presents a number of questions for naval architects and marine engineers, from 'What does net zero mean?' to 'What technologies are needed to make this transition?' This chapter aims to explain the terminology and technologies associated with net zero shipping, and aims to give the reader some useful support in a rapidly evolving field.

## Energy

All vessels are reliant on some form of energy to propel them. There has been a historical evolution from human power, through wind power, to coal and then oil. The demands to make shipping faster or more reliable led the evolution towards having stored energy, or fuel, on board a vessel to power its propulsion. As this evolution continues, and we look to use energy sources with a reduced environmental impact, some of the key questions for naval architects and marine engineers to address are: what energy sources are available, and how do we store enough energy on the vessel to meet the requirements we have?

Figure 10.1 shows a comparison of the energy density of a variety of fuel options, many of them considered more environmentally friendly than existing marine diesel. Unfortunately, marine diesel is one of the most energy-dense fuels available, with around 15 times the amount of energy in a given volume than Lithium Ion Batteries can store. Any ship powered by Lithium Ion Batteries will need significantly more space within it to achieve the same performance criteria for speed and range as a vessel powered by diesel. This clearly isn't possible in many cases, and solutions to these problems will be outlined later, but the important point to note from figure 10.1 is that transitions towards more environmentally friendly fuels will require either more space or a reduction in performance criteria for vessels.

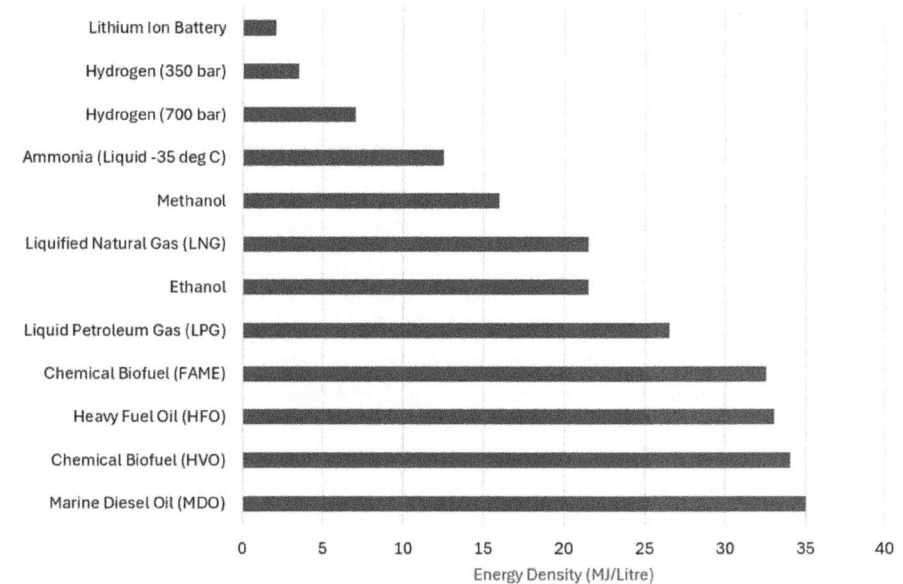

▲ **Figure 10.1** *Energy density comparison for a variety of fuel types*

It is not only the energy density that makes marine diesel a popular fuel choice; numerous other factors are at play. Short-term exposure to small quantities is not particularly damaging to human health. It is not volatile at room temperature and yet it does not require an excessively high temperature to start combusting and thereby release its energy.

The unfortunate side is that for every tonne of diesel burned, around 3 tonnes of carbon dioxide gas is produced, which in excessive quantities in the atmosphere has been shown to lead to global warming. Whilst other gases generated in combustion may absorb and radiate more heat than carbon dioxide, the quantity of carbon dioxide produced is the reason it has become the focal gas in terms of global warming.

As stated previously, a number of the fuels listed in figure 10.1 are considered more environmentally friendly than marine diesel oil, and are mentioned with regards to shipping moving towards having net-zero greenhouse gas emissions. The concept of net-zero emissions is that the greenhouse gases produced by using a fuel are balanced by some other means. For example, if a vegetable oil is used within a diesel engine, the carbon dioxide produced in combustion will be balanced by the amount of carbon dioxide the plants absorbed during their lifespan before becoming a fuel. Other gases produced, such as nitrogen oxides, would need to be dealt with by an exhaust scrubbing system for this to be considered a net-zero process.

These distinctions are important for marine engineers to be aware of, as a number of alternative energy sources for vessels are not producing zero emissions at the exhaust of the engine, but are considered to be a net-zero means of propulsion. In addition to this, when battery propulsion is considered, it is important to remember that just because there is no exhaust, there are emissions associated with both the production of the battery and the electricity generated that is stored in them.

These examples show that selecting environmentally-friendly alternatives to diesel is complex, and when done properly requires a lifecycle assessment of the vessel to be made, a study that considers the environmental impact of a vessel from production through to end of life.

Figure 10.2 shows a schematic of a propulsion system, which could be applied to any vessel.

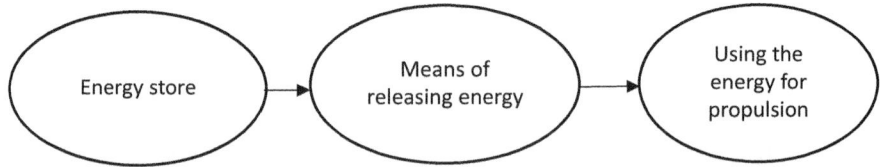

▲ **Figure 10.2** *Schematic of a propulsion system*

The combination of the significant differences in energy densities, along with multiple varying practical constraints of using different fuels for different applications, make it highly unlikely that in the long term there will be one single replacement for diesel engines. The propulsion means and fuel selected will be highly dependent on the usage profile of a particular vessel, which will lead to craft being more specific.

To add to the complexity, as shown in figure 10.3, the energy stored within different fuel types can be released in different ways, with various advantages and disadvantages associated with them.

As such, marine engineers of the future will require knowledge of a wider range of technologies than was previously the case. The remainder of this chapter will outline some of the key technologies and fuel types that will play a part in decarbonizing shipping.

| Energy store | Means of releasing Energy | Using energy for propulsion |
| --- | --- | --- |
| Marine Diesel Oil | Combustion within ICE | ICE connected to propeller shaft |
| Hydrogen | Combustion within ICE | ICE connected to propeller shaft |
| Hydrogen | Hydrogen fuel cell | Electricity produced by fuel cell powering electric motor |
| Ammonia | Combustion within ICE | ICE connected to propeller shaft |
| Ammonia | Ammonia passed via splitter to produce nitrogen (vented to air) and hydrogen (to fuel cell) | Electricity produced by fuel cell powering electric motor |
| Lithium Ion Battery | Electrical circuit | Electric motor |

▲ **Figure 10.3** *Comparison of different fuel types with associated propulsion technology*

# Energy Stores

The following section lists a number of the key fuels or stores of energy that are expected to be seen in larger quantities on vessels in the future.

**HVO (Hydrotreated Vegetable Oil):** HVO is made by using hydrogen to chemically change plant-based oils so that they can be used as a diesel replacement. Modern diesel engines need very little modification in order to be able to run on HVO, although there is a slight reduction in the energy density in comparison with diesel. As stated previously, the carbon dioxide produced in burning the fuel is offset by that which the plant matter absorbed as it grew, and this represents one of the problems for HVO as a replacement for diesel, in that its sustainability is dependent on where the raw feedstock plant material comes from. Restrictions on land available for agriculture, and whether it should be used for growing food instead, are reasons why the use of HVO is not seen as the single replacement fuel for diesel.

**Methanol:** Methanol is one of the fuels that has been identified a suitable replacement fuel, with large container ships having been built with methanol engines in place. The chemical formulation of methanol is $CH_3OH$, and so due to it containing carbon, carbon dioxide will be produced when burned. As it is possible to manufacture methanol from biological sources, the carbon footprint of the fuel can be considered to be significantly lower than that of fossil fuels. Not all sources of methanol are of this type; a significant percentage of methanol production is from chemically processing coal or natural gas, with an associated larger carbon footprint than bio-methanol.

Methanol can either be used in fuel cells or in internal combustion engines. However, the engines do need modification in comparison to fuel oil engines. There are a number of safety considerations with methanol, in terms of how it is handled as a fuel. It can burn without a flame, and it has a low flash point. Often methanol tanks will have nitrogen pumped into them, to ensure that any vapours that form at the top of the tank cannot combust.

As can be seen in figure 10.1, methanol has less than half the energy density of diesel, and so a greater proportion of the ship has to be allocated to fuel tanks than with diesel. Methanol mixes readily with water, and is not considered to be problematic in the event of a spill.

**Ammonia**: Ammonia has the chemical formulation $NH_3$, which means as there is no carbon present, no carbon dioxide will be produced when it is burned. This makes it attractive as an alternative fuel, and it has a greater energy density than hydrogen alone, as the nitrogen molecule serves the role of an efficient carrier of hydrogen. Unfortunately, there are a number of disadvantages, which in part is why ammonia has only previously been used as a fuel in periods of fuel shortage (such as the Second World War).

The process of manufacturing ammoni, can be heavily energy intensive, and so the overall emissions footprint of the fuel is very dependent on where this energy came from. Additionally, if ammonia is not burned at the correct temperature, nitrous oxides may form, though these can be dealt with using exhaust after treatment. The temperature at which it will start to combust is higher than diesel, which leads to additional loads being placed on the engine. It is corrosive and toxic, meaning special systems have to be in place for handling and storage.

Despite this, engines have been developed to run on ammonia, and an ever increasing number of ships that are fuelled by ammonia are being built.

**Hydrogen**: The appeal of using hydrogen as an alternative fuel is that on combustion, the only byproduct is water. Unfortunately, hydrogen will burn rapidly and not produce the sustained heat that other fuels will. This means that if it is used within internal combustion engines, the efficiency of the engine is about half of that of a diesel fueled one. This, coupled with the significantly lower energy density of hydrogen, means its role as a fuel is more likely to be for use in fuel cells, which have a similar efficiency to a diesel-fueled engine.

Figure 10.1 shows that the energy density of hydrogen is low, and it needs to be stored at high pressure to make it viable as a fuel. Tanks capable of withstanding this pressure are costly to build, and vessels travelling further offshore often look to store hydrogen at a higher pressure than is becoming standard within the automotive sector (350 bar). The application of hydrogen as a fuel is likely to be in vessels that can use automotive tanks, as these will be cheaper due to the economies of scale.

Hydrogen can be produced in a variety of ways, but some are dependent on the fossil fuel industry. A more sustainable supply is hydrogen produced from water electrolysis powered by renewable energy sources. This is, however, currently the minority of the global hydrogen supply.

**Batteries**: The development of handheld electronics has seen lithium ion batteries become sufficiently energy dense to be adopted in the automotive world. This has led to the maritime sector looking at batteries as a form of delivering zero emissions propulsion. Whilst at first glance, the suitable applications would be for smaller vessels, there has been extensive use of batteries in large ferries, often having charging installations capable of delivering power in the megawatt range. There are future energy density increases expected from battery chemistries currently in development, but there will always be limitations to the scale batteries are able to power oceanic vessels.

**Nuclear:** Whilst there are numerous military ships operating using nuclear fuel sources and reactors, the technology has only been applied to merchant shipping in a handful of vessels. One such example is the NS *Savannah*, which operated from 1962 until 1971. Russia has used nuclear power in 13 of its ice breakers, with more under construction. Widespread adoption has not been seen for a variety of reasons, such as the complexities of decommissioning, general operating expenses, and environmental concerns. In addition, nuclear power can be a politically contentious matter in different countries around the world. The barriers that have prevented its adoption in the past are still present, and are likely to be in the future. The economics of application in land-based power generation are likely to be more favourable than in ship-based power generation.

# Means of Releasing Energy

**Internal Combustion Engines (ICE):** The name Diesel has become associated with a particular fuel type, but originally, Rudolph Diesel's first patent in 1892 was intended for an engine that would run on vegetable oil. Regardless of which fuel is used, diesel engines work by compressing air to get it hot, then adding a fuel, which burns rapidly, releasing its energy to move the pistons that rotate the shaft. As with many inventions, the diesel engine took time to be adopted, with the first ocean-going diesel ship being built in 1912. Having more than a century's worth of development behind it, modern diesel engines are very refined pieces of precision technology, with well-established supply chains and support networks.

It is in part this reliability, established supply chains and known cost implications that make internal combustion engines suitable for alternative fuel sources, provided minor modifications are made. These are typically changes to the injection systems or fuel lines to account for the different properties of the fuel. Due to the existing large engine manufacturers putting considerable research into alternative fuels, internal combustion engines are likely to continue to play a major role within shipping for a long time to come.

**Fuel Cells:** A fuel cell is similar to a battery in that it provides electricity through electrochemical reaction. Where they differ is that they need a continuous supply of fuel to do that, and do not act as a storage device. They consist of two electrodes (a negative anode and a positive cathode) which are either side of an electrolyte. Fuel is supplied (typically in the form of hydrogen) to the anode and air (containing oxygen) is supplied to the cathode. Typically at the anode there is a catalyst, which separates the hydrogen into electrons and hydrogen ions. The hydrogen ions pass through the electrolyte to react with oxygen at the cathode, in the process generating water ($H_2O$) and heat. The electrons' path to the cathode is through the electrical load connected to the fuel cell, as it would be in a battery.

Whilst the chemical reaction is between hydrogen and oxygen, fuel cells can operate with different fuels rather than just pure hydrogen. Both methanol and ammonia can be used as fuel supplies, but this will lead to different outputs, and may require additional equipment to take the hydrogen out of the molecule prior to entering the fuel cell. In addition to water and heat, methanol will yield carbon dioxide and ammonia will yield nitrogen. This may add complication to the system, methanol and ammonia both have higher energy density than hydrogen on its own, and the storage of them is more straightforward.

Fuel cells have had a long history, being conceived in 1838, but the relatively high cost per kW of power produced in comparison to internal combustion engines has meant they have only had use in specific applications where they have advantages, such as in space or submarines.

# Using Energy for Propulsion

Modern diesel development has led to an increase in diesel-electric vessels (such as cruise ships or offshore supply vessels), and so there are already a large number of ships that have electric motors. Transitioning to alternative powering solutions will increase this number, particularly for smaller vessels.

There are different characteristics in terms of how an electric motor delivers its torque compared to an internal combustion engine, as can be seen in figure 10.4.

An internal combustion engine typically shows an increase in torque reaching an optimum value (often between 2,000–3,000 rpm), and then decreasing. Contrastingly, an electric motor can deliver its maximum torque instantaneously, with a reduction in torque with rpm as the motor reaches higher speeds. This difference can allow electric motors to turn slightly larger diameter propellers, at low rpm, thereby gaining in efficiency. Care has to be taken with regards the motor control, so that the master of the vessel has an increase in the power available at a predictable rate, rather than high levels of thrust at low rpm. This could potentially be dangerous in close-quarter manoeuvring situations.

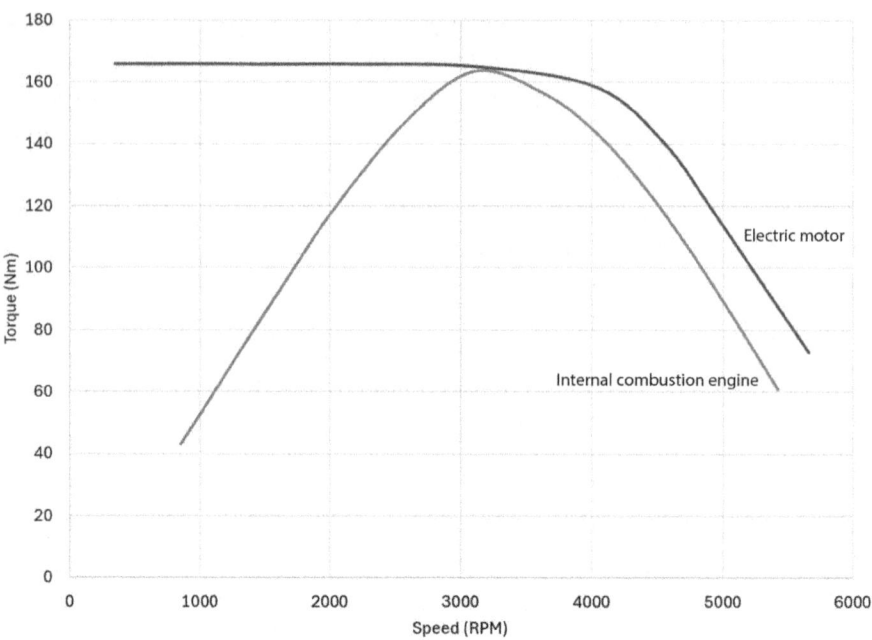

▲ **Figure 10.4** *Comparison of different fuel types with associated propulsion technology*

# Exhaust Treatment Technologies

Existing exhaust treatment technologies have made significant reductions in reducing the pollutants emitted from marine exhausts, and have been necessary for engines to meet the requirements placed on them by the IMO. Systems such as Selective Catalytic Reduction (SCR), or scrubbers as they are often known, have been used to reduce particulate matter along with sulphur and nitrous oxides. Whilst there are different forms of scrubbers, typically they work by reacting the acidic components within the exhaust gas with an alkaline material. The remaining exhaust can then leave, and any by-products be treated before safe disposal. The passing of the exhaust gas through another system leads to a decrease in the efficiency of the engine, and there is a balance between how many acidic gases can be removed without being too detrimental to engine performance. As both sulphur and nitrous oxides are more acidic than carbon dioxide, it is these gases that react first, and carbon dioxide levels are typically much the same post treatment.

Research is being conducted into how the exhaust of marine engines can be treated to remove the carbon dioxide, and store it in a format that can be more easily dealt with. Such methods are particularly attractive when used in conjunction with a fuel such as bio methanol, as it would then improve the tail pipe emissions, and reduce the carbon footprint of the fuel even further.

# Summary

As this chapter has highlighted, the complexity of vessel operations and requirements mean it is highly unlikely a long-term single replacement for diesel will emerge. It is also very likely that the technologies mentioned in this chapter are used in conjunction with one another, in hybrid systems, to utilise the various advantages they individually offer. This should also allow for greater overall efficiencies, as the most appropriate fuel will be used at a particular time. Whilst many vessels are already highly efficient, future energy prices are likely to increase, thus changing the cost benefit point of energy saving devices, which are likely to become more commonplace than has been the case with diesel.

It is hoped this chapter has served as an introduction to the technological revolution that shipping is at the start of, and supports future Marine Engineering Officers in understanding technologies they are likely to meet during their careers.

# SOLUTIONS TO CHAPTER 2 TEST EXAMPLES

**A. 2.1:**

**a)**
$$\text{Mass of steel} = 300 \text{ g}$$
$$= 0.300 \text{ kg}$$
$$\text{Volume of steel} = 42 \text{ cm}^3$$
$$= 42 \times 10^{-6} \text{ m}^3$$
$$\text{Density of steel} = 0.300 \div (42 \times 10^{-6})$$
$$= 7143 \text{ kg/m}^3$$

**b)**
$$\text{Mass of equal volume of water} = 42 \times 10^{-3} \text{ kg}$$
$$\text{Relative density of steel} = 0.300 \div (42 \times 10^{-3})$$
$$= 7.143$$

Alternatively,

$$\text{Density of water} = 1000 \text{ kg/m}^3$$
$$\text{Relative density of steel} = \text{density of steel} \div \text{density of water}$$
$$= 7143 \div 1000$$
$$= 7.143$$

**c)**
$$\text{Volume of steel} = 100 \text{ cm}^3$$
$$= 100 \times 10^{-6} \text{ m}^3$$
$$\text{Mass of steel} = 7143 \times 100 \times 10^{-6} \text{ kg/m}^3 \times \text{m}^3$$
$$= 0.7143 \text{ kg}$$

**A. 2.2:**
$$\text{Volume of wood and metal immersed} = 3.5 \times 10^3 + 250 - 100$$
$$= 3650 \text{ cm}^3$$
$$\text{Mass of wood and metal} = 1.000 \times 3650 \text{ g/cm}^3 \times \text{cm}^3$$
$$= 3650 \text{ g}$$
$$\text{Mass of wood} = 3.5 \times 10^3 \times 1.000 \times 0.60$$
$$= 2100 \text{ g}$$
$$\text{Mass of metal} = 3650 - 2100$$
$$= 1550 \text{ g}$$
$$\text{Mass of equal volume of fresh water} = 250 \times 1.000$$
$$= 250 \text{ g}$$
$$\text{Relative density} = 1550 \div 250$$
$$= 6.20$$

**A. 2.3:**
$$\text{Mass of raft} = 1000 \times 0.7 \times 3 \times 2 \times 0.25$$
$$= 1.05 \times 10^3 \text{ kg}$$
$$\text{Mass of raft when completely submerged} = 1018 \times 3 \times 2 \times 0.25$$
$$= 1.527 \times 10^3 \text{ kg}$$
$$\text{Mass required to submerge raft} = 1.527 \times 10^3 - 1.05 \times 10^3$$
$$= 0.477 \times 10^3 \text{ kg}$$
$$= 477 \text{ kg}$$

**A. 2.4:**
**a)**
$$\text{Displacement of barge} = 1025 \times 65 \times 12 \times 5.5$$
$$= 4397 \times 10^3 \text{ kg}$$
$$= 4397 \text{ tonne}$$

**b)**
$$\text{Draught in fresh water} = 5.5 \times 1.025 \div 1.000$$
$$= 5.638 \text{ m}$$

**A. 2.5:**

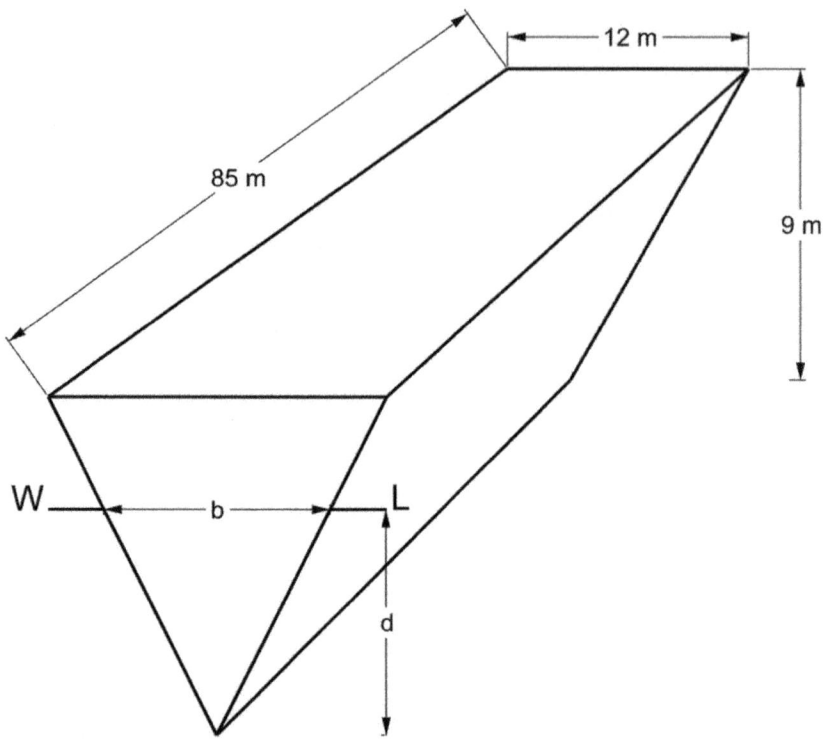

▲ Figure A. 2.1

$$\text{Let } d = \text{draught}$$
$$b = \text{breadth at waterline}$$
$$\text{By similar triangles } b = 12 \div 9 \times d$$
$$= 4 \div 3 \times d$$
$$\text{At draught d, displacement} = 1.025 \times 85 \times b \times d \div 2$$
$$= 1.025 \times 85 \times 4 \times d \div 3 \times d \div 2$$
$$= 58.08d^2 \text{ tonne}$$

Tabulating

| Draught d | $d^2$ | Displacement tonne |
|-----------|-------|--------------------|
| 0 | 0 | 0 |
| 1.25 | 1.563 | 91 |
| 2.50 | 6.250 | 363 |

| Draught d | d² | Displacement tonne |
|-----------|--------|--------------------|
| 3.75      | 14.062 | 817                |
| 5.00      | 25.000 | 1452               |
| 6.25      | 39.062 | 2269               |
| 7.50      | 56.250 | 3267               |

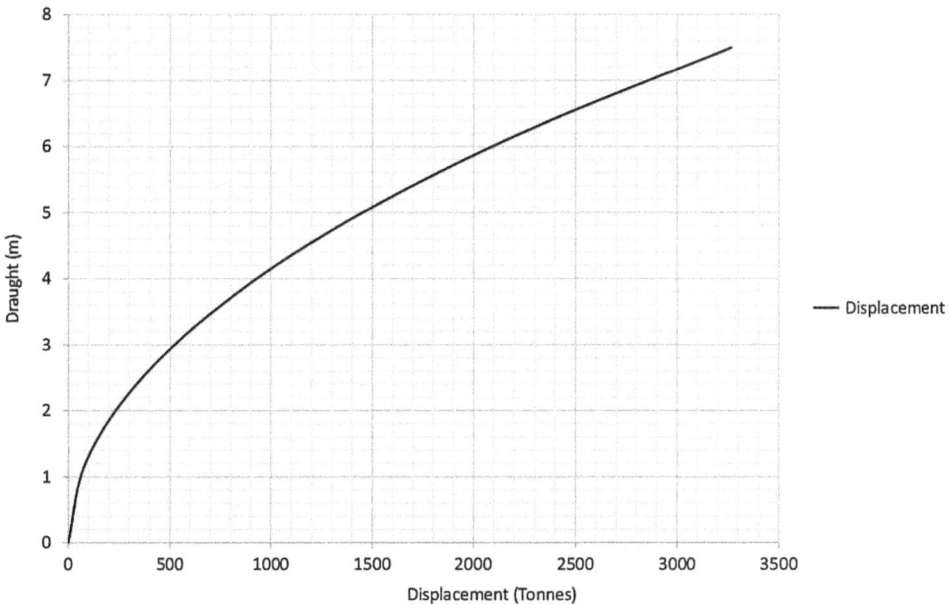

▲ Figure A. 2.2

At 6.5 m draught, displacement in sea water is 2450 tonne.

$$\text{Displacement in fresh water} = 2450 \times 1.000 \div 1.025$$
$$= 2390 \text{ tonne}$$

**A. 2.6:** 

$$\text{Immersed volume of cylinder} = 1 \div 2 \times 15 \times \pi \div 4 \times 4^2$$
$$= 30\,\pi \text{ m}^3$$
$$\text{Mass of cylinder} = 1.025 \times 30\,\pi$$
$$= 96.62 \text{ tonne}$$

**A. 2.7:**
$$\text{Mass of water displaced} = 1.025 \times 22$$
$$= 22.55 \text{ tonne}$$

Therefore:

$$\text{Apparent mass of bilge keels} = 36 - 22.55$$
$$= 13.45 \text{ tonne}$$
$$\text{Increase in draught} = 13.45 \div 20$$
$$= 0.673 \text{ cm}$$

**A. 2.8:**

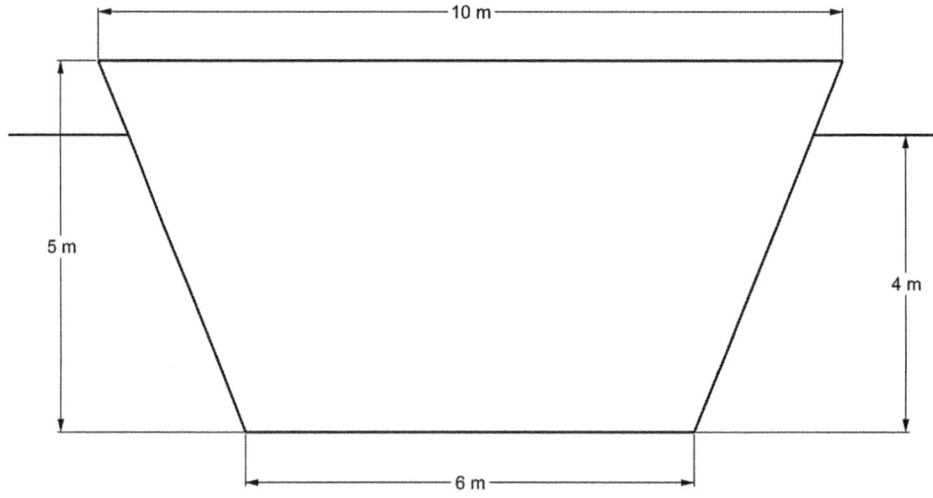

▲ Figure A. 2.3

$$\text{Breadth at waterline} = 6 + (4 \div 5 \times 4)$$
$$= 9.2 \text{ m}$$
$$\text{Displacement} = 1.025 \times 40 \times (6 + 9.2) \div 2 \times 4$$
$$= 1246 \text{ tonne}$$

**A. 2.9:**
$$\text{TPC} = 0.01025 \times A_w$$

| Draught | Waterplane area | TPC |
|---------|-----------------|-------|
| 7.5 | 1845 | 18.91 |
| 6.25 | 1690 | 17.32 |

| Draught | Waterplane area | TPC |
|---------|-----------------|------|
| 5.00 | 1535 | 15.73 |
| 3.75 | 1355 | 13.89 |
| 2.50 | 1120 | 11.48 |

$$\text{Increase in draught required} = 0.2\,\text{m}$$
$$= 20\,\text{cm}$$
$$\text{Mean TPC at } 6.2\,\text{m} = 17.25$$
$$\text{Mass required} = 17.25 \times 20$$
$$= 345\,\text{tonne}$$

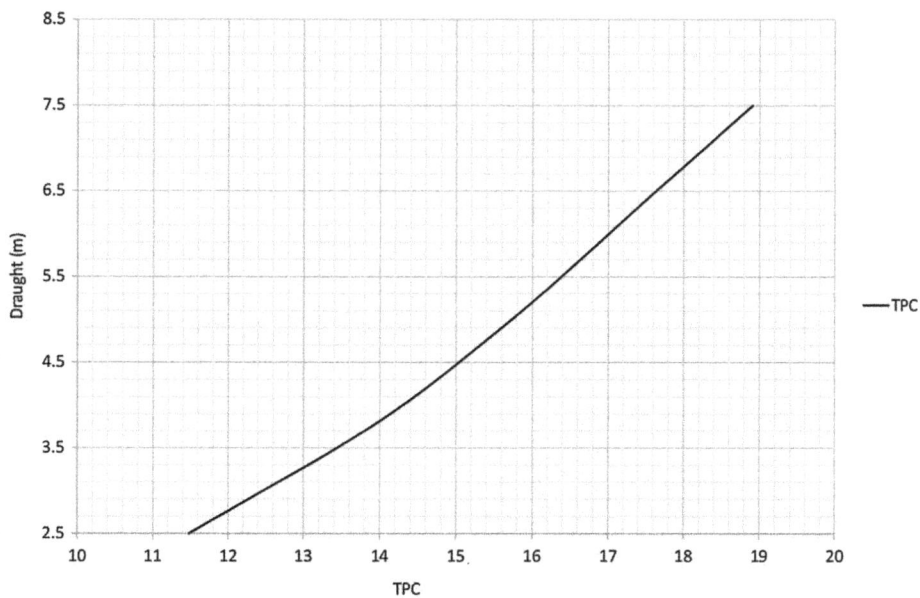

▲ Figure A. 2.4

**A. 2.10:**

$$\text{Volume of displacement} = 19500 \div 1.025$$
$$= 19024 \text{ m}^3$$
$$\text{Waterplane area} = \text{TPC} \div 0.01025$$
$$= 26.5 \div 0.01025$$
$$= 2585 \text{ m}^2$$
$$\text{Block coefficient} = 19024 \div (150 \times 20.5 \times 8)$$
$$= 0.773$$
$$\text{Prismatic coefficient} = C_b \div C_m$$
$$= 0.773 \div 0.94$$
$$= 0.822$$
$$\text{Waterplane area coefficient} = 2585 \div (150 \times 20.5)$$
$$= 0.841$$

**A. 2.11:**

$$\text{Let length of ship} = L$$
$$\text{Then breadth} = 0.13L$$
$$\text{and draught} = 0.13L \div 2.1$$
$$= 0.0619L$$
$$\text{Volume of displacement} = 9450 \div 1.025$$
$$= 9219.5 \text{m}^3$$
$$\text{Then } C_b = \nabla \div (L \times B \times d)$$
$$0.7 = 9219.5 \div (L \times 0.13L \times 0.0619L)$$
$$L^3 = 9219.5 \div (0.7 \times 0.13 \times 0.0619)$$
$$\text{Length of ship } L = 117.9 \text{ m}$$
$$C_p = 9219.5 \div (117.9 \times 106)$$
$$\text{Prismatic coefficient} = 0.738$$

**A. 2.12:**

$$\text{Let length of ship} = L$$
$$\text{Then draught} = L \div 18$$
$$\text{and breadth} = 2.1 \times \text{draught}$$
$$= 2.1 \div 18 \times L$$
$$\text{TPC sea water} = 0.01025 A_w$$
$$\text{TPC fresh water} = 0.0100 A_w$$
$$\text{TPC sea water} - \text{TPC fresh water} = 0.7$$
$$0.01025 A_w - 0.0100 A_w = 0.7$$
$$0.00025 A_w = 0.7$$
$$A_w = 0.7 \div 0.00025$$
$$= 2800 \text{m}^2$$

$$\text{But } Aw = Cw \times L \times B$$
$$2800 = 0.83 \times L \times (2.1 \div 18)L$$
$$L^2 = 2800 \times 18 \div (0.83 \times 2.1)$$
$$\text{From which } L = 170\,\text{m}$$
$$\text{TPC fresh water} = 0.0100 \times 2800$$
$$= 28$$

**A. 2.13:**

| ½ girth | SM | Product |
|---------|-----|---------|
| 2.1 | 1 | 2.1 |
| 6.6 | 4 | 26.4 |
| 9.3 | 2 | 18.6 |
| 10.5 | 4 | 42.0 |
| 11.0 | 2 | 22.0 |
| 11.0 | 4 | 44.0 |
| 11.0 | 2 | 22.0 |
| 9.9 | 4 | 39.6 |
| 7.5 | 2 | 15.0 |
| 3.9 | 4 | 15.6 |
| 0 | 1 | 0 |
|  |  | 247.3 |

$$\text{Common interval} = 9\,\text{m}$$
$$\text{Wetted surface area} = \frac{2}{3} \times 9 \times 247.3$$
$$= 1483.8\,\text{m}^2$$
$$\tfrac{1}{2}\% = 7.42\,\text{m}^2$$
$$\text{Appendages} = 30\,\text{m}^2$$
$$\text{Total wetted surface area} = 1521.22\,\text{m}^2$$

**A. 2.14:**

**a)**

$$\text{Volume of displacement} = 14000 \div 1.025$$
$$= 13659\,\text{m}^3$$
$$S = 1.7Ld + \text{volume} \div d$$
$$= 1.7 \times 130 \times 8 + 13653 \div 8$$
$$= 1768.0 + 1707.3$$
$$= 3475.3\,\text{m}^2$$

**b)**
$$S = c\sqrt{\Delta L}$$
$$= 2.58\sqrt{14000 \times 130}$$
$$= 3480\,\text{m}^2$$

**A. 2.15:**
$$\text{Volume of existing barge} = 75 \times 9 \times 6$$
$$= 4050\,\text{m}^3$$

Volumes of similar ships are proportional to length[3]

$$ie \quad V_1 \div V_2 = (L_1 \div L_2)^3$$
$$L_2 = L_1(V_2 \div V_1)^{1/3}$$
$$L_2 = 75 \times (3200 \div 4050)^{1/3}$$
$$\text{New length } L_2 = 69.34\,\text{m}$$
$$\text{New beam } B_2 = 9 \times (3200 \div 4050)^{1/3}$$
$$= 8.32\,\text{m}$$
$$\text{New depth } D_2 = 5.55\,\text{m}$$

**A. 2.16:**
$$\text{Let } S = \text{wetted surface area of small ship}$$
$$2S = \text{wetted surface area of large ship}$$
$$\Delta = \text{displacement of small ship}$$
$$\Delta + 2000 = \text{displacement of large ship}$$

It was shown that for similar ships:

$$\Delta \text{ is proportional to } S^{3/2}$$

Hence:
$$\Delta \div (\Delta + 2000) = (S \div 2S)^{3/2}$$
$$2^{3/2} \times \Delta = \Delta + 2000$$
$$2.828\Delta = \Delta + 2000$$
$$1.828\Delta = 2000$$
$$\Delta = 2000 \div 1.828$$

$$\text{Displacement of smaller ship } \Delta = 1094\,\text{tonne}$$

**A. 2.17:**

For similar ships : $\Delta$ is proportional to $L^3$

$$\Delta_1 \div \Delta_2 = (L_1 \div L_2)^3$$

Displacement of model $\Delta_2 = 11000 \times (6 \div 120)^3$

$$= 1.375 \text{ tonne}$$

$S$ is proportional to $L^2$

$$S_1 \div S_2 = (L_1 \div L_2)^2$$

Wetted surface area of model

$$S_2 = 2500 \times (6 \div 120)^2$$
$$= 6.25 \text{ m}^2$$

**A. 2.18:**

| ½ width | SM | Product for area |
|---------|----|------------------|
| 1 | 1 | 1 |
| 7.5 | 4 | 30 |
| 12 | 2 | 24 |
| 13.5 | 4 | 54 |
| 14 | 2 | 28 |
| 14 | 4 | 56 |
| 14 | 2 | 28 |
| 13.5 | 4 | 54 |
| 12 | 2 | 24 |
| 7 | 4 | 28 |
| 0 | 1 | 0 |
| | | $327 = \Sigma_A$ |

Common interval $h = 180 \div 10$

$$= 18\text{m}$$

**a)**

Waterplane area $= h/3\Sigma_A \times 2 = (18 \div 3) \times 327 \times 2 = 3924 \text{ m}^2$

Waterplane area $= h/3\Sigma_A \times 2$

$$(18 \div 3) \times 327 \times 2$$

$$= 3924 \text{ m}^2$$

**b)**
$$\text{TPC} = (\text{waterplane area} \times \text{density}) \div 100$$
$$= 3924 \times 1.025 \div 100$$
$$= 40.22$$

**c)**
$$\text{Waterplane area coefficient} = \text{waterplane area} \div (\text{length} \times \text{breadth})$$
$$= 3924 \div (180 \times 28)$$
$$= 0.779$$

**A. 2.19:**

| Waterplane area | SM | Product for volume |
|---|---|---|
| 865 | 1 | 865 |
| 1735 | 4 | 6940 |
| 1965 | 2 | 3930 |
| 2040 | 4 | 8160 |
| 2100 | 2 | 4200 |
| 2145 | 4 | 8580 |
| 2215 | 1 | 2215 |
| | | $34890 = \Sigma_\nabla$ |

$$\text{Common interval } h = 180 \div 10$$
$$= 18\,\text{m}$$
$$\text{Volume of displacement} = h/3\Sigma_\nabla$$
$$\text{displacement} = h/3\Sigma_\nabla \times \rho$$
$$= (1.5 \div 3) \times 34890 \times 1.025$$
$$= 17881\,\text{tonne}$$

**A. 2.20:**

| Cross-sectional area | SM | Product for volume |
|---|---|---|
| 5 | 1 | 5 |
| 60 | 4 | 240 |
| 116 | 2 | 232 |
| 145 | 4 | 580 |
| 152 | 2 | 304 |

| Cross-sectional area | SM | Product for volume |
|---|---|---|
| 153 | 4 | 612 |
| 153 | 2 | 306 |
| 151 | 4 | 604 |
| 142 | 2 | 284 |
| 85 | 4 | 340 |
| 0 | 1 | 0 |
| | | $3507 = \Sigma_\nabla$ |

$$\text{Common interval} = 140 \div 10$$
$$= 14 \, \text{m}$$
$$\text{Volume of displacement} = h/3 \Sigma_\nabla$$
$$= (14 \div 3) \times 3507$$
$$= 16.366 \, \text{m}^3$$

**a)**
$$\text{Displacement} = \text{volume} \times \text{density}$$
$$= 16366 \times 1.025$$
$$= 16775 \, \text{tonne}$$

**b)** Block coefficient $= \text{volume of displacement} \div (\text{length} \times \text{breadth} \times \text{draught})$
$$= 16366 \div 140 \times 18 \times 9$$
$$= 0.722$$

**c)** Midship section area coefficient $= \text{midship section area} \div (\text{breadth} \times \text{draught})$
$$= 153 \div (18 \times 9)$$
$$= 0.944$$

**d)** Prismatic coefficient $= \text{volume of displacement} \div (\text{length} \times \text{midship section area})$
$$= 16366 \div (140 \times 153)$$
$$= 0.764$$

Alternatively

$$\text{Prismatic coefficient} = \text{block coefficient} \div \text{midship section area coefficient}$$
$$= 0.722 \div 0.944$$
$$= 0.764$$

**A. 2.21:**

| Section | ½ ordinate | SM | Product for area | Lever | Product for 1st moment |
|---------|-----------|-----|---------|-------|----------------|
| AP | 1.2 | ½ | 0.6 | +5 | + 3.0 |
| ½ | 3.5 | 2 | 7.0 | +4½ | +31.5 |
| 1 | 5.3 | 1 | 5.3 | +4 | +21.2 |
| 1½ | 6.8 | 2 | 13.6 | +3½ | +47.6 |
| 2 | 8.0 | 1½ | 12.0 | +3 | +36.0 |
| 3 | 8.3 | 4 | 33.2 | +2 | +66.4 |
| 4 | 8.5 | 2 | 17.0 | +1 | + 17.0 |
| 5 | 8.5 | 4 | 34.0 | 0 | +222.7 = $\Sigma_{MA}$ |
| 6 | 8.5 | 2 | 17.0 | −1 | −17.0 |
| 7 | 8.4 | 4 | 33.6 | −2 | −67.2 |
| 8 | 8.2 | 1½ | 12.3 | −3 | −36.9 |
| 8½ | 7.9 | 2 | 15.8 | −3½ | −55.3 |
| 9 | 6.2 | 1 | 6.2 | −4 | −24.8 |
| 9½ | 3.5 | 2 | 7.0 | −4½ | −31.5 |
| FP | 0 | ½ | 0 | −5 | −0 |
| | | | 214.6 = $\Sigma_A$ | | −232.7 = $\Sigma_{MF}$ |

$$\text{Common interval} = 120 \div 10$$
$$= 12\,\text{m}$$

**a)**
$$\text{Waterplane area} = h/3\,\Sigma_A \times 2$$
$$= (12 \div 3) \times 214.6 \times 2$$
$$= 1716.8\,\text{m}^2$$

Since $\Sigma_{MF}$ exceeds $\Sigma_{MA}$, the centroid will be forward of midships

**b)**     Distance of centroid from midships $= h \times (\sum_{MF} + \sum_{MA}) \div \sum_A$
$$= 12 \times (222.7 - 232.7) \div 214.6$$
$$= 0.559 \, \text{m forward}$$

**A. 2.22:**

| TPC | SM | Product for displacement | Lever | Product for 1st moment |
|-----|-----|-----|-----|-----|
| 4.0 | 1 | 4.0 | 0 | 0 |
| 6.1 | 4 | 24.4 | 1 | 24.4 |
| 7.8 | 2 | 15.6 | 2 | 31.2 |
| 9.1 | 4 | 36.4 | 3 | 109.2 |
| 10.3 | 2 | 20.6 | 4 | 82.4 |
| 11.4 | 4 | 45.6 | 5 | 228.0 |
| 12.0 | 1 | 12.0 | 6 | 72.0 |
| | | $158.6 = \sum_\Delta$ | | $547.2 = \sum_M$ |

$$\text{Common interval} = 1.5 \, \text{m}$$
$$= 150 \, \text{cm}$$

**a)**     $\text{Displacement} = h/3 \sum_\Delta (h \text{ in cm})$
$$= 150 \div 3 \times 158.6$$
$$= 7930 \, \text{tonne}$$

**b)**     $KB = h \sum_M \div \sum_\Delta (h \text{ in m})$
$$= 1.5 \times 547.2 \div 158.6$$
$$= 5.175 \, \text{m}$$

**A. 2.23:**

| ½ breadths | SM | Product for area | Lever | Product for 1st moment | Lever | Product for 2nd moment |
|-----|-----|-----|-----|-----|-----|-----|
| 0.3 | 1 | 0.3 | +5 | + 1.5 | +5 | + 7.5 |
| 3.8 | 4 | 15.2 | +4 | +60.8 | +4 | +243.2 |
| 6.0 | 2 | 12.0 | +3 | +36.0 | +3 | + 108.0 |
| 7.7 | 4 | 30.8 | +2 | +61.6 | +2 | + 123.2 |

| ½ breadths | SM | Product for area | Lever | Product for 1st moment | Lever | Product for 2nd moment |
|---|---|---|---|---|---|---|
| 8.3 | 2 | 16.6 | +1 | +16.6 | +1 | + 16.6 |
| 9.0 | 4 | 36.0 | 0 | $+176.5 = \Sigma_{MA}$ | 0 | 0 |
| 8.4 | 2 | 16.8 | −1 | −16.8 | −1 | + 16.8 |
| 7.8 | 4 | 31.2 | −2 | −62.4 | −2 | + 124.8 |
| 6.9 | 2 | 13.8 | −3 | −41.4 | −3 | + 124.2 |
| 4.7 | 4 | 18.8 | −4 | −75.2 | −4 | + 300.8 |
| 0 | 1 | 0 | −5 | 0 | −5 | 0 |
| | | $191.5 = \Sigma_A$ | | $-195.8 = \Sigma_{MF}$ | | $+ 1065.1 = \Sigma_I$ |

Common interval = 15 m

**a)**
$$\text{Waterplane area} = h/3\,\Sigma_A \times 2$$
$$= 15 \div 3 \times 191.5 \times 2$$
$$= 1915\,m^2$$

**b)**
$$\text{Distance of centroid from midships} = h \times (\Sigma_{MF} + \Sigma_{MA}) \div \Sigma_A$$
$$= 15 \times (176.5 - 195.8) \div 191.5$$
$$= 1.512\,m\ \text{forward}$$

**c)**

Second moment of area about midships

$$= h^3/3\,\Sigma_I \times 2$$
$$= 15^3 \div 3 \times 1065.1 \times 2$$
$$= 2396475\,m^4$$

By parallel axis theorem, second moment of area about centroid

$$= 2396475 - 1915 \times 1.512^2$$
$$= 2396475 - 4378$$
$$= 2392097\,m^4$$

**A. 2.24:**

| Draught | Displacement | SM | Product for moment |
|---------|--------------|-----|--------------------|
| 0 | 0 | 1 | 0 |
| 1 | 189 | 4 | 756 |
| 2 | 430 | 2 | 860 |
| 3 | 692 | 4 | 2768 |
| 4 | 977 | 1 | 977 |
| | | | $5361 = \Sigma_M$ |

$$\text{Common interval} = 1\,\text{m}$$
$$\text{Area of curve} = h/3\,\Sigma_M$$
$$= \tfrac{1}{3} \times 5361$$
$$= 1787\,\text{tonne m}$$

$$\text{VCB below waterline} = \text{area of curve} \div \text{displacement}$$
$$= 1787 \div 977$$
$$= 1.829\,\text{m}$$
$$KB = 4 - 1.829$$
$$= 2.171\,\text{m}$$

**A. 2.25:**

| ½ ordinate | (½ ordinate)³ | SM | Product for 2nd moment |
|------------|---------------|-----|------------------------|
| 1.6 | 4.1 | 1 | 4.1 |
| 5.7 | 185.2 | 4 | 740.8 |
| 8.8 | 681.5 | 2 | 1363.0 |
| 10.2 | 1061.2 | 4 | 4244.8 |
| 10.5 | 1157.6 | 2 | 2315.2 |
| 10.5 | 1157.6 | 4 | 4630.4 |
| 10.5 | 1157.6 | 2 | 2315.2 |
| 10.0 | 1000.0 | 4 | 4000.0 |
| 8.0 | 512.0 | 2 | 1024.0 |
| 5.0 | 125.0 | 4 | 500.0 |
| 0 | 0 | 1 | 0 |
| | | | $21137.5 = \Sigma_I$ |

$$\text{Common interval} = 16\,\text{m}$$

Second moment of area about centreline

$$= h/9 \times \Sigma_i \times 2$$
$$= 16 \div 9 \times 21137.5 \times 2$$
$$= 75155\,\text{m}^4$$

**A. 2.26:**

| Cross-sectional area | SM | Product for volume | Lever | Product for 1st moment |
|---|---|---|---|---|
| 2 | 1 | 2 | +5 | + 10 |
| 40 | 4 | 160 | +4 | + 640 |
| 79 | 2 | 158 | +3 | + 474 |
| 100 | 4 | 400 | +2 | + 800 |
| 103 | 2 | 206 | +1 | + 206 |
| 104 | 4 | 416 | 0 | $+2130 = \Sigma_{MA}$ |
| 104 | 2 | 208 | −1 | −208 |
| 103 | 4 | 412 | −2 | −824 |
| 97 | 2 | 194 | −3 | −582 |
| 58 | 4 | 232 | −4 | −928 |
| 0 | 1 | 0 | −5 | 0 |
| | | $2388 = \Sigma_\nabla$ | | $-2542 = \Sigma_{MF}$ |

Common interval = 12 m

**a)** $\quad \text{Displacement} = h/3\Sigma_\nabla \times \rho$
$$= 12 \div 3 \times 2388 \times 1025$$
$$= 9790.8\,\text{tonne}$$

**b)** Centre of buoyancy from midships

$$= h \times (\Sigma_{MF} + \Sigma_{MA}) \div \Sigma_\nabla$$
$$= 12 \times (2130 - 2542) \div 2388$$
$$= 2.07\,\text{m forward}$$

# SOLUTIONS TO CHAPTER 3 TEST EXAMPLES

**A. 3.1:**

| Mass | Kg | Vertical moment |
|------|-----|-----------------|
| 4000 | 6.0 | 24000 |
| 1000 | 0.8 | 800 |
| 200 | 1.0 | 200 |
| 5000 | 5.0 | 25000 |
| 3000 | 9.5 | 28500 |
| 13200 | | 78500 |

Thus displacement = 13200 tonne
Centre of gravity above keel = 78500 ÷ 13200
= 5.947 m

**A. 3.2:**

| Mass | Kg | Vertical moment | LCG from midships | Longitudinal moment | Longitudinal moment |
|------|-----|-----------------|-------------------|---------------------|---------------------|
| 5000 | 6.0 | 30000 | 1.5 for'd | 1.5 for'd | 7500 for'd |
| 500 | 10.0 | 5000 | 36.0 aft | 36.0 aft | 18000 aft |
| 5500 | | 35000 | | | 10500 aft |

$$\text{Centre of gravity above keel} = 35000 \div 5500$$
$$= 6.364 \text{ m}$$
$$\text{Centre of gravity from midships} = 10500 \div 5500$$
$$= 1.909 \text{ m aft}$$

**A. 3.3:**

$$\text{Shift in centre of gravity} = 300 \times (24 + 40) \div 6000$$
$$= 3.2 \text{ m aft}$$
$$\text{New position of centre of gravity} = 1.2 - 3.2$$
$$= -2 \text{ m}$$
$$\text{or} = 2 \text{ m aft of midships}$$

**A. 3.4:**

a) Shift in centre of gravity due to transfer of oil

$$= 250 \times (75 + 50) \div 17000$$
$$= 1.838 \text{ m aft}$$
$$\text{Therefore new position of centre of gravity} = 1.0 + 1.838$$
$$= 2.838 \text{ m aft of midships}$$

b) Taking moments about midships:

$$\text{new position of centre of gravity} = (17000 \times 2.838 - 200 \times 50) \div (17000 - 200)$$
$$= (48246 - 10000) \div 16800$$
$$= 2.277 \text{ m aft of midships}$$

**A. 3.5:** Shift in centre of gravity due to lowered cargo

$$= 500 \times 3 \div 3000$$
$$= 0.5 \text{ m down}$$

Taking moments about the new position of the centre of gravity: shift in centre of gravity due to added cargo

$$= (3000 \times 0 + 500 \times 3.5) \div (3000 + 500)$$
$$= 1750 \div 3500$$
$$= 0.5 \text{ m up}$$

ie the position of the centre of gravity does not change.

**A. 3.6**:

| | Mass | Kg | Vertical moment |
|---|---|---|---|
| | 2000 | 1.5 | 3000 |
| | 300 | 4.5 | 1350 |
| | 50 | 6.0 | 300 |
| Total removed | 2350 | | 4650 |
| Original | 10000 | 3.0 | 30000 |
| Final | 7650 | | 25350 |

$$\text{New displacement} = 7650 \text{ tonne}$$
$$\text{New centre of gravity} = 25350 \div 7650$$
$$= 3.314 \text{ m above keel}$$

**A. 3.7:**

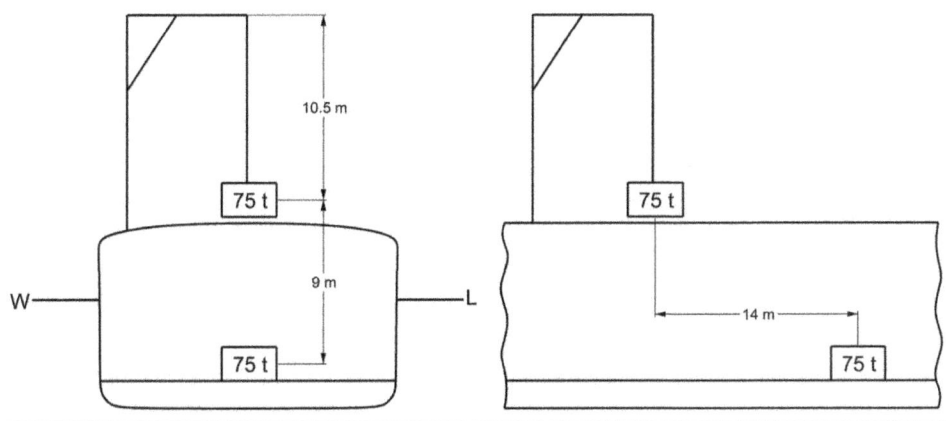

▲ Figure A. 3.1

a)  Since the centre of gravity of a suspended mass is at the point of suspension, the mass is virtually raised to the derrick head.

$$\text{Rise in centre of gravity} = 75 \times 10.5 \div 8000$$
$$= 0.0984 \text{ m}$$

b)  When the mass is at the derrick head, there is no further movement of the centre of gravity.

$$\text{ie rise in centre of gravity} = 0.0984 \text{ m}$$

c) When the mass is in its final position, the centre of gravity moves down and forwards.

$$\text{Vertical shift in centre of gravity} = 75 \times 9 \div 8000$$
$$= 0.0844 \text{ m down}$$
$$\text{Longitudinal shift in centre of gravity} = 75 \times 14 \div 8000$$
$$= 0.131 \text{ m forward}$$

# SOLUTIONS TO CHAPTER 4 TEST EXAMPLES

**A. 4.1:**

$$GM = KB + BM - KG$$
$$BM = I \div volume$$
$$= 425 \times 10^3 \times 1.025 \div 12000$$
$$= 3.63\,m$$
$$GM = 3.60 + 3.63 - 6.50$$
$$= 0.73\,m$$

**A. 4.2:**

$$BM = I \div volume$$
$$= 60 \times 10^3 \times 1.025 \div 10000$$
$$= 6.15\,m$$
$$KG = (4000 \times 6.30 + 2000 \times 7.50 + 4000 \times 9.15) \div (4000 + 2000 + 4000)$$
$$= (25200 + 15000 + 36000) \div 10000$$
$$= 76800 \div 10000$$
$$= 7.68\,m$$
$$GM = 2.15 + 6.15 - 7.68$$
$$= 0.62\,m$$

**A. 4.3:**

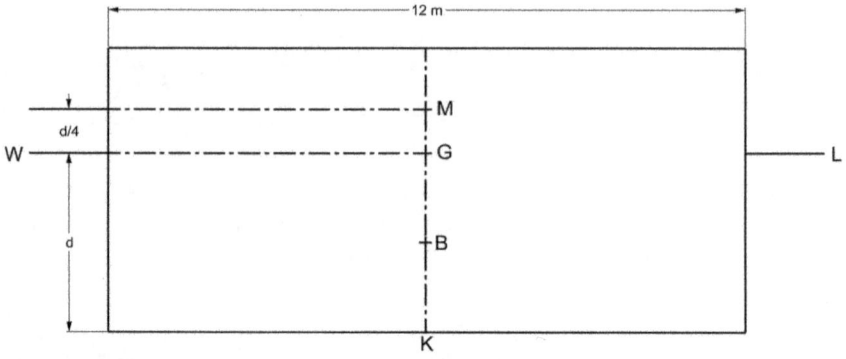

▲ Figure A. 4.1

$$\text{Let } d = \text{draught}$$
$$\text{Then } KB = d \div 2$$
$$BM = B^2 \div (12d)$$
$$KG = d$$
$$GM = d \div 4$$
$$KG + GM = KB + BM$$
$$d + d \div 4 = d \div 2 + 12 \div d$$
$$\tfrac{3}{4}d \div 4 = 12 \div d$$
$$d^2 = 4 \div 3 \times 12$$
$$= 16$$
$$\text{Therefore: } d = 4 \text{ m}$$

**A. 4.4:**

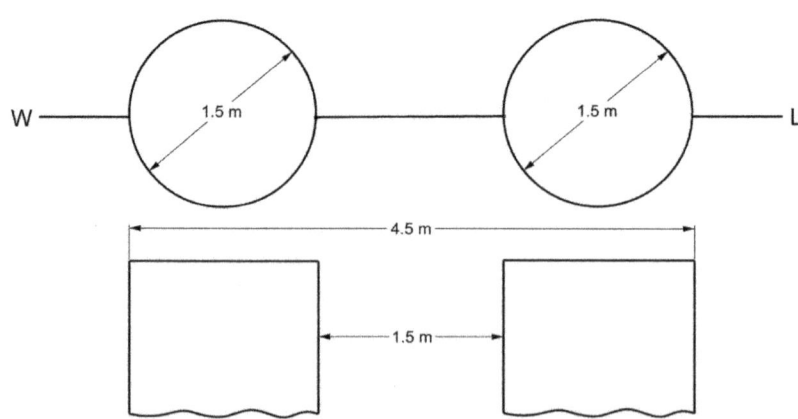

▲ Figure A. 4.2

Although the raft is formed by cylinders, the waterplane consists of two rectangles, a distance of 1.5 m apart. Second moment of area of waterplane about centreline

$$I = 1 \div 12 \times 6 \times 4.5^3 - 1 \div 12 \times 6 \times 1.5^3$$
$$= 1 \div 12 \times 6(4.5^3 - 1.5^3)$$
$$= \frac{1}{2} \times 87.75$$
$$= 43.875 m^4$$

Volume of displacement $V = 2 \times 6 \times \pi \div 4 \times 1.5^2 \times \frac{1}{2}$
$$= 10.603 \, m^3$$
$$BM = I \div volume$$
$$= 43.875 \div 10.603$$
$$= 4.138 \, m$$

**A. 4.5: a)**

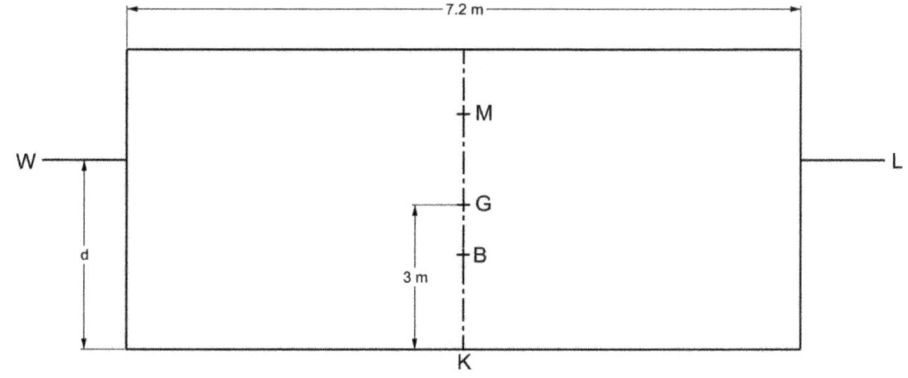

▲ Figure A. 4.3

$$\text{let } d = \text{draught}$$
$$KB = d \div 2$$
$$BM = B^2 \div 12d$$
$$= 4.32 \div d$$

| D | KB | BM | KM |
|---|----|----|----|
| 0 | 0 | ∞ | ∞ |
| 0.5 | 0.25 | 8.640 | 8.890 |
| 1.0 | 0.50 | 4.320 | 4.820 |
| 1.5 | 0.75 | 2.880 | 3.630 |
| 2.0 | 1.00 | 2.160 | 3.160 |

| D | KB | BM | KM |
|---|---|---|---|
| 2.5 | 1.25 | 1.728 | 2.978 |
| 3.0 | 1.50 | 1.440 | 2.940 |
| 3.5 | 1.75 | 1.234 | 2.984 |
| 4.0 | 2.00 | 1.080 | 3.080 |

**b)** Between draughts of 2.4 m and 3.6 m, the vessel will be unstable.

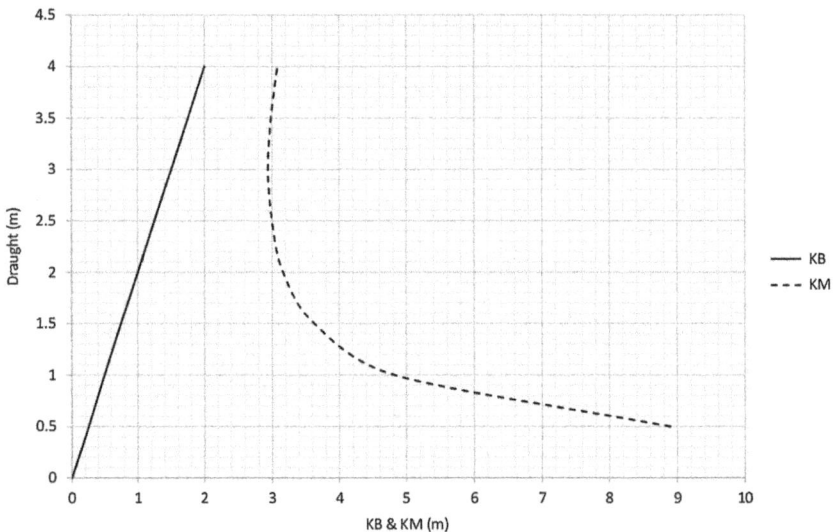

▲ Figure A. 4.4

**A. 4.6:**

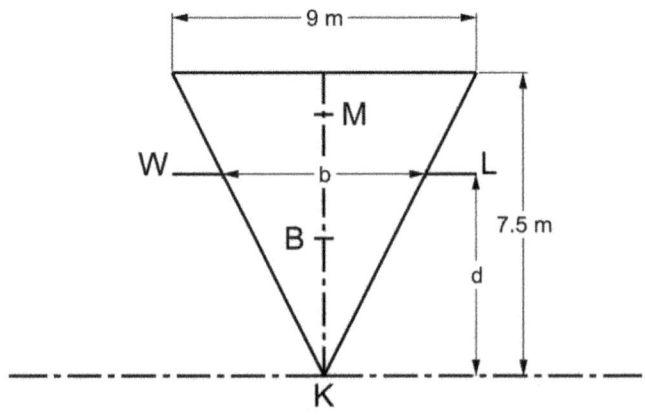

▲ Figure A. 4.5

Let $d$ = draught

$b$ = breadth at waterline

By similar triangles:

$$b = 9 \div 7.5 \times d$$
$$= 1.2d$$
$$KB = \tfrac{2}{3}d$$
$$BM = b^2 \div 6d$$
$$= (1.2d)^2 \div 6d$$
$$= 0.24d$$

| D | KB | BM | KM |
|---|---|---|---|
| 0 | 0 | 0 | 0 |
| 0.5 | 0.333 | 0.120 | 0.453 |
| 1.0 | 0.667 | 0.240 | 0.907 |
| 1.5 | 1.000 | 0.360 | 1.360 |
| 2.0 | 1.333 | 0.480 | 1.813 |
| 2.5 | 1.667 | 0.600 | 2.267 |
| 3.0 | 2.000 | 0.720 | 2.720 |

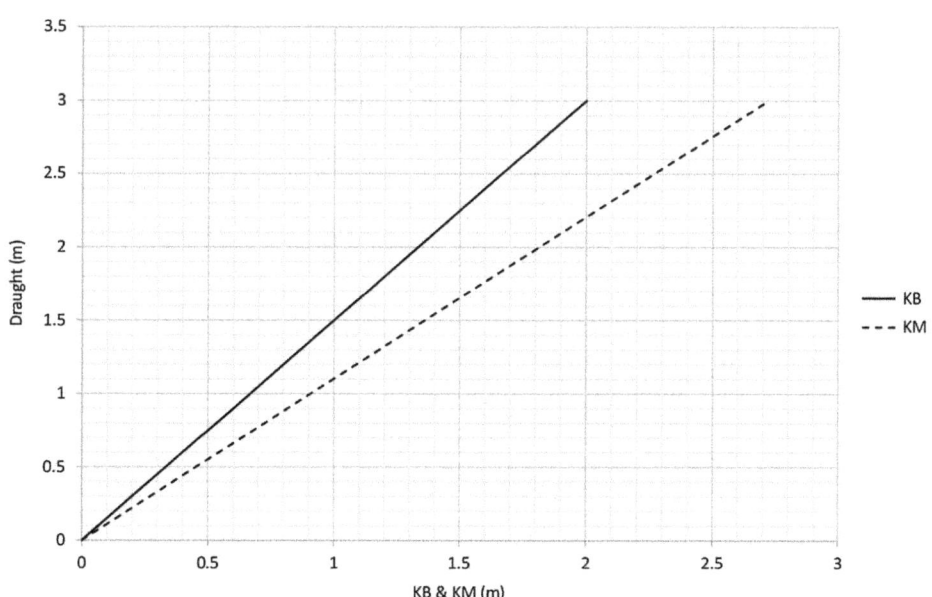

▲ Figure A. 4.6

**A. 4.7:**

$$GM = m \times d \div (\text{displacement} \times \tan \theta)$$
$$\tan \theta = \text{deflection of pendulum} \div \text{length of pendulum}$$
$$= 0.110 \div 8.5$$
$$GM = 10 \times 14 \div (8000 \times 0.110 \div 8.5)$$
$$= 1.352 \, \text{m}$$
$$KG = KM - GM$$
$$= 7.15 - 1.352$$
$$= 5.798 \, \text{m}$$

**A. 4.8:** Mean deflection

$$= (81 + 78 + 85 + 83 + 79 + 82 + 84 + 80) \div 8$$
$$= 81.5 \, \text{mm}$$
$$GM = 6 \times 13.5 \times 7.5 \div (4000 \times 0.0815)$$
$$= 1.863 \, \text{m}$$

**A. 4.9:** Virtual reduction in *GM* due to free surface

$$= \rho \times i \div (\rho \times \text{volume})$$
$$= i \div \text{volume}$$
$$i = 1 \div 12 \times 12 \times 9^3$$
$$= 729$$
$$\text{Volume} = 5000 \div 1.025$$
$$\text{Free surface effect} = 729 \times 1.025 \div 5000$$
$$= 0.149 \, \text{m}$$

**A. 4.10: a)**

$$GM = KM - KG$$
$$= 5.00 - 4.50$$
$$= 0.5 \, \text{m}$$

But
$$GM = (m \times d) \div (\text{displacement} \times \tan \theta)$$
$$\tan \theta = (m \times d) \div (\text{displacement} \times GM)$$
$$= 10 \times 12 \div (8000 \times 0.50)$$
$$= 0.030$$

Angle of heel $\theta = 1.718°$

**b)**
$$\text{Free surface effect} = i \div \text{volume}$$

$$\text{Free surface effect} = i \div \text{volume}$$
$$= 75 \times 15^3 \times 1.025 \div (12 \times 8000)$$
$$= 0.270 \text{ m}$$
$$\text{Virtual } GM = 0.507 - 0.27$$
$$= 0.24 \text{ m}$$
$$\tan \theta = 10 \times 12 \div (8000 \times 0.23)$$
$$= 0.0652$$
$$\text{Angle of heel } \theta = 3.73°$$

**A. 4.11:**

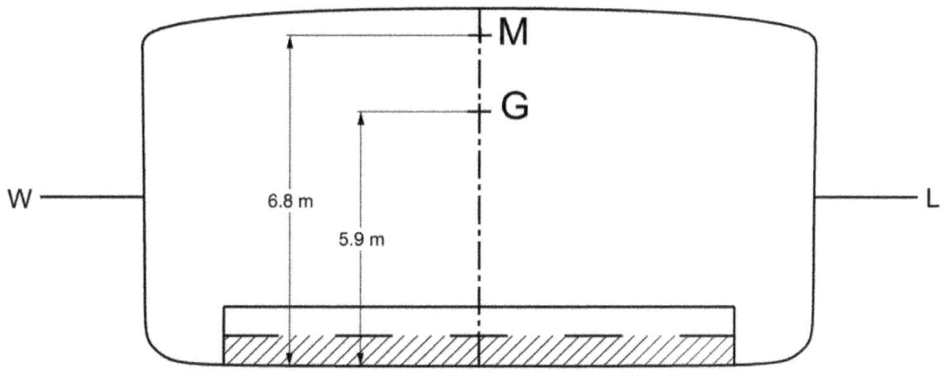

▲ Figure A. 4.7

$$\text{Mass of water added} = 10.5 \times 12 \times 0.6 \times 1.025$$
$$= 77.49 \text{ tonne}$$
$$\text{New } KG = (6000 \times 5.9 + 77.49 \times 0.3) \div (6000 + 77.49)$$
$$= (35400 + 23.25) \div 6077.49$$
$$= 5.829 \text{ m}$$
$$\text{Free surface effect} = (\rho \times i) \div \text{displacement}$$
$$= 1.025 \times 10.5 \times 12^3 \div (12 \times 6077.49)$$
$$= 0.255 \text{ m}$$
$$\text{Virtual } KG = 5.829 + 0.255$$
$$= 6.084 \text{ m}$$
$$GM = 6.80 - 6.084$$
$$= 0.716 \text{ m}$$

**A. 4.12:**  **a)**  Free surface effect $= (\rho \times i) \div$ displacement

$$= (0.8 \times 9 \times 10 \times 24^3 \div 12) \div 25000$$

$$= 3.318\,m$$

**b)** With centreline bulkhead, free surface effect is reduced to one-quarter of (a)

$$\text{Free surface effect} = \tfrac{1}{4} \times 3.226$$

$$= 0.807\,m$$

**c)** With twin longitudinal bulkheads and equal tanks, free surface effect is reduced to one-ninth of a)

$$\text{Free surface effect} = 1/9 \times 3.226$$

$$= 0.358\,m$$

**d)** The only change is in the second moment of area.

$$\text{For centre tank } i = 9 \times 10 \times 12^3 \div 12$$

$$= 9 \times 10 \times 1728 \div 12$$

$$= 12960\,m^4$$

$$\text{For wing tanks } i = (9 \times 10 \times 6^3 \div 12) \times 2$$

$$= (9 \times 10 \times 216 \div 12) \times 2$$

$$= 3240\,m^4$$

$$\text{Thus for both tanks } i = 12960 + 3240$$

$$= 16200\,m^4$$

$$\text{Total free surface effect} = 0.8 \times 16200 \div 25000$$

$$= 0.518\,m$$

**A. 4.13:**

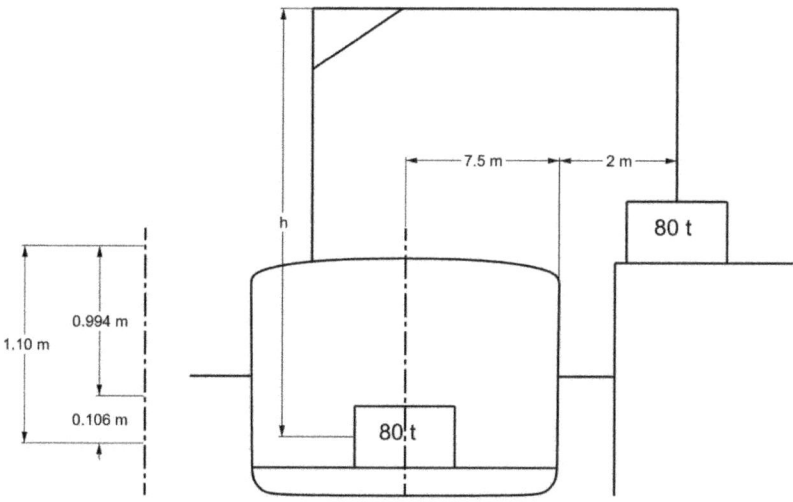

▲ Figure A. 4.8

$$GM = (m \times d) \div (\text{displacement} \times \tan \theta)$$
$$= 80 \times 9.5 \div (12500 \times \tan 3.5°)$$
$$= 0.094 \text{ m}$$

Thus the metacentric height is reduced from 1.1 m to 0.994 m when the mass is suspended over the quay. Since the draught does not alter, and hence the transverse metacentre remains in the same position, this reduction in metacentric height must be due to a rise in the centre of gravity. This rise is due to the effect of the suspended mass.

$$\text{Rise in centre of gravity} = 1.10 - 0.994$$
$$= 0.106 \text{ m}$$

Let $h$ be the distance from the centre of gravity of the mass to the derrick head.

$$\text{Then rise in centre of gravity} = \text{mass} \times h \div \text{displacement}$$
$$0.106 = 80 \times h \div 12500$$
$$h = 0.106 \times 12500 \div 80$$
$$= 16.56 \text{ m}$$

**A. 4.14:**

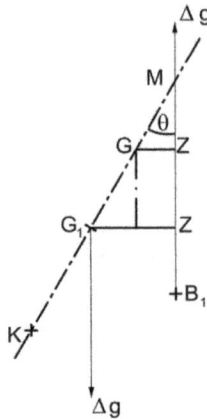

Since the actual centre of gravity $G_1$ is *below* the assumed centre of gravity $G$, the ship is *more* stable, and

$$G_1Z = GZ + GG_1\sin\theta$$
$$GG_1 = 3.50 - 3.00$$
$$= 0.5\,\text{m}$$

| $\theta$ | $\sin\theta$ | $GG_1\sin\theta$ | $GZ$ | $G_1Z$ |
|---|---|---|---|---|
| 0 | 0 | 0 | 0 | 0 |
| 15° | 0.2588 | 0.129 | 0.25 | 0.379 |
| 30° | 0.500 | 0.250 | 0.46 | 0.710 |
| 45° | 0.7071 | 0.353 | 0.51 | 0.864 |
| 60° | 0.8660 | 0.433 | 0.39 | 0.823 |
| 75° | 0.9659 | 0.483 | 0.10 | 0.583 |
| 90° | 1.000 | 0.500 | −0.38 | 0.120 |

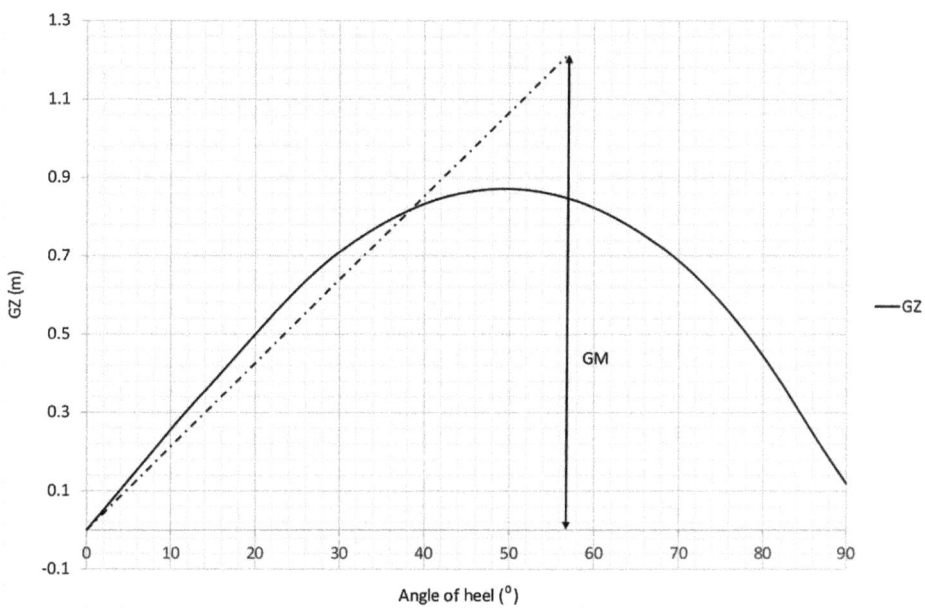

▲ Figure A. 4.10

**A. 4.15:** The dynamical stability of a ship to any given angle is represented by the area under the righting moment curve to that angle.

| θ | Righting moment | SM | Product |
|---|---|---|---|
| 0 | 0 | 1 | 0 |
| 15° | 1690 | 4 | 6760 |
| 30° | 5430 | 2 | 10 860 |
| 45° | 9360 | 4 | 37 440 |
| 60° | 9140 | 1 | 9140 |
| | | | 64 200 |

$$\text{Common interval} = 15°$$
$$= 15 \div 57.3 \text{ radians}$$
$$\text{Dynamical stability} = \tfrac{1}{3} \times 15 \div 57.3 \times 64200$$
$$= 5602 \text{ kJ}$$

**A. 4.16:**

Wall-sided formula $GZ = \sin\theta\,(GM + \tfrac{1}{2}BM\,\tan^2\theta)$

$$BM = I \div \text{volume}$$
$$= 82 \times 10^3 \times 1.025 \div 18000$$
$$= 4.67\,\text{m}$$
$$GM = KB + BM - KG$$
$$= 5.25 + 4.67 - 9.24$$
$$= 0.68\,\text{m}$$

| θ | tan θ | tan² θ | ½ BM tan² θ | GM + ½ BM tan² θ | sin θ | GZ |
|---|-------|--------|-------------|------------------|-------|-----|
| 0° | 0.0 | 0.0 | 0.0 | 0.68 | 0.0 | 0 |
| 5° | 0.087 | 0.008 | 0.018 | 0.698 | 0.087 | 0.061 |
| 10° | 0.176 | 0.031 | 0.073 | 0.753 | 0.174 | 0.131 |
| 15° | 0.268 | 0.072 | 0.168 | 0.848 | 0.259 | 0.219 |
| 20° | 0.364 | 0.132 | 0.309 | 0.989 | 0.342 | 0.338 |

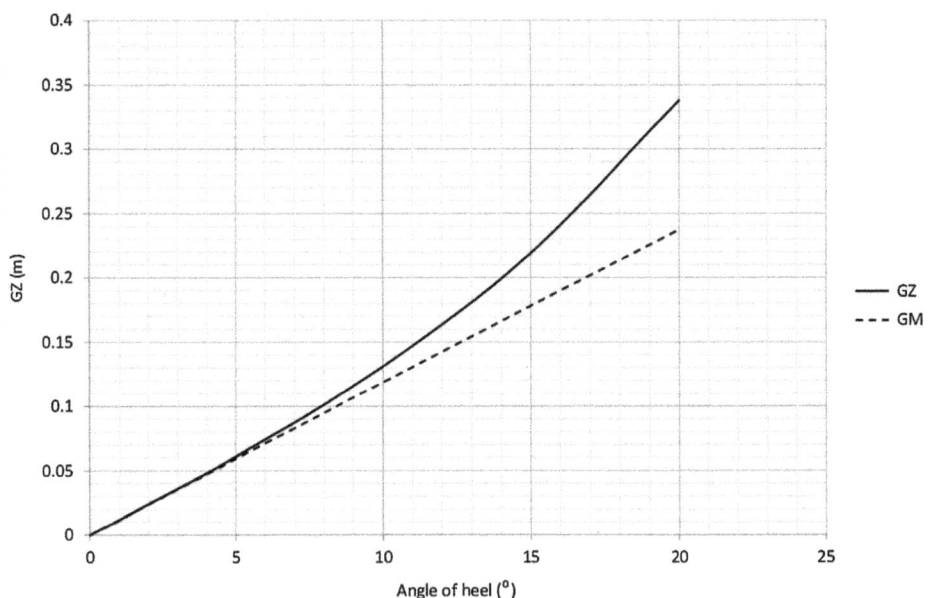

▲ Figure A. 4.11

**A. 4.17:**

$$\text{New } KG = (7200 \times 5.20 - 300 \times 0.60) \div (7200 - 300)$$
$$= 37260 \div 6900$$
$$= 5.4 \text{ m}$$
$$GM = KM - KG$$
$$= 5.35 - 5.40$$
$$= -0.05 \text{ m}$$
$$BM = KM - KB$$
$$= 5.35 - 3.12$$
$$= 2.23 \text{ m}$$
$$\tan \theta = \pm \sqrt{(-2GM \div BM)}$$
$$= \pm \sqrt{(0.10 \div 2.23)}$$
$$= \pm 0.2118$$

Angle of loll $\theta = \pm 11.956°$

# SOLUTIONS TO CHAPTER 5 TEST EXAMPLES

**A. 5.1:**

**a)**

$$\text{Change in trim} = \text{trimming moment} \div \text{MCT 1 cm}$$
$$\text{Therefore: MCT 1 cm} = 100 \times 75 \div 65$$
$$= 115.4 \text{ tonne m}$$

**b)**

$$\text{MCT 1 cm} = \text{displacement} \times GM_L \div (100\,L)$$
$$\text{Therefore: } GM_L = 115.4 \times 100 \times 125 \div 12000$$
$$= 120.2 \text{ m}$$

**c)**

$$GG_1 = m \times d \div \text{displacement}$$
$$= 100 \times 75 \div 12000$$
$$= 0.625 \text{ m}$$

**A. 5.2:**

$$\text{Bodily sinkage} = \text{mass added} \div \text{TPC}$$
$$= 110 \div 13$$
$$= 8.46 \text{ cm}$$
$$\text{Change in trim} = 110 \times (24 + 2.5) \div 80$$
$$= 36.44 \text{ cm by the stern}$$
$$\text{Change forward} = -36.44 \div 120(120 \div 2 - 2.5)$$
$$= -17.46 \text{ cm}$$
$$\text{Change aft} = +36.44 \div 120(120 \div 2 + 2.5)$$
$$= +18.98 \text{ cm}$$

$$\text{New draught forward} = 5.50 + 0.085 - 0.175$$
$$= 5.41\,\text{m}$$
$$\text{New draught aft} = 5.80 + 0.085 + 0.190$$
$$= 6.075\,\text{m}$$

**A. 5.3:**

$$\text{Bodyrise} = 180 \div 18$$
$$= 10\,\text{cm}$$
$$\text{MCT 1 cm} = (14000 \times 125) \div (100 \times 130)$$
$$= 134.6\,\text{tonne m}$$
$$\text{Change in trim} = 180 \times (40 - 3) \div 134.6$$
$$= -49.48\,\text{cm by the stern}$$
$$= +49.48\,\text{cm by the head}$$
$$\text{Change forward} = +(49.48 \div 130) \times (130 \div 2 + 3)$$
$$= +25.88\,\text{cm}$$
$$\text{Change aft} = -(49.48 \div 130) \times (130 \div 2 - 3)$$
$$= -23.6\,\text{cm}$$
$$\text{New draught forward} = 7.50 - 0.10 + 0.259$$
$$= 7.659\,\text{m}$$
$$\text{New draught aft} = 8.10 - 0.10 - 0.236$$
$$= 7.764\,\text{m}$$

**A. 5.4:** If the draught aft remains constant, the reduction in draught aft due to change in trim must equal the bodily sinkage.

$$\text{Bodily sinkage} = 180 \div 11$$
$$= 16.36\,\text{cm}$$
$$\text{Change in trim aft} = -(t \div L) \times \text{WF}$$
$$16.36 = t \div 90 \times (90 \div 2 - 2)$$
$$t = 16.36 \times 90 \div 43$$
$$= 34.24\,\text{cm by the head}$$
$$\text{But: change in trim } t = m \times d \div \text{MCT 1 cm}$$
$$34.24 = 180 \times d \div 50$$
$$d = 34.24 \times 50 \div 180$$
$$= 9.511\,\text{m}$$

Thus the mass must be placed 9.511 m forward of the centre of flotation or 7.511 m forward of midships.

$$\text{Change in trim forward} = +34.24 \div 90(90 \div 2 + 2)$$
$$= 17.88 \text{ cm}$$
$$\text{New draught forward} = 5.80 + 0.164 + 0.179$$
$$= 6.143 \text{ m}$$

**A. 5.5:**

$$\text{Change in trim required} = 8.90 - 8.20$$
$$= 0.7 \text{ m}$$
$$= 70 \text{ cm by the head}$$
$$\text{Therefore } 70 = m \times d \div \text{MCT 1 cm}$$
$$= m \times (60 + 1.5) \div 260$$
$$m = 70 \times 260 \div 61.5$$
$$= 296 \text{ tonne}$$
$$\text{Bodily sinkage} = 296 \div 28$$
$$= 10.57 \text{ cm}$$
$$\text{Change forward} = +70 \div 150(150 \div 2 + 1.5)$$
$$= +35.70 \text{ cm}$$
$$\text{Change aft} = -70 \div 150(150 \div 2 - 1.5)$$
$$= -34.30 \text{ cm}$$
$$\text{New draught forward} = 8.20 + 0.106 + 0.357$$
$$= 8.663 \text{ m}$$
$$\text{New draught aft} = 8.90 + 0.106 - 0.343$$
$$= 8.663 \text{ m}$$

(It is not necessary to calculate both draughts but this method checks the calculation.)

**A.5.6:**

| Mass | Distance from F | Moment forward | Moment aft |
|------|------|------|------|
| 20 | 41.5A | — | 830 |
| 50 | 24.5A | — | 1225 |
| 30 | 3.5A | — | 105 |
| 70 | 4.5F | 315 | — |
| 15 | 28.5F | 427.5 | — |
| 185 | | 742.5 | 2160 |

$$\text{Bodily sinkage} = 185 \div 9$$
$$= 20.55 \, \text{cm}$$
$$\text{Excess moment aft} = 2160 - 742.5$$
$$= 1417.5 \, \text{tonne m}$$
$$\text{Change in trim} = 1417.5 \div 55$$
$$= 25.77 \, \text{cm by the stern}$$
$$\text{Change forward} = -25.77 \div 110 \times (110 \div 2 - 1.5)$$
$$= 12.53 \, \text{cm}$$
$$\text{Change aft} = +25.77 \div 110 \times (110 \div 2 + 1.5)$$
$$= +13.24 \, \text{cm}$$
$$\text{New draught forward} = 4.20 + 0.205 - 0.125$$
$$= 4.28 \, \text{m}$$
$$\text{New draught aft} = 4.45 + 0.205 + 0.132$$
$$= 4.787 \, \text{m}$$

**A. 5.7:**

| Mass | Distance from F | Moment forward | Moment aft |
|------|-----------------|----------------|------------|
| + 160 | 66.5A | — | +10640 |
| +200 | 23.5F | +4700 | — |
| – 120 | 78.5A | — | – 9420 |
| – 70 | 19.5A | — | – 1365 |
| + 170 | | +4700 | – 145 |

$$\text{Bodily sinkage} = 170 \div 28$$
$$= 6.07 \, \text{cm}$$
$$\text{Excess moment forward} = 4700 - (-145)$$
$$= 4845 \, \text{tonne m}$$
$$\text{Change in trim} = 4845 \div 300$$
$$= 16.15 \, \text{cm by the head}$$
$$\text{Change forward} = (16.15 \div 170) \times (170 \div 2 - 3.5)$$
$$= +7.74 \, \text{cm}$$
$$\text{Change aft} = -(16.15 \div 170) \times (170 \div 2 + 3.5)$$
$$= -8.41 \, \text{cm}$$
$$\text{New draught forward} = 6.85 + 0.061 + 0.077$$
$$= 6.988 \, \text{m}$$
$$\text{New draught aft} = 7.50 + 0.061 - 0.084$$
$$= 7.477 \, \text{m}$$

**A. 5.8:**

| Item | Mass | Lcg | Moment forward | Moment aft |
|------|------|-----|----------------|------------|
| Lightweight | 1050 | 4.64A | — | 4872.0 |
| Cargo | 2150 | 4.71F | 10126.5 | — |
| Fuel | 80 | 32.55A | — | 2604.0 |
| Water | 15 | 32.90A | — | 493.5 |
| Stores | 5 | 33.60F | 168.0 | — |
| | 3300 | | 10294.5 | 7969.5 |

$$\text{Excess moment forward} = 10294.5 - 7969.5$$
$$= 2325 \text{ tonne m}$$
$$\text{LCG of loaded ship} = 2325 \div 3300$$
$$= 0.705 \text{ m forward of midships.}$$

The mean draught, MCT 1 cm, LCB and LCF may be found by interpolation from the tabulated values.

$$\text{Displacement difference 4.5 m to 5 m} = 3533 - 3172$$
$$= 361 \text{ tonne}$$
$$\text{Actual difference required} = 3300 - 3172$$
$$= 128 \text{ tonne}$$
$$\text{Proportion of draught difference} = 128 \div 361$$
$$\text{Actual draught difference} = 0.50 \times 128 \div 361$$
$$= 0.50 \times 0.355$$
$$= 0.177 \text{ m}$$
$$\text{Therefore: mean draught} = 4.50 + 0.177$$
$$= 4.677 \text{ m}$$
$$\text{MCT 1 cm difference 4.5 m to 5 m} = 43.10 - 41.26$$
$$= 1.84 \text{ tonne m}$$
$$\text{Actual difference} = 1.84 \times 0.355$$
$$= 0.65 \text{ tonne m}$$
$$\text{MCT 1 cm} = 41.26 + 0.65$$
$$= 41.91 \text{ tonne m}$$

LCB difference 4.5 m to 5 m = −0.24 m

$$\text{Actual difference} = -0.24 \times 0.355$$
$$= -0.085$$
$$\text{LCB} = 1.24 - 0.085$$
$$= 1.155 \text{ m forward of midships}$$

LCF difference 4.5 m to 5 m = 0.43 m

$$\text{Actual difference} = 0.43 \times 0.355 = 0.153$$
$$\text{Therefore: LCF} = 0.84 + 0.153$$
$$= 0.993 \text{ m aft of midships}$$

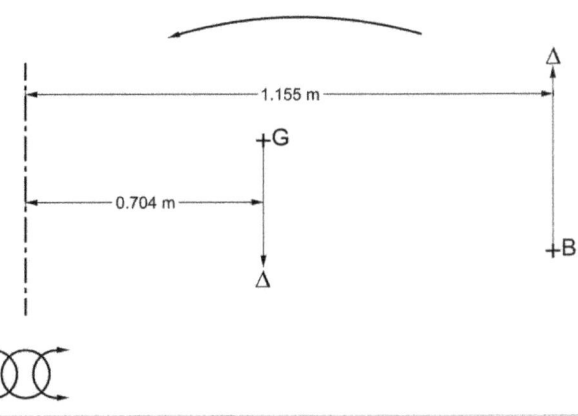

▲ Figure A. 5.1

$$\text{Trimming lever} = 1.155 - 0.704$$
$$= 0.451 \text{ m by the stern}$$
$$\text{Trimming moment} = 3300 \times 0.451$$
$$\text{Trim} = 3300 \times 0.451 \div 41.91$$
$$= 35.51 \text{ cm by the stern}$$
$$\text{Change forward} = -(35.51 \div 80) \times (80 \div 2 + 0.993)$$
$$= -18.20 \text{ cm}$$
$$\text{Change aft} = +(35.51 \div 80) \times (80 \div 2 - 0.993)$$
$$= +17.31 \text{cm}$$
$$\text{Draught forward} = 4.677 - 0.182$$
$$= 4.495 \text{ m}$$
$$\text{Draught aft} = 4.677 + 0.173$$
$$= 4.85 \text{ m}$$

**A. 5.9:**

Change in mean draught $= (100 \times \text{displacement} \div A_W) \times (\rho_s - \rho_r) \div (\rho_s \times \rho_r)$

$$= (100 \times 15000 \div 1950) \times (1.022 - 1.005) \div (1.005 \times 1.022)$$
$$= 12.73 \text{ cm}$$

Since the vessel moves from river water into sea water, the draught will be reduced.

**A. 5.10:**

$$\text{Let } \rho_s = \text{density of sea water in kg/m}^3$$

Change in mean draught $= (100 \times \text{displacement} \div A_W) \times (\rho_s - \rho_r) \div (\rho_s \times \rho_r)$

$$10 = (7000 \times 100 \div 1500) \times (\rho_s - 1005) \times 1000 \div (\rho_s \times 1005)$$
$$(\rho_s - 1005) \div (\rho_s \times 1005) = (10 \times 1500) \div (7000 \times 100 \times 1000)$$
$$(\rho_s - 1005) = (10 \times 1500) \div (7000 \times 100 \times 1000) \times (\rho_s \times 1005)$$
$$(\rho_s - 1005) = 0.02153\rho_s$$
$$\rho_s(1 - 0.02153) = 1005$$
$$\rho_s = 1005 \div 0.97847$$
$$\rho_{42} = 1027 \text{ kg/m}^3$$

**A. 5.11:**

| $\frac{1}{2}$ ordinate | SM | Product |
|---|---|---|
| 0 | 1 | 0 |
| 2.61 | 4 | 10.44 |
| 3.68 | 2 | 7.36 |
| 4.74 | 4 | 18.96 |
| 5.84 | 2 | 11.68 |
| 7.00 | 4 | 28.00 |
| 7.30 | 2 | 14.60 |
| 6.47 | 4 | 25.88 |
| 5.35 | 2 | 10.70 |
| 4.26 | 4 | 17.04 |
| 3.16 | 2 | 6.32 |
| 1.88 | 4 | 7.52 |
| 0 | 1 | 0 |
| | | 158.50 |

$$\text{Common interval} = 90 \div 12\,\text{m}$$
$$\text{Waterplane area} = \tfrac{2}{3} \times (90 \div 12) \times 158.50$$
$$= 792.5\,\text{m}^2$$

**a)** $\quad$ TPC $= 792.5 \times 1.024 \div 100$
$$= 8.115$$

**b)** $\quad$ Mass required $= 12 \times 8.115$
$$= 97.38\,\text{tonne}$$

**c)** $\quad$ Change in mean draught $= (8200 \times 100 \div 792.5) \times (1.024 - 1.005) \div (1.005 \times 1.024)$
$$= 19.10\,\text{cm increase}$$

**A. 5.12:** $\qquad\qquad$ Let $\Delta = $ displacement in sea water
Then $(\Delta - 360) = $ displacement in fresh water

Volume of displacement in sea water $= \Delta \div 1.025\,\text{m}^3$
Volume of displacement in fresh water $= (\Delta - 360) \div 1.000\,\text{m}^3$

Since the draught remains constant, these two volumes must be equal

$$(\Delta - 360) \div 1.000 = \Delta \div 1.025$$
$$1.025\Delta - 1.025 \times 360 = \Delta$$
$$0.025\Delta = 1.025 \times 360$$
$$\Delta = 1.025 \times 360 \div 0.025$$
Displacement in sea water $= 14760\,\text{tonne}$

**A. 5.13:**

Change in mean draught $= (5200 \times 100 \div 1100) \times (1.023 - 1.002) \div (1.002 \times 1.023)$
$$= 9.68\,\text{cm increase}$$
MCT 1 cm $= 5200 \times 95 \div 100 \times 90$
$$= 54.88\,\text{tonne m}$$
FB $= 0.6 + 2.2$
$$= 2.8\,\text{m}$$
Change in trim $= (\Delta \times \text{FB})(\rho_s - \rho_r) \div (\text{MCT 1 cm} \times \rho_s)$
$$= 5200 \times 2.8 \times (1.023 - 1.002) \div (54.88 \times 1.023)$$
$$= 5.45\,\text{cm by the head}$$

$$\text{Change forward} = +(5.45 \div 90) \times (90 \div 2 + 2.2)$$
$$= +2.86 \text{ cm}$$
$$\text{Change aft} = -(5.45 \div 90) \times (90 \div 2 - 2.2)$$
$$= -2.59 \text{ cm}$$
$$\text{New draught forward} = 4.95 + 0.097 + 0.029$$
$$= 5.076 \text{ m}$$
$$\text{New draught aft} = 5.35 + 0.097 - 0.026$$
$$= 5.421 \text{m}$$

**A. 5.14:**

$$\text{Change in mean draught} = (22000 \times 100 \div 3060) \times (1.026 - 1.007) \div (1.007 \times 1.026)$$
$$= 13.22 \text{ cm reduction}$$
$$\text{Change in trim} = 22000 \times 3 \times (1.026 - 1.007) \div 280 \times 1.007$$
$$= 4.5 \text{ cm by the stern}$$
$$\text{Change forward} = -(4.5 \div 160) \times (160 \div 2 + 4)$$
$$= -2.36 \text{ cm}$$
$$\text{Change aft} = +(4.5 \div 160) \times (160 \div 2 - 4)$$
$$= +2.14 \text{ cm}$$
$$\text{New draught forward} = 8.15 - 0.132 - 0.024$$
$$= 7.994 \text{ m}$$
$$\text{New draught aft} = 8.75 - 0.132 + 0.021$$
$$= 8.639 \text{ m}$$

**A. 5.15:**   **a)**   $$\text{Volume of lost buoyancy} = 12 \times 10 \times 4 \text{m}^3$$
$$\text{Area of intact waterplane} = (60 - 12) \times 10$$
$$= 48 \times 10 \text{m}^2$$
$$\text{Increase in draught} = (12 \times 10 \times 4) \div (48 \times 10)$$
$$= 1 \text{m}$$
$$\text{Therefore: new draught} = 4 + 1$$
$$= 5 \text{ m}$$

**b)**   $$\text{Volume of lost buoyancy} = 0.85 \times 12 \times 10 \times 4 \text{m}^3$$
$$\text{Area of intact waterplane} = (60 - 0.85 \times 12) \times 10$$
$$= 49.8 \times 10 \text{ m}^2$$
$$\text{Increase in draught} = (0.85 \times 12 \times 10 \times 4) \div (49.8 \times 10)$$
$$= 0.819 \text{ m}$$
$$\text{Therefore: new draught} = 4 + 0.819$$
$$= 4.819 \text{ m}$$

**c)**      Volume of lost buoyancy $= 0.60 \times 12 \times 10 \times 4 \, m^3$
Area of intact waterplane $= (60 - 0.60 \times 12) \times 10$
$$= 52.8 \times 10 \, m^2$$
$$= \text{Increase in draught } (0.60 \times 12 \times 10 \times 4) \div (52.8 \times 10)$$
$$= 0.545 \, m$$
Therefore: new draught $= 4 + 0.545$
$$= 4.545 \, m$$

**A. 5.16:**                    Density of coal $= 1.000 \times 1.28$
$$= 1.28 \, \text{tonne/m}^3$$
Volume of 1 tonne of solid coal $= 1 \div 1.28$
$$= 0.781 m^3$$
Volume of 1 tonne of stowed coal $= 1.22 \, m^3$

Therefore: in every 1.22 $m^3$ of volume, 0.439 $m^3$ is available for water

Hence permeability $\mu = 0.439 \div 1.22$
$$= 0.3598$$
Increase in draught $= 0.3598 \times 9 \times 8 \times 3 \div ((50 - 0.3598 \times 9) \times 8)$
$$= 0.208 \, m$$
Therefore: new draught $= 3 + 0.208$
$$= 3.208 \, m$$

**A. 5.17:**

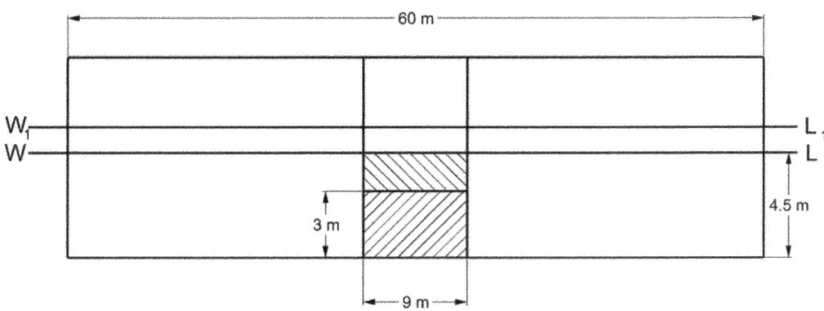

▲ **Figure A. 5.2**

**a)**      Volume of lost buoyancy $= 9 \times 12 \times 3 \, m^3$
Area of intact waterplane $= 60 \times 12 \, m^2$

(Since the water is restricted at the flat, the whole of the waterplane is intact.)

$$\text{Increase in draught} = (9 \times 12 \times 3) \div (60 \times 12)$$
$$= 0.45\,\text{m}$$
$$\text{New draught} = 4.5 + 0.45$$
$$= 4.95\,\text{m}$$

**b)**  $\quad$ Volume of lost buoyancy $= 9 \times 12 \times (4.5 - 3)\,\text{m}^3$
$$\text{Area of intact waterplane} = (60 - 9) \times 12\,\text{m}^2$$
$$\text{Increase in draught} = 9 \times 12 \times 1.5 \div (51 \times 12)$$
$$= 0.265\,\text{m}$$
$$\text{New draught} = 4.5 + 0.265$$
$$= 4.765\,\text{m}$$

**A. 5.18:** $\quad$ Mass of barge and teak $= 25 \times 4 \times 1.2 \times 1.000$
$$= 120\,\text{tonne}$$
$$\text{Mass of teak} = 25 \times 4 \times 0.120 \times 0.805$$
$$= 9.66\,\text{tonne}$$
$$\text{Mass of barge} = 120 - 9.66$$
$$= 110.34\,\text{tonne}$$

If it is assumed that the teak is first removed and then the compartment bilged:

$$\text{Draught when teak removed} = 110.34 \div (1.000 \times 25 \times 4)$$
$$= 1.103\,\text{m}$$
$$\text{Increase in draught due to bilging} = (5 \times 4 \times 1.1034) \div ((25 - 5) \times 4)$$
$$= 0.276\,\text{m}$$
$$\text{Final draught} = 1.103 + 0.276$$
$$= 1.379\,\text{m}$$

**A. 5.19:** It has been convenient to consider this as a three-part question, but any one part could constitute an examination question.

**a)**

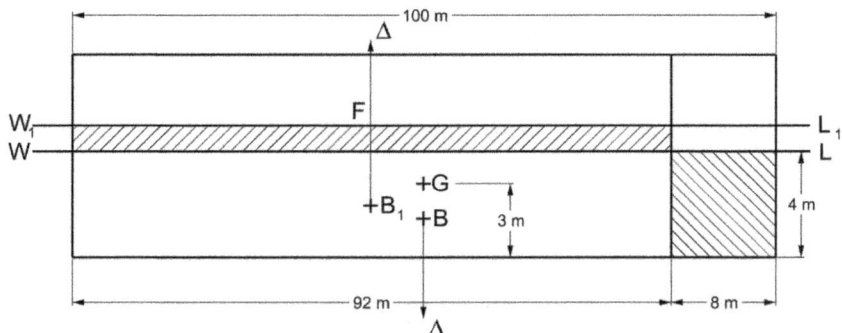

▲ Figure A. 5.3

$$\text{Increase in mean draught} = (8 \times 12 \times 4) \div ((100 - 8) \times 12)$$
$$= 0.348 \text{ m}$$
$$\text{New mean draught} = 4 + 0.348$$
$$= 4.348$$
$$KB_1 = 4.348 \div 2$$
$$= 2.174$$
$$B_1 M_L = 92^2 \div (12 \times 4.348)$$
$$= 162.22$$
$$GM_L = 2.17 + 162.22 - 3.00$$
$$= 161.39$$
$$\text{Change in trim} = (50 \times L \times l) \div GM_L$$
$$= (50 \times 100 \times 8) \div 161.39$$
$$= 247.8 \text{ cm by the head}$$
$$\text{Change forward} = +(247.8 \div 100) \times (100 \div 2 + 4)$$
$$= +133.8 \text{ cm}$$
$$\text{Change aft} = -(247.8 \div 100) \times (100 \div 2 - 4)$$
$$= -114.0 \text{ cm}$$
$$\text{New draught forward} = 4.348 + 1.338$$
$$= 5.686 \text{ m}$$
$$\text{New draught aft} = 4.348 - 1.140$$
$$= 3.208 \text{ m}$$

**b)**

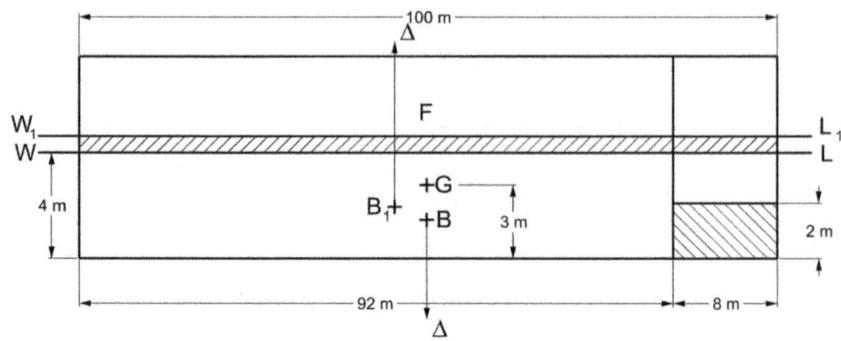

▲ Figure A. 5.4

$$\text{Volume of lost buoyancy} = 8 \times 12 \times 2 \text{ m}^3$$
$$\text{Area of intact waterplane} = 100 \times 12 \text{ m}^2$$
$$\text{Increase in draught} = (8 \times 12 \times 2) \div (100 \times 12)$$
$$= 0.16 \text{ m}$$
$$\text{New draught} = 4.0 + 0.16$$
$$= 4.16 \text{ m}$$
$$KB_1 = 2.08 \text{ m (approx)}$$
$$I_F = (1/12) \times L^3 \times B$$
$$= (1/12) \times 100^3 \times 12$$
$$= 1.0 \times 10^6 \text{ m}^4$$
$$\text{Volume} = L \, Bd$$
$$= 100 \times 12 \times 4$$
$$B_1M_L = 1.0 \times 10^6 \div (100 \times 12 \times 4)$$
$$= 208.33 \text{ m}$$
$$GM_L = 2.08 + 208.33 - 3.00$$
$$= 207.41 \text{ m}$$

To find the longitudinal shift in the centre of buoyancy, consider the volume of lost buoyancy moved to the centre of the intact waterplane, which in this case is midships.

$$BB_1 = (8 \times 12 \times 2 \times (50-4)) \div (100 \times 12 \times 4)$$
$$= 1.84 \text{ m}$$

Trimming moment $= \Delta BB_1$

$$\text{MCT 1 cm} = \Delta \times GM_L \div (100L)$$
$$= (\Delta \times 207.41) \div (100 \times 100)$$

$$\text{Change in trim} = 1.84 \times 100 \times 100 \div 207.41$$
$$= 88.71 \text{ cm by the head}$$

$$\text{Change forward} = +88.71 \div 100 \times 50$$
$$= +44.36 \text{ cm}$$

$$\text{Change aft} = -88.71 \div 100 \times 50$$
$$= -44.36 \text{ cm}$$

$$\text{New draught forward} = 4.16 + 0.444$$
$$= 4.604 \text{ m}$$

$$\text{New draught aft} = 4.16 - 0.444$$
$$= 3.716 \text{ m}$$

**c)**

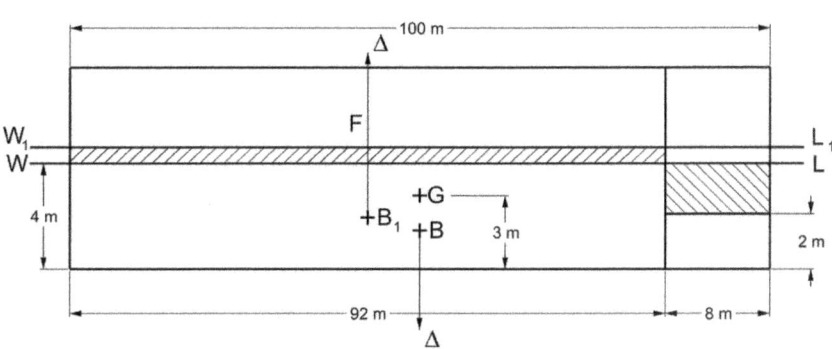

$$\text{Volume of lost buoyancy} = 8 \times 12 \times 2\ \text{m}^3$$
$$\text{Area of intact waterplane} = (100 - 8) \times 12\ \text{m}^2$$
$$\text{Increase in draught} = 8 \times 12 \times 2 \div ((100 - 8) \times 12)$$
$$= 0.174\ \text{m}$$
$$\text{New draught} = 4.0 + 0.174$$
$$= 4.174\ \text{m}$$
$$KB_1 = 2.087\ \text{m}$$
$$I_F = 92^3 \times 12 \div 12$$
$$= 778688\ \text{m}^4$$
$$\text{Volume} = 100 \times 12 \times 4$$
$$B_1 M_L = 778688 \div 4800$$
$$= 162.23\ \text{m}$$
$$GM_L = 2.09 + 162.22 - 3.00$$
$$= 161.31\text{m}$$
$$\text{Shift in centre of buoyancy } BB_1 = (8 \times 12 \times 2 \times (46 + 4)) \div (100 \times 12 \times 4)$$
$$= 2\ \text{m}$$
$$\text{Change in trim} = (2 \times 100 \times 100) \div 161.31$$
$$= 124\ \text{cm by the head}$$
$$\text{Change forward} = +(124 \div 100) \times (100 \div 2 + 4)$$
$$= +67\ \text{cm}$$
$$\text{Change aft} = -(124 \div 100) \times (100 \div 2 - 4)$$
$$= -57\ \text{cm}$$
$$\text{New draught forward} = 4.174 + 0.670$$
$$= 4.844\ \text{m}$$
$$\text{New draught aft} = 4.174 - 0.570$$
$$= 3.604\ \text{m}$$

# SOLUTIONS TO CHAPTER 6 TEST EXAMPLES

**A. 6.1:**

| Width | SM | Product for area | (Width)$^2$ | SM | Product for 1st moment | (Width)$^3$ | SM | Product for 2nd moment |
|-------|----|------------------|-------------|----|------------------------|-------------|----|------------------------|
| 10 | 1 | 10 | 100 | 1 | 100 | 1000 | 1 | 1000 |
| 9 | 4 | 36 | 81 | 4 | 324 | 729 | 4 | 2916 |
| 7 | 2 | 14 | 49 | 2 | 98 | 343 | 2 | 686 |
| 4 | 4 | 16 | 16 | 4 | 64 | 64 | 4 | 256 |
| 1 | 1 | 1 | 1 | 1 | 1 | 1 | 1 | 1 |
| | | 77 = Σa | | | 587 = Σm | | | 4859 = Σi |

$$\text{Common interval} = 12 \div 4$$
$$= 3\,\text{m}$$
$$\text{Area of surface } a = h/3\,\Sigma_A$$
$$= 3 \div 3 \times 77$$
$$= 77\,\text{m}^2$$

Distance of centroid from longitudinal bulkhead

$$= \Sigma_M \div (2 \times \Sigma_A)$$
$$= 587 \div (2 \times 77)$$
$$= 3.812\,\text{m}$$

Second moment of area about longitudinal bulkhead

$$i_b = h/9\Sigma_i$$
$$= 3 \div 9 \times 4859$$
$$= 1619.7 \, m^4$$

By parallel axis theorem, second moment of area about centroid

$$i_g = i_b - ay^2$$
$$= 1619.7 - 77 \times 3.812^2$$
$$= 1619.7 - 1118.7$$
$$= 501 m^4$$

**A. 6.2:**          Load on tank top $= \rho gAh$
$$9.6 \times 10^6 = 1025 \times 9.81 \times 12 \times 10 \times h$$
$$h = (9.6 \times 10^6) \div 1025 \div 9.81 \div 12 \div 10$$
$$= 7.96 \, m$$

**A. 6.3: a)** Load on top $= \rho gAh$

Since H is zero

$$\text{Load on top} = 0$$
$$\text{Load on short side} = 1000 \times 9.81 \times 12 \times 1.4 \times 0.7$$
$$= 115.3 \times 10^3 \, N$$
$$= 115.3 \, kN$$

**b)**          Load on top $= 1000 \times 9.81 \times 15 \times 12 \times 7$
$$= 12.36 \times 10^6 \, N$$
$$= 12.36 \, MN$$
$$\text{Load on short side} = 1000 \times 9.81 \times 12 \times 1.4 \times (7 + 0.7)$$
$$= 1.269 \times 10^6 \, N$$
$$= 1.269 \, MN$$

**A. 6.4:**          Pressure at bottom $= \rho gh$
$$= 1025 \times 9.81 \times 6$$
$$= 60.33 \times 10^3 \, N/m^2$$
$$= 60.34 \, kN/m^2$$
$$\text{Load on bulkhead} = \rho gAh$
$$= 1025 \times 9.81 \times 9 \times 6 \times 3$$
$$= 1.629 \times 10^6 \, N$$
$$= 1.629 \, MN$$

**A. 6.5:**

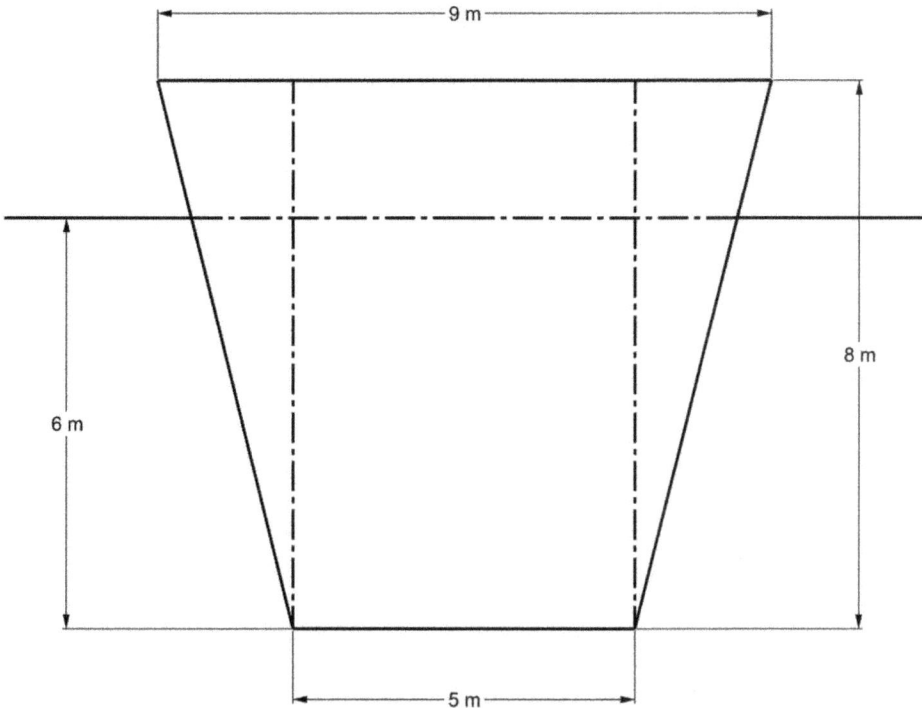

▲ Figure A. 6.1

**a)** Width of bulkhead at a depth of 6 m

$$= 5 + (4 \times 6 \div 8)$$
$$= 8\,m$$

Divide the bulkhead into a rectangle and two triangles as shown in the above figure.

$$\text{Load on rectangle} = 0.85 \times 1000 \times 9.81 \times 5 \times 6 \times 3$$
$$= 750.4 \times 10^3\,N$$
$$= 750.4\,kN$$
$$\text{Load on two triangles} = 2 \times 0.85 \times 1000 \times 9.81 \times (1.5 \times 6 \div 2) \times (6 \div 3)$$
$$= 150.1 \times 10^3\,N$$
$$= 150.1\ kN$$
$$\text{Total load on bulkhead} = 750.4 + 150.1$$
$$= 900.5\ kN$$

**b)**

$$\text{Load on rectangle} = 0.85 \times 1000 \times 9.81 \times 5 \times 8 \times (8 \div 2 + 4)$$
$$= 2.669 \times 10^6 \text{ N}$$
$$= 2.669 \text{ MN}$$
$$\text{Load on two triangles} = 2 \times 0.85 \times 1000 \times 9.81 \times (2 \times 8 \div 2) \times (8 \div 3 + 4)$$
$$= 889.0 \times 10^3 \text{ N}$$
$$= 0.890 \text{ MN}$$
$$\text{Total load on bulkhead} = 2.669 + 0.890$$
$$= 3.559 \text{ MN}$$

**A. 6.6: a)**

$$\text{Load on bulkhead} = \rho g A h$$
$$= 0.9 \times 1000 \times 9.81 \times 10 \times 12 \times 6$$
$$= 6.357 \times 10^6 \text{ N}$$
$$= 6.357 \text{ MN}$$
$$\text{Distance of centre of pressure from top of bulkhead} = 2/3\,D$$
$$= 2/3 \times 12$$
$$= 8 \text{ m}$$

**b)**

$$\text{Load on bulkhead} = 0.9 \times 1000 \times 9.81 \times 10 \times 12$$
$$\times (12 \div 2 + 3)$$
$$= 9.535 \times 10^6 \text{ N}$$
$$= 9.535 \text{ MN}$$
$$\text{Distance of centre of pressure from surface of oil} = (I_{NA} \div AH) + H$$
$$= (1 \div 12) \times 10 \times 12^3$$
$$\div (10 \times 12 \times 9) + 9$$
$$= 1.333 + 9$$
$$= 10.333 \text{ m}$$
$$\text{Distance of centre of pressure from top of bulkhead} = 10.333 - 3$$
$$= 7.333 \text{ m}$$

**A. 6.7:**

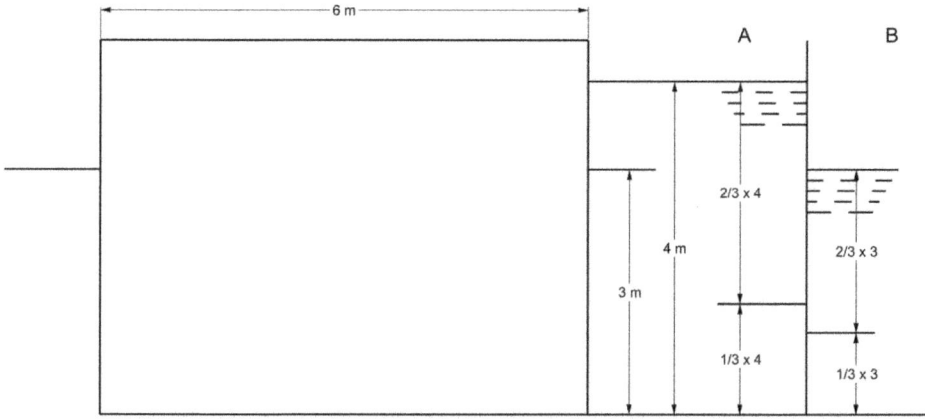

▲ Figure A. 6.2

**a)**

$$\text{Load on side A} = 1025 \times 9.81 \times 6 \times 4 \times 2$$
$$= 482.7 \times 10^3 \text{ N}$$
$$= 482.7 \text{ kN}$$
$$\text{Centre of pressure on side A} = 2 \div 3 \times 4$$
$$= 2.667 \text{ m from top}$$
$$= 1.333 \text{ m from bottom}$$
$$\text{Load on side B} = 1000 \times 9.81 \times 6 \times 3 \times 1.5$$
$$= 265.0 \times 103 \text{ N}$$
$$= 265.0 \text{ kN}$$
$$\text{Resultant load} = 482.7 - 265.0$$
$$= 217.7 \text{ kN}$$

**b)**

$$\text{Centre of pressure on side B} = 2 \div 3 \times 3$$
$$= 2 \text{ m from top}$$
$$= 1 \text{ m from bottom}$$
Taking moments of load about bottom of gate :
$$\text{Resultant centre of pressure from bottom} = (482.6 \times 1.333 - 265.0 \times 1) \div$$
$$(482.7 - 265.0)$$
$$= (643.47 - 265.0) \div (217.7)$$
$$= 1.74 \text{ m}$$

**A. 6.8:**

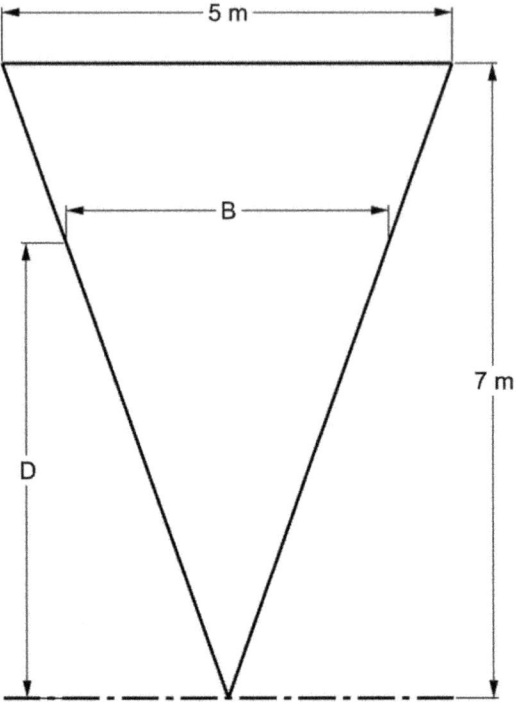

▲ Figure A. 6.3

There are two types of solution possible, one if the water is above the top of the bulkhead and one if the water is below the top of the bulkhead. Assume water at top edge:

$$\text{Load on bulkhead} = 1025 \times 9.81 \times (5 \times 7) \div 2 \times 7 \div 3$$
$$= 410.6 \times 10^3 \text{ N}$$
$$= 410.6 \text{ kN}$$

Since the load on the bulkhead is only 190 kN, the water must be below the top of the bulkhead.

$$\text{Width at water level } B = 5 \div 7 \times D$$
$$190 \times 10^3 = 1025 \times 9.81 \times (D \div 2) \times (5D \div 7) \times (D \div 3)$$
$$D^3 = 190 \times 10^3 \times 2 \times 7 \times 3 \div (1025 \times 9.81 \times 5)$$
$$\text{from which } D = 5.414 \text{ m}$$
$$\text{Centre of pressure below surface of water} = 1 \div 2 \times D$$
$$= 2.707 \text{ m}$$
$$\text{Centre of pressure below top of bulkhead} = 2.707 + (7.00 - 5.414)$$
$$= 4.293 \text{ m}$$

**A. 6.9:**

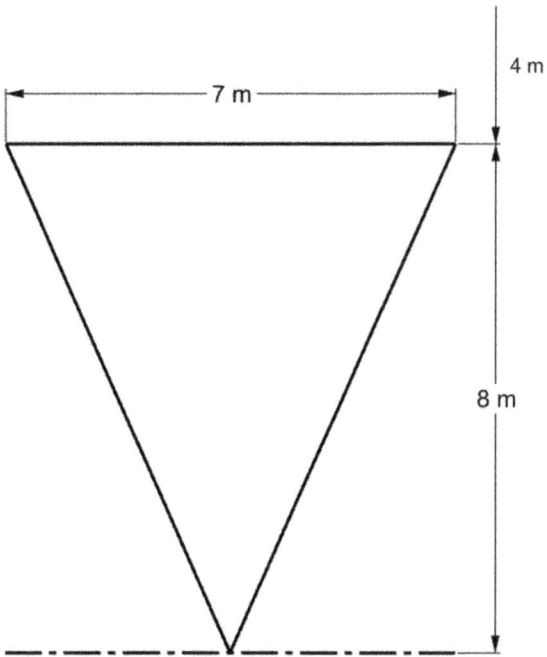

▲ Figure A. 6.4

**a)**

$$\text{Load on bulkhead} = 1025 \times 9.81 \times (7 \times 8 \div 2) \times (8 \div 3)$$
$$= 750.8 \times 10^3 \text{ N}$$
$$= 750.8 \text{ kN}$$
$$\text{Centre of pressure from top} = 1 \div 2 \times D$$
$$= 4 \text{ m}$$

**b)**

$$\text{Load on bulkhead} = 1025 \times 9.81 \times (7 \times 8 \div 2) \times (8 \div 3 + 4)$$
$$= 1.877 \times 10^6 \text{ N}$$
$$= 1.877 \text{ MN}$$
$$\text{For triangle } I_{NA} = 1 \div 36 \times BD^3$$
$$\text{Centre of pressure from surface of water} = I_{NA} \div (AH) + H$$
$$= (1 \div 36) \times 7 \times 8^3 \div ((1 \div 2) \times 7 \times 8 \times$$
$$(1 \div 3) \times 8 \times 4) + ((1 \div 3) \times 8 \times 4)$$
$$= 2 \times 7 \times 8^3 \div (36 \times 7 \times 8 \times 6.667) + 6.667$$
$$= 7.2 \text{ m}$$
$$\text{Centre of pressure from top of bulkhead} = 7.200 - 4.000$$
$$= 3.2 \text{ m}$$

**A. 6.10:** **a)**                                    Let $l$ = height of bulkhead

Maximum shearing force in stiffeners = 200 kN

$$= 2/3 \times load$$

Therefore:        $load = 200 \times 3 \div 2$

$$= 300 \text{ kN}$$

But load on stiffener $= 1025 \times 9.81 \times l \times (l \div 9) \times (l \div 2)$

$$l^3 = 300 \times 10^3 \times 9 \times 2 \div (1025 \times 9.81)$$

from which        $l = 8.128 \text{ m}$

**b)**                          Shearing force at top $= 1/3 \times load$

$$= 1/3 \times 300$$

$$= 100 \text{ kN}$$

**c)**                          Position of zero shear $= l/\sqrt{3}$

$$= 8.128/\sqrt{3}$$

$$= 4.693 \text{ m from top}$$

**A. 6.11:**

| Width | SM | Product for area | Lever | Product for 1st moment | Lever | Product for 2nd moment |
|-------|-----|------------------|-------|------------------------|-------|------------------------|
| 8.0 | 1 | 8.0 | 0 | 0 | 0 | 0 |
| 7.5 | 4 | 30.0 | 1 | 30.0 | 1 | 30.0 |
| 6.5 | 2 | 13.0 | 2 | 26.0 | 2 | 52.0 |
| 5.7 | 4 | 22.8 | 3 | 68.4 | 3 | 205.2 |
| 4.7 | 2 | 9.4 | 4 | 37.6 | 4 | 150.4 |
| 3.8 | 4 | 15.2 | 5 | 76.0 | 5 | 380.0 |
| 3.0 | 1 | 3.0 | 6 | 18.0 | 6 | 108.0 |
|  |  |  |  | $256.0 = \Sigma_M$ |  | $925.6 = \Sigma_I$ |

Common interval $= 1.2$ m

Load on bulkhead $=$ density $\times g \times$ 1st moment

$$= \rho g\, h^2/3\Sigma_M$$

$$= 1025 \times 9.81 \times (1.2^2 \div 3) \times 256.0$$

$$= 1236 \text{ kN}$$

Centre of pressure from top $= 2^{nd}$ moment about top $\div$ 1st moment about top

$$= h \times \Sigma_I \div \Sigma_M$$

$$= 1.2 \times 925.6 \div 256.0$$

$$= 4.339 \text{ m}$$

# SOLUTIONS TO CHAPTER 7 TEST EXAMPLES

**A. 7.1:**

$$R_f = f\,S\,V^2$$
$$= 0.424 \times 3200 \times 17^{1.825}$$
$$= 238900\,\text{N}$$
$$= 238.9\,\text{kN}$$
$$\text{Power} = R_f \times V$$
$$= 238.9 \times (17 \times 0.5144)$$
$$= 2089.1\,\text{kW}$$

**A. 7.2:**

$$\text{At } 3\,\text{m/s } R_f = 13\,\text{N/m}^2$$
$$\text{At } 15\,\text{knots } R_f = 13 \times (15 \div 3 \times 0.5144)^{1.97}$$
$$= 83.605\,\text{N/m}^2$$
$$\text{Therefore}: R_f = 83.605 \times 3800$$
$$= 317700\,\text{N}$$
$$\text{Power} = 317700 \times 15 \times 0.5144$$
$$= 2451400\,\text{W}$$
$$= 2451.4\,\text{kW}$$

**A. 7.3:**

$$At\ 180\ m/min\ R_f = 12\,N/m^2$$

$$At\ 14\ knots\ R_f = 12 \times [(14 \times 0.5144) \div (180 \div 60)]^{1.9}$$

$$= 63.35\,N/m^2$$

$$R_f = 63.35 \times 4000\ N$$

$$= 253411\ N$$

$$R_t = R_f \div 0.7$$

$$R_t = 253411 \div 0.7$$

$$= 362016\,N$$

$$Effective\ power = R_t \times V$$

$$= 362016 \times 14 \times 0.5144$$

$$= 2607095\ W$$

$$= 2607.1\,kW$$

**A. 7.4:**

$$Wetted\ surface\ area\ S = c\sqrt{\Delta L}$$

$$= 2.55\sqrt{(125 \times 16 \times 7.8 \times 0.72 \times 1.025 \times 125)}$$

$$= 3059\,m^2$$

$$R_f = f\,SV^2$$

$$= 0.423 \times 3059 \times 17.5^{1.825}$$

$$= 240140\,N$$

$$Power = 240140 \times 17.5 \times 0.5144$$

$$= 2161739\ W$$

$$= 2161.7\,kW$$

**A. 7.5:**

$$R_f\ is\ proportional\ to\ L^3$$

$$Therefore:\ R_f = 36 \times (20 \div 1)^3$$

$$= 288000\,N$$

$$V\ is\ proportional\ to\ \sqrt{L}$$

$$V = 3 \times \sqrt{(20 \div 1)}$$

$$= 13.417\,knots$$

$$Power = 288000 \times 13.417 \times 0.5144$$

$$= 1987691\ W$$

$$= 1987.7\ kW$$

**A. 7.6:**

$V$ is proportional to $\sqrt{L}$

$V$ is proportional to $\Delta^{\frac{1}{3}}$

$$\text{Therefore: } V = 16 \times (24000 \div 14000)^{\frac{1}{3}}$$
$$= 17.503 \text{ knots}$$

$R_f$ is proportional to $\Delta$

$$\text{Therefore: } R_f = 113 \times (24000 \div 14000)$$
$$= 193.7 \text{ kN}$$

**A. 7.7:**

$$\text{Let } V = \text{speed in m/s}$$
$$\text{At 3 m/s } R_f = 11 \times 1.025$$
$$= 11.275 \text{ N/m}^2 \text{ in sea water}$$
$$R_f = 11.275 \times 2500$$
$$= 28188 \text{ N}$$
$$\text{At } V \text{ m/s } R_f = 28.19(V \div 3)^{1.92} \text{ kN}$$
$$\text{and } R_t = R_f \div 0.72$$
$$\text{effective power} = R_t \times V$$
$$= 1100 \text{ kW}$$
$$(R_f \div 0.72) \times V = 1100$$
$$(28.19 \div 0.72) \times (V \div 3)^{1.92} \times V = 1100$$
$$V^{2.92} = 1100 \times (0.72 \div 28.19) \times 3^{1.92}$$
$$V = 6.454 \text{ m/s}$$
$$V = 6.454 \div 0.5144$$
$$V = 12.547 \text{ knots}$$

**A. 7.8:**

Model:

$$R_f = 35 \text{ N in fresh water}$$
$$= 35 \times 1.025$$
$$= 35.875 \text{ N}$$
$$f = 0.417 + (0.773 \div (6 + 2.862))$$
$$= 0.417 + 0.0872$$
$$= 0.5042$$
$$R_f = 0.5042 \times 7 \times 3^{1.825}$$
$$= 26.208 \text{ N}$$
$$R_r = R_t - R_f$$
$$= 35.875 - 26.208$$
$$= 9.667 \text{ N}$$

Ship:

$R_r$ is proportional to $L^3$

$$\text{Therefore: } R_r = 9.667 \times (120 \div 6)^3$$
$$= 77336 \text{ N}$$

$S$ is proportional to $L^2$

$$S = 7 \times (120 \div 6)^2$$
$$= 2800 \text{ m}^2$$

$V$ is proportional to $\sqrt{L}$

$$\text{Therefore: } V = 3\sqrt{(120 \div 6)}$$
$$= 13.416 \text{ knots}$$
$$f = 0.417 + (0.773 \div (120 + 2.862))$$
$$= 0.417 + 0.0063$$
$$= 0.4233$$
$$R_f = 0.4233 \times 2800 \times 13.416^{1.825}$$
$$= 135400 \text{ N}$$
$$R_t = 77336 + 135400$$
$$= 212736 \text{ N}$$
$$\text{Effective power (naked)} = 212736 \times 13.416 \times 0.5144$$
$$= 1468.1 \text{ kW}$$
$$\text{Effective power} = \text{ep} \times \text{SCF}$$
$$= 1468.1 \times 1.15$$
$$= 1688.4 \text{ kW}$$

**A. 7.9:**
$$\text{sp} = \Delta^{\frac{2}{3}} \times V^3 \div 550$$
$$= 12000^{\frac{2}{3}} \times 16^3 \div 550$$
$$= 3903 \text{ kW}$$

**A.7.10: a)**
$$2800 = \Delta^{\frac{2}{3}} \times 14^3 \div 520$$
$$\Delta^{\frac{2}{3}} = 2800 \times 520 \div 14^3$$
$$\Delta = 12223 \text{ tonne}$$

**b)**
$$\text{New speed} = 0.85\,V$$
$$= 0.85 \times 14$$

sp is proportional to $V^3$

Therefore:

$$sp_1 \div sp_2 = (V_1 \div V_2)^3$$
$$sp_2 = 2800 \times (0.85 \times 14 \div 14)^3$$
$$= 1720\,\text{kW}$$

**A. 7.11:**
$$2100 = 8000^{\frac{2}{3}} \times V^3 \div 470$$
$$V^3 = 2100 \times 470 \div 8000^{\frac{2}{3}}$$
$$V = 13.51\,\text{knots}$$

**A. 7.12:**
**a)** $\Delta = 150 \times 19 \times 8 \times 0.68 \times 1.025$
$$= 15890\,\text{tonne}$$
$$sp = 15890^{\frac{2}{3}} \times 18^3 \div 600$$
$$= 6143\,\text{kW}$$

**b)** sp is proportional to $R_t \times V$

sp is proportional to $V^3 \times V$

sp is proportional to $V^4$

Therefore:

$$sp = 6143 \times (21 \div 18)^4$$
$$= 11382\,\text{kW}$$

*Note:* In practice, there will be a gradual increase in the index of speed.

**A. 7.13:**
$$\text{Fuel cons/day} = \Delta^{\frac{2}{3}} \times V^3 \div \text{fuel coefficient tonne}$$
$$= 15000^{\frac{2}{3}} \times 14.5^3 \div 62500$$
$$= 29.67\,\text{tonne}$$

**A. 7.14:**
$$25 = 9000^{\frac{2}{3}} \times V^3 \div 53500$$
$$V^3 = 25 \times 53500 \div 9000^{\frac{2}{3}}$$
$$V = 14.57 \text{ knots}$$

**A. 7.15:**
At 16 knots: time taken $= 2000 \div (16 \times 24)$
$$= 5.208 \text{ days}$$
total consumption $= 28 \times 5.208$
$$= 145.8 \text{ tonne}$$

But total consumption is proportional to $V^2$

Therefore: at 14 knots, total consumption $= 145.8 \times (14 \div 16)^2$
$$= 111.6 \text{ tonne}$$
Saving in fuel $= 145.8 - 111.6$
$$= 34.2 \text{ tonne}$$

**A. 7.16:**

Total fuel used $= 115 - 20$
$$= 95 \text{ tonne}$$
At 15 knots, total consumption for 1100 nautical miles $= 40 \times 1100 \div (15 \times 24)$
$$= 122.2 \text{ tonne}$$

But total consumption is proportional to $V^2$

$$95 \div 122.3 = V^2 \div 15^2$$
$$V = 15\sqrt{(95 \div 122.3)}$$
$$= 13.22 \text{ knots}$$
Time taken $= 1100 \div 13.22$
$$= 83.21 \text{ hours}$$

**A. 7.17:**
**a)** $\text{cons}_1 \div \text{cons}_2 = (V_1 \div V_2)^3$
$$V_2^3 = V_1^3 (\text{cons}_2 \div \text{cons}_1)$$
$$V_2 = V_1 \times (\text{cons}_2 \div \text{cons}_1)^{\frac{1}{3}}$$
$$V_2 = 14_\times (1.15 \times 23 \div 23)^{\frac{1}{3}}$$
$$V_2 = 14.67 \text{ knots}$$

**b)**
$$V_2 = 14 \times (0.88 \times 23 \div 23)^{\frac{1}{3}}$$
$$= 13.42 \text{ knots}$$

**c)**
$$V_2 = 14 \times (18 \div 23)^{\frac{1}{3}}$$
$$= 12.90 \text{ knots}$$
$$\text{Fuel consumption/hour} = 0.12 + 0.001 V^3 \text{ tonne}$$
$$= 0.12 + 0.001 \times 14^3$$
$$= 2.864 \text{ tonne}$$

**A. 7.18:**

**a)** Over 1700 nautical miles:
$$\text{total fuel consumption} = 2.864 \times 1700 \div 14$$
$$= 347.8 \text{ tonne}$$

**b)**
$$\text{Saving in fuel} = 10 \text{ t/day}$$
$$= 10 \div 24 \text{ t/h}$$
$$\text{Therefore: new fuel consumption} = 2.864 - 10 \div 24$$
$$= 2.447 \text{l/h}$$
$$2.447 = 0.12 + 0.001 V^3$$
$$0.001 V^3 = 2.447 - 0.12$$
$$V^3 = 2327$$
$$V = 13.25 \text{ knots}$$

**A. 7.19:**
$$\text{Let } C = \text{normal consumption per hour}$$
$$V = \text{normal speed}$$

For first 8 hours:
$$\text{speed} = 1.2 V$$
$$\text{cons/h} = C \times (1.2V \div V)^3$$
$$= 1.728 C$$
$$\text{cons for 8 hours} = 8 \times 1.728 C$$
$$= 13.824 C$$

For next 10 hours:

$$speed = 0.9\,V$$
$$cons/h = C \times (0.9V \div V)^3$$
$$= 0.729\,C$$
$$cons \text{ for } 10 \text{ hours} = 10 \times 0.729\,C$$
$$= 7.29\,C$$

For remaining 6 hours:

$$speed = V$$
$$cons \text{ for } 6 \text{ hours} = 6\,C$$
$$\text{Total cons for } 24 \text{ hours} = 13.824\,C + 7.29\,C + 6\,C$$
$$= 27.114\,C$$
$$\text{Normal cons for } 24 \text{ hours} = 24\,C$$
$$\text{Therefore: increase in cons} = 27.114\,C - 24\,C$$
$$= 3.114\,C$$
$$\% \text{ increase in cons} = 3.114\,C \div 24\,C \times 100$$
$$= 12.98\%$$

**A. 7.20:**

$$\text{Let } C = \text{consumption per day at 18 knots}$$
$$\text{Then } C - 22 = \text{consumption per day at 14.5 knots}$$
$$C \div (C - 22) = (18 \div 14.5)^3$$
$$= 1.913$$
$$C = 1.913C - 1.913 \times 22$$
$$0.913\,C = 1.913 \times 22$$
$$C = 1.913 \times 22 \div 0.913$$
$$= 46.09 \text{ tonne/day}$$

**A. 7.21:**

$$\text{At 17 knots: cons/day} = 42 \text{ tonne}$$
$$\text{At } V \text{ knots: cons/day} = 28 \text{ tonne}$$
$$\text{But at } V \text{ knots: cons/day} = 1.18\,C$$
$$\text{Where } C = 42 \times (V \div 17)^3$$
$$1.18\,C = 28$$

Therefore:

$$28 = 1.18 \times 42 \times (V \div 17)^3$$
$$V^3 = 17^3 \times 28 \div (1.18 \times 42)$$
$$V = 17(28 \div (1.18 \times 42))^{\frac{1}{3}}$$
$$= 14.06 \text{ knots}$$

# SOLUTIONS TO CHAPTER 8 TEST EXAMPLES

**A. 8.1:**

$$\text{Theoretical speed } V_T = P \times N \times 60 \div 1852$$
$$= 5 \times 105 \times 60 \div 1852$$
$$= 17.01 \text{ knots}$$
$$\text{Apparent slip} = (V_T - V) \div V_T \times 100$$
$$\text{Speed of advance } V_a = V(1-w)$$
$$= 14 \times (1-0.35)$$
$$= 9.10 \text{ knots}$$
$$\text{Real slip} = (V_T - V_a) \div V_T \times 100$$
$$= (17.01 - 9.10) \div 17.01 \times 100$$
$$= 46.50\%$$

**A. 8.2:**

$$p = P \div D$$
$$\text{Therefore: pitch } P = p \times D$$
$$= 0.8 \times 5.5$$
$$= 4.4 \text{ m}$$
$$\text{Theoretical speed } V_T = (4.4 \times 120 \times 60) \div 1852$$
$$= 17.11 \text{ knots}$$
$$\text{Real slip} = (V_T - V_a) \div V_T \times 100$$
$$0.35 = (17.11 - V_a) \div 17.11$$
$$\text{Therefore: speed of advance } V_a = 17.11 \times (1 - 0.35)$$
$$= 11.12 \text{ knots}$$
$$\text{But } V_a = V(1 - w_t)$$
$$\text{Therefore: ship speed } V = V_a \div (1 - w_t)$$
$$= 11.12 \div (1 - 0.32)$$
$$= 16.35 \text{ knots}$$
$$\text{Apparent slip} = (V_T - V) \div V_T \times 100$$
$$= (17.11 - 16.35) \div 17.11 \times 100$$
$$= 4.44\%$$

**A. 8.3:**

$$C_b = \Delta \div (L \times B \times d \times \rho)$$
$$= 12400 \div (120 \times 17.5 \times 7.5 \times 1.025)$$
$$= 0.768$$
$$w = 0.5 \times 0.768 - 0.05$$
$$= 0.334$$
$$\text{Speed of advance } V_a = V(1 - w)$$
$$= 12 \times (1 - 0.334)$$
$$= 7.992 \text{ knots}$$
$$\text{Real slip} = (V_T - V_a) \div V_T \times 100$$
$$0.30 \, V_T = V_T - 7.992$$
$$V_T = 7.992 \div 0.7$$
$$= 11.42 \text{ knots}$$
$$\text{But } V_T = (P \times N \times 60) \div 1852$$
$$\text{Pitch } P = 11.42 \times 1852 \div (100 \times 60)$$
$$= 3.52 \text{ m}$$
$$\text{Diameter } D = P \div p$$
$$= 3.52 \div 0.75$$
$$= 4.7 \text{ m}$$
$$\text{Apparent slip} = (11.42 - 12) \div 11.42 \times 100$$
$$= -5.08\%$$

**A. 8.4:**

$$V_T = 4.8 \times 110 \times 60 \div 1852$$

Theoretical speed

$$= 17.11\,\text{knots}$$

Apparent slip $-s = (V_T - V) \div V_T \times 100$

Real slip $+1.5s = (V_T - V_a) \div V_T \times 100$   (1)

$$V_a = V(1-w)$$

$$= 0.75\,V$$

Therefore: real slip $+1.5s = (V_T - 0.75\,V) \div V_T \times 100$   (2)

Multiply (1) by 1.5

$$-1.5s = 1.5 \times (V_T - V) \div V_T \times 100$$   (3)

Adding (2) and (3)

$$(V_T - 0.75\,V) \div V_T \times 100 + 1.5 \times (V_T - V) \div V_T \times 100 = 0$$

Hence:

$$V_T - 0.75\,V + 1.5\,V_T - 1.5\,V = 0$$

$$2.25\,V = 2.5\,V_T$$

$$V = 2.5 \div 2.25 \times 17.11$$

$$= 19.01\,\text{knots}$$

Substitute for V in (1)

$$-s = (17.11 - 19.01) \div 17.11 \times 100$$

Apparent slip $= -11.10\%$

Real slip $= -1.5 \times (-11.10)$

$$= +16.66\%$$

**A. 8.5: a)**

Theoretical speed $V_T = (4.3 \times 95 \times 60) \div 1852$

$$= 13.23\,\text{knots}$$

Real slip : $0.28 = (13.23 - V_a) \div 13.23$

$$V_a = 13.23 \times (1 - 0.28)$$

$$= 9.53\,\text{knots}$$

**b)**

$$\text{Effective disc area } A = (\pi \div 4) \times (D^2 - d^2)$$
$$= (\pi \div 4) \times (4.6^2 - 0.75^2)$$
$$= 16.18 \, m^2$$
$$\text{Thrust} = \rho \, A \, s \, P^2 n^2$$
$$= 1.025 \times 16.18 \times 0.28 \times 4.3^2 \times (95 \div 60)^2$$
$$= 215.2 \, kN$$

**c)**

$$\text{Thrust power} = T \times V_a$$
$$= 215.2 \times 9.53 \times 0.5144$$
$$= 1055 \, kW$$

**A. 8.6:**

$$T_1 N_1 = T_2 N_2$$
$$17.5 \times 115 = T_2 \times 90$$
$$T_2 = 17.5 \times 115 \div 90$$
$$\text{Thrust pressure} = 22.36 \, bar$$

**A. 8.7:**

$$tp_1 \div tp_2 = T_1 V_1 \div T_2 V_2$$
$$T_2 = (T_1 V_1 tp_2) \div (V_2 tp_1)$$
$$= 19.5 \times (V_1 \div 0.88 V_1) \times (2900 \div 3400)$$
$$\text{New thrust pressure} = 18.9 \, bar$$

**A. 8.8:**

$$tp = 2550 \, kW$$
$$dp = tp \div \text{propeller efficiency}$$
$$dp = 2550 \div 0.65$$
$$sp = dp \div \text{transmission efficiency}$$
$$sp = 2550 \div (0.65 \times 0.94)$$
$$ip = sp \div \text{mechanical efficiency}$$
$$ip = 2550 \div (0.65 \times 0.94 \times 0.83)$$

**a)**

$$\text{Indicated power} = 5028 \, kW$$
$$ep = dp \times QPC$$
$$= (2550 \div 0.65) \times 0.71$$

**b)**

$$\text{Effective power} = 2785 \text{ kW}$$
$$sp = (\Delta^{\frac{2}{3}}V^3) \div C$$
$$V^3 = (sp \times C) \div \Delta^{\frac{2}{3}}$$
$$V^3 = 2550 \div (0.65 \times 0.94) \times (420 \div 15000^{\frac{2}{3}})$$

**c)** Ship speed $V = 14.23$ knots

**A. 8.9:**

$$\text{Theoretical speed } V_T = 4 \times 125 \div 60$$
$$= 8.33 \text{ m/s}$$
$$\text{Real slip} = (V_T - V_a) \div V_T$$
$$0.36 = (8.33 - V_a) \div 8.33$$
$$V_a = 8.33 \times (1 - 0.36)$$
$$= 5.33 \text{ m/s}$$
$$tp = dp \times \text{propeller efficiency}$$
$$= 2800 \times 0.67$$
$$= 1876 \text{ kW}$$
$$\text{But } tp = T \times V_a$$
$$1876 = T \times 5.33$$
$$T = 1876 \div 5.33$$
$$\text{Propeller thrust} = 351.97 \text{ kN}$$

**A. 8.10:**

$$\text{Pitch} = \tan\theta \times 2\pi R$$
$$= \tan 21.5° \times 2\pi \times 2$$
$$= 4.95 \text{ m}$$

**A. 8.11:**

$$\tan\theta = 40 \div 115$$
$$= 0.3478$$
$$\theta = 19.179°$$
$$\text{Pitch} = 0.3478 \times 2\pi \times 2.6$$
$$= 5.682 \text{ m}$$
$$\sin\theta = \text{horizontal ordinate} \div \text{blade width}$$
$$\text{Width} = 40 \div \sin 19.179°$$
$$= 121.8 \text{ cm}$$
$$= 1.218 \text{ m}$$

# SOLUTIONS TO CHAPTER 9 TEST EXAMPLES

**A. 9.1:**

$$\text{Ship speed} = 18 \text{ knots}$$
$$= 18 \times 0.5144 = 9.26 \text{ m/s}$$
$$\text{Torque } T = 580 AV^2 \sin \alpha \times b$$
$$= 580 \times 25 \times 9.26^2 \times 0.5736 \times 1.2$$
$$= 855816 \text{ Nm}$$

But
$$= T \div J = q \div r$$

and
$$J = \pi r^4 \div 2$$

Therefore: $T = \pi r^4 q \div (2r)$
$$r^3 = (T \times 2) \div (\pi q)$$
$$= 855816 \times 2 \div (\pi \times 85 \times 10^6)$$
$$r = 0.1858 \text{ m}$$
$$\text{Diameter} = 0.3716 \text{ m}$$
$$= 372 \text{ mm}$$

**A. 9.2:**

$$\text{Torque } T = 580 AV^2 \sin \alpha \times b$$
$$= 580 \times 13 \times (14 \times 0.5144)^2 \times 1.1 \times \sin \alpha$$
$$= 430226 \sin \alpha$$

| Angle α | sin α | T | SM | Product |
|---------|-------|------|-----|---------|
| 0° | 0 | 0 | 1 | 0 |
| 10° | 0. 1736 | 74690 | 4 | 298760 |
| 20° | 0.3420 | 147140 | 2 | 294280 |
| 30° | 0.5000 | 215110 | 4 | 860440 |
| 40° | 0.6428 | 276550 | 1 | 276550 |
| | | | | 1730030 |

$$\text{Common interval} = 10 \div 573$$
$$\text{Work done} = \tfrac{1}{3} \times 10 \div 573 \times 1730030$$
$$= 100640 \text{ J}$$
$$= 100.64 \text{ kJ}$$

**A. 9.3:**

$$\text{Rudder area} = L \times d \div 60$$
$$= 150 \times 8.5 \div 60$$
$$= 21.25 \text{ m}^2$$
$$\text{Torque } T = Jq \div r$$
$$= \pi r^4 q \div (2r)$$
$$= \pi r^3 q \div 2$$
$$= \pi \div 2 \times 0.16^3 \times 70 \times 10^6$$
$$= 0.4504 \times 10^6 \text{ Nm}$$

Therefore:

$$0.4504 \times 10^6 = 580 \times 21.25 \times V^2 \times 0.9 \times \sin 35°$$
$$V^2 = 0.4504 \times 10^6 \div (580 \times 21.25 \times 0.9 \times 0.5736)$$
$$V = 8.414 \text{ m/s}$$
$$\text{Maximum ship speed } V = 8.414 \div 0.5144$$
$$= 16.35 \text{ knots}$$

**A. 9.4:**

Normal force on rudder $F_n = 580AV^2\sin\alpha$

Transverse force on rudder $F_t = 580AV^2\sin\alpha\cos\alpha$

$$F_t = 580\times12\times(16\times0.5144)^2\times\sin 35°\times\cos 35°$$
$$= 221600\,\text{N}$$
$$= 221.6\,\text{kN}$$
$$\tan\theta = (F_t\times NL)\div(\Delta\times g\times GM)$$
$$= (221.6\times1.6)\div(5000\times9.81\times0.24)$$
$$= 0.0301$$

Angle of heel $\theta = 1.724°$

**A. 9.5:** Speed of ship $V = 17$ knots

$$V = 17\times0.5144$$
$$= 8.745\,\text{m/s}$$
$$\tan\theta = (V^2\times GL)\div(\Delta\times g\times GM)$$
$$= 8.746^2\times(7.00-4.00)\div(9.81\times450\times(7.45-7.00))$$
$$= 0.1155$$

Angle of heel $\theta = 6.588°$

# SELECTION OF EXAMINATION QUESTIONS

## SECTION 1

**Q. E1.1:** A box barge is 15 m long, 6 m wide and floats in water of 1.016 t/m³ at a draught of 3 m. 150 tonne cargo is now added. Calculate the load exerted by the water on the sides, ends and bottom.

**Q. E1.2:** A ship has a load displacement of 6000 tonne and centre of gravity 5.3 m above the keel. 1000 tonne are then removed 2.1 m above the keel, 300 tonne moved down 2.4 m, 180 tonne placed on board 3 m above the keel and 460 tonne placed on board 2.2 m above the keel. Find the new position of the centre of gravity.

**Q. E1.3:** A ship of 8000 tonne displacement floats in sea water of 1.025 t/m³ and has a TPC of 14. The vessel moves into fresh water of 1.000 t/m³ and loads 300 tonne of oil fuel. Calculate the change in mean draught.

**Q. E1.4:** The wetted surface area of one ship is 400% that of a similar ship. The displacement of the latter is 4750 tonne more than the former. Calculate the displacement of the smaller ship.

**Q. E1.5:** A ship of 9000 tonne displacement floats in fresh water of 1.000 t/m³ at a draught 50 mm below the sea water line. The waterplane area is 1650 m². Calculate the mass of cargo that must be added so that when entering sea water of 1.025 t/m³ it floats at the sea water line.

**Q. E1.6:** The effective power of ship is 1400 kW at 12 knots, the propulsive efficiency 65% and the fuel consumption 0.3 kg/kWh, based on shaft power. Calculate the fuel required to travel 10000 nautical miles at 10 knots.

**Q. E1.7:** A ship is 60 m long, 16 m beam and has a draught of 5 m in sea water, block coefficient 0.7 and waterplane area coefficient 0.8. Calculate the draught at which it will float in fresh water.

**Q. E1.8:** A tank top manhole 0.5 m wide and 0.65 m long has semicircular ends. The studs are pitched 30 mm outside the line of hole and 100 mm apart. The cross-sectional area of the studs between the threads is 350 mm². The tank is filled with salt water to a height of 7.5 m up the sounding pipe. Calculate the stress in the studs.

**Q. E1.9:** A small vessel has the following particulars before modifications are carried out. Displacement 150 tonne, *GM* 0.45 m, *KG* 1.98 m, *KB* 0.9 m, TPC 2.0 and draught 1.65 m.

After modification, 20 tonne has been added, *Kg* 3.6 m. Calculate the new *GM*, assuming constant waterplane area over the change in draught.

**Q. E1.10:** A ship displacing 10000 tonne and travelling at 16 knots has a fuel consumption of 41 tonne per day. Calculate the consumption per day if the displacement is increased to 13750 tonne and the speed is increased to 17 knots. Within this speed range, fuel consumption per day varies as (speed)$^{3.8}$.

**Q. E1.11** The ½ ordinates of a waterplane 320 m long are 0, 9, 16, 23, 25, 25, 22, 18 and 0 m respectively. Calculate:
   **a)** waterplane area
   **b)** TPC
   **c)** waterplane area coefficient.

**Q. E1.12:** A double bottom tank is filled with sea water to the top of the air pipe. The pressure on the outer bottom is found to be 1.20 bar while the pressure on the inner bottom is found to be 1.05 bar. Calculate the height of the air pipe above the inner bottom and the depth of the tank.

**Q. E1.13:** A ship 125 m long and 17.5 m beam floats in sea water of 1.025 t/m³ at a draught of 8 m. The waterplane area coefficient is 0.83, block coefficient 0.759 and midship section area coefficient 0.98. Calculate:
   **a)** prismatic coefficient
   **b)** TPC
   **c)** change in mean draught if the vessel moves into river water of 1.016 t/m³.

**Q. E1.14:** At 90 rev/min, a propeller of 5 m pitch has an apparent slip of 15% and wake fraction 0.10. Calculate the real slip.

**Q. E1.15:** A hopper barge of box form 50 m long and 10 m wide floats at a draught of 2 m in sea water when the hopper, which is 15 m long and 5 m wide, is loaded with mud having relative density twice that of the sea water, to the level of the waterline.

Doors in the bottom of the hopper are now opened, allowing the mud to be discharged. Calculate the new draught.

**Q. E1.16:** It is found that by reducing the fuel consumption of a vessel to 43 tonne/day, the speed is reduced 2.2 knots and the saving in fuel for a voyage of 3500 nautical miles is 23%. Determine:

**a)** the original daily fuel consumption, and

**b)** the original speed.

**Q. E1.17:** A ship of 7000 tonne displacement, having *KG* 6 m and TPC 21, floats at a draught of 6 m. 300 tonne of cargo is now added at *Kg* 1 m and 130 tonne removed at *Kg* 5 m. The final draught is to be 6.5 m and *KG* 5.8 m. Two holds are available for additional cargo, one having *Kg* 5 m and the other *Kg* 7 m. Calculate the mass of cargo to be added to each hold.

**Q. E1.18:** A block of wood of uniform density has a constant cross section in the form of a triangle, apex down. The width is 0.5 m and the depth 0.5 m. It floats at a draught of 0.45 m. Calculate the metacentric height.

**Q. E1.19:** The waterplane area of a ship at 8.4 m draught is 1670 m². The areas of successive waterplanes at 1.4 m intervals below this are 1600, 1540, 1420, 1270, 1080 and 690 m² respectively. Calculate the displacement in fresh water at 8.4 m draught and the draught at which the ship would lie in sea water with the same displacement.

**Q. E1.20:** A floating dock 150 m long, 24 m overall width and 9 m draught consists of a rectangular bottom compartment 3 m deep and rectangular wing compartments 2.5 m wide. A ship with a draught of 5.5 m is floated in. 4000 tonne of ballast is pumped out of the dock to raise the ship 1.2 m. Calculate the mean TPC of the ship.

**Q. E1.21:** A ship of 7500 tonne displacement has its centre of gravity 6.5 m above the keel. Structural alterations are made, when 300 tonne are added 4.8 m above the keel; 1000 tonne of oil fuel are then added 0.7 m above the keel. Calculate:

**a)** the new position of the centre of gravity

**b)** the final centre of gravity when 500 tonne of oil fuel are used.

**Q. E1.22:** A ship of 15000 tonne displacement floats at a draught of 7 m in water of 1000 t/m³. It is required to load the maximum amount of oil to give the ship a draught of 7 m in sea water of 1.025 t/m³. If the waterplane area is 2150 m², calculate the mass of oil required.

**Q. E1.23:** A historic vessel has a bulkhead 12 m wide and 9 m high. It is secured at the base by an angle bar having 20 mm diameter rivets on a pitch of 80 mm. The bulkhead is loaded on one side only to the top edge with sea water. Calculate the stress in the rivets.

**Q. E1.24:** The ½ ordinates of a waterplane 96 m long are 1.2, 3.9, 5.4, 6.0, 6.3, 6.3, 6.3, 5.7, 4.5, 2.7 and 0 m respectively. A rectangular double bottom tank with parallel sides is 7.2 m wide, 6 m long and 1.2 m deep. When the tank is completely filled with oil of 1.15 m³/tonne, the ship's draught is 4.5 m. Calculate the draught when the sounding in the tank is 0.6 m.

**Q. E1.25:** A ship enters harbour and discharges 6% of its displacement. It then travels upriver to a berth and the total change in draught is found to be 20 cm. The densities of the harbour and berth water are respectively 1.023 t/m³ and 1.006 t/m³ and the TPC in the harbour water is 19. Calculate the original displacement and state whether the draught has been increased or reduced.

**Q. E1.26:** The fuel consumption of a vessel varies within certain limits as (speed)$^{2.95}$. If, at 1.5 knots above and below the normal speed, the power is 9200 kW and 5710 kW respectively, find the normal speed.

**Q. E1.27:** The length of a ship is 7.6 times the breadth, while the breadth is 2.85 times the draught. The block coefficient is 0.69, prismatic coefficient 0.735, waterplane area coefficient 0.81 and the wetted surface area 7000 m². The wetted surface area $S$ is given by:

$$S = 1.74\,Ld + \nabla \div d$$

Calculate:

**a)** displacement in tonne

**b)** area of immersed midship section

**c)** waterplane area.

**Q. E1.28:** A ship 120 m long, 17 m beam and 7.2 m draught has a block coefficient of 0.76. A parallel section 6 m long is added to the ship amidships. The midship sectional area coefficient is 0.96. Find the new displacement and block coefficient.

**Q. E1.29:** State what is meant by the Admiralty Coefficient and what its limitations are. A ship has an Admiralty Coefficient of 355, a speed of 15 knots and shaft power 7200 kW. Calculate its displacement.

If the speed is now reduced by 16%, calculate the new power required.

**Q. E1.30:** A ship of 8100 tonne displacement, 120 m long and 16 m beam floats in water of density 1025 kg/m³. In this condition, the ship has the following hydrostatic data:

Prismatic coefficient = 0.70

Midship area coefficient = 0.98

Waterplane area coefficient = 0.82

A full depth midship compartment, which extends the full breadth of the ship, is flooded. Calculate the length of the compartment if, after flooding, the mean draught is 7.5 m.

**Q. E1.31:** A ship of 22000 tonne displacement has a draught of 9 m in river water of 1.008 t/m³. The waterplane area is 3200 m². The vessel then enters sea water of 1.026 t/m³. Calculate the change in displacement as a percentage of the original displacement in order to:
a) keep the draught the same
b) give a draught of 8.55 m.

**Q. E1.32:** A ship has a fuel consumption of 60 tonne per 24 hours when the displacement is 15500 tonne and the ship speed 14 knots. Determine the ship speed during a passage of 640 nautical miles if the displacement is 14500 tonne and the total fuel consumption is 175 tonne.

**Q. E1.33:** The TPC values of a ship at 1.2 m intervals of draught, commencing at the load waterline, are 19, 18.4, 17.4, 16.0, 13.8, 11.0 and 6.6 respectively. If 6.6 represents the value at the keel, calculate the displacement in tonne and state the load draught.

**Q. E1.34:** If the density of sea water is 1.025 t/m³ and the density of fresh water is 1.000 t/m³, prove that the Statutory Fresh Water Allowance is

$$\Delta \div (40 \text{ TPC}) \text{ cm.}$$

A ship of 12000 tonne displacement loads in water of 1.012 t/m³. By how much will the Summer Load Line be submerged if it is known that 130 tonne must be removed before sailing? The TPC in sea water is 17.7.

**Q. E1.35:** A collision bulkhead is in the form of an isosceles triangle and has a depth of 7 m and a width at the deck of 6 m. The bulkhead is flooded on one side with water of density 1025 kg/m³ and the resultant load on the bulkhead is estimated at 195 kN. Calculate the depth of water to which the bulkhead is flooded.

**Q. E1.36:** A ship of 8000 tonne displacement has a metacentric height of 0.46 m, centre of gravity 6.6 m above the keel and centre of buoyancy 3.6 m above the keel. Calculate the second moment of area of the waterplane about the centreline of the ship.

**Q. E1.37:** A ship has a displacement of 9800 tonne. 120 tonne of oil fuel are moved from an after tank to a tank forward. The centre of gravity of the ship moves 0.75 m forward.
The forward tank already contains 320 tonne of oil fuel and after the transfer 420 tonne of fuel are used. The centre of gravity of the ship now moves to a new position 0.45 m aft of the vessel's original centre of gravity. Find the

distance from the ship's original centre of gravity to the centre of gravity of each tank.

**Q. E1.38:** A ship of 18000 tonne displacement carries 500 tonne of fuel. If the fuel coefficient is 47250 when the ship's speed is 12 knots, calculate:
**a)** the range of operation of the ship
**b)** the percentage difference in range of the ship's operation if the speed of the ship is increased by 5%.

**Q. E1.39:** Before bunkering in harbour, the draught of a vessel of 12000 tonne displacement is 8.16 m, the waterplane area being 1625 m². After loading 1650 tonne of fuel and entering sea water of 1.024 t/m³, the draught is 9.08 m. Assuming that the waterplane area remains constant and neglecting any fuel, etc, expended in moving the vessel, calculate the density of the harbour water.

**Q. E1.40:** Describe how an inclining experiment is carried out.
A vessel of 8000 tonne displacement was inclined by moving 5 tonne through 12 m. The recorded deflections of a 6 m pendulum were 73, 80, 78 and 75 mm. If the *KM* for this displacement was 5.1 m, calculate *KG*.

**Q. E1.41:** A rectangular bulkhead 17 m wide and 6 m deep has a head of sea water, on one side only, of 2.5 m above the top of the bulkhead. Calculate:
**a)** the load on the bulkhead
**b)** the pressure at the top and bottom of the bulkhead.

**Q. E1.42:** For a ship of 4600 tonne displacement, the metacentric height (*GM*) is 0.77 m. A 200 tonne container is moved from the hold to the upper deck.
Determine the angle of heel developed if, during this process, the centre of mass of the container is moved 8 m vertically and 1.1 m transversely.

**Q. E1.43:** The load draught of a ship is 7.5 m in sea water and the corresponding waterplane area is 2100 m². The areas of parallel waterplanes at intervals of 1.5 m below the load waterplane are 1930, 1720, 1428 and 966 m² respectively.
Draw the TPC curve. Assuming that the displacement of the portion below the lowest given waterplane is 711 tonne, calculate the displacement of the vessel when:
**a)** fully loaded, and
**b)** floating at a draught of 4.5 m in sea water.

**Q. E1.44:** Define *centre of buoyancy* and show with the aid of sketches how a vessel that is stable will return to the upright after being heeled by an external force.
A vessel displacing 8000 tonne has its centre of gravity 1 m above the centre of buoyancy when in the upright condition. If the moment tending to right

the vessel is 570 tonne m when the vessel is heeled over 7°, calculate the horizontal distance the centre of buoyancy has moved from its original position.

**Q. E1.45:** A propeller rotates at 2 rev/s with a speed of advance of 12 knots and a real slip of 0.30. The torque absorbed by the propeller is 250 kNm and the thrust delivered is 300 kN. Calculate:
a) the pitch of the propeller
b) the thrust power
c) the delivered power.

**Q. E1.46:** A box barge 40 m long and 7.5 m wide floats in sea water with draughts forward and aft of 1.2 m and 2.4 m respectively. Where should a mass of 90 tonne be added to obtain a level keel draught?

**Q. E1.47:** For a box-shaped barge of 16 m beam floating at an even keel draught of 6 m in water of density 1025 kg/m³, the tonne per centimetre immersion (TPC) is 17. A full-depth midship compartment 20 m in length and 16 m breadth has a permeability of 0.80.
If the compartment is bilged, determine:
a) the draught
b) the position of the metacentre above the keel, if the second moment of area of the intact waterplane about the centreline is 75000 m⁴.

**Q. E1.48:** A box barge is 7.2 m wide and 6 m deep. Draw the metacentric diagram using 1 m intervals of draught up to the deck line.

**Q. E1.49:** A propeller has a diameter of 4.28 m, pitch ratio of 1.1 and rotates at a speed of 2 rev/s. If the apparent and true slip are 0.7% and 12% respectively, calculate the wake speed.

Pitch ratio = propeller pitch ÷ propeller diameter

**Q. E1.50:** At a ship speed of 12 knots, the shaft power for a vessel is 1710 kW and the fuel consumption is 0.55 kg/kWh.
Determine for a speed of 10 knots:
a) the quantity of fuel required for a voyage of 7500 miles
b) the fuel coefficient if the ship's displacement is 6000 tonne.

**Q. E1.51:** A ship of 12000 tonne displacement has a metacentric height of 0.6 m and a centre of buoyancy 4.5 m above the keel. The second moment of area of the waterplane about the centreline is 42.5 × 10³ m⁴. Calculate the height of the centre of gravity above the keel.

**Q. E1.52:** A box-shaped barge 37 m long, 6.4 m beam, floats at an even keel draught of 2.5 m in water of density 1025 kg/m³. If a mass is added and the vessel moves into water of density 1000 kg/m³, determine the magnitude of this mass if the forward and aft draughts are 2.4 m and 3.8 m respectively.

**Q. E1.53:** A ship of 7200 tonne displacement has two similar bunkers adjacent to each other, the capacity of each being 495 tonne and their depth 9.9 m. If one of the bunkers is completely full and the other completely empty, find how much fuel must be transferred to lower the ship's centre of gravity by 120 mm.

**Q. E1.54:** A ship of 7000 tonne displacement has *KM* 7.3 m. Masses of 150 tonne at a centre 3 m above and 60 tonne at a centre 5.5 m below the original centre of gravity of the ship are placed on board. A ballast tank containing 76 tonne of water at *Kg* 0.6 m is then discharged.

Calculate the original height of the ship's centre of gravity above the keel if the final metacentric height is 0.5 m and *KM* is assumed to remain constant.

**Q. E1.55:** A wall-sided ship 120 m long floats at a draught of 3.5 m in sea water. The waterplane area has the following ½ ordinates:

| Station | 0 | 1 | 2 | 3 | 4 | 5 | 6 | 7 | 8 | 9 | 10 |
|---|---|---|---|---|---|---|---|---|---|---|---|
| Half breadth (m) | 0.1 | 2.4 | 5.1 | 7.4 | 8.4 | 8.4 | 8.4 | 7.4 | 5.1 | 2.4 | 0.1 |

A central midship compartment 24 m long extending to the full breadth of the ship, and having a permeability of 50%, is bilged. Calculate the new draught.

**Q. E1.56: a)** State FOUR precautions to be taken when carrying out an inclining experiment.

**b)** A vessel is inclined in the following condition:

| | |
|---|---|
| Displacement | 5500 tonne |
| Transverse metacentre above keel | 7.5 m |
| Transverse shift of inclining ballast | 12 m |
| Mass of inclining ballast | 15 tonne |
| Length of inclining pendulum | 8 m |
| Deflection of pendulum | 0.25 m |

The following changes are then made to the loading condition of the ship:

i)   mass removed: 150 tonne, *Kg* 9.75 m

ii)  mass added: 220 tonne, *Kg* 9 m.

Calculate the final distance of the centre of gravity above the keel (*KG*).

**Q. E1.57:** For a ship of displacement 15500 tonne, 138 m long and 18.5 m beam, the even keel draught is 8.5 m in sea water of density 1025 kg/m³. The propeller pitch ratio is 0.83, and at 1.92 rev/s the speed of the ship is 15.5 knots with a real slip ratio of 0.35. The Taylor wake fraction $W_t = 0.5\,C_b - 0.048$, and also $W_t = 1$ - speed of advance ÷ ship speed.

Calculate the pitch and diameter of the propeller.

**Q. E1.58: a)** A barge of constant triangular cross section floats apex down at an even keel draught of 5 m in sea water of density 1025 kg/m³. The deck is 40 m long, 14 m wide and is 7 m above the keel. Find the displacement of the barge.

**b)** If an empty midship compartment 6 m long extending to the full width and depth of the barge is bilged, find the new draught.

**Q. E1.59:** For a propeller of diameter 5.4 m, pitch ratio 0.875 and blade area ratio (BAR) 0.46, the developed thrust was 860 kN at 1.87 rev/s, when the real slip and propeller efficiency were found to be 28% and 68% respectively. Calculate for this condition of loading:

**a)**  the thrust power (*tp*)

**b)**  the delivered power (*dp*)

**c)**  the shaft torque

**d)**  the mean pressure on the blades due to the thrust load.

*Note:* Propeller efficiency = *tp* ÷ *dp*

**Q. E1.60:** A ship 150 m long, 18 m beam, floats at an even keel draught of 7 m in sea water of density 1025 kg/m³. The half areas of immersed sections commencing from the after perpendicular (AP) are:

| Station | 0 | 1 | 2 | 3 | 4 | 5 | 6 | 7 | 8 | 9 | 10 |
|---|---|---|---|---|---|---|---|---|---|---|---|
| Half areas of immersed sections (m²) | 10 | 26 | 40 | 59 | 60 | 60 | 60 | 56 | 38 | 14 | 4 |

Calculate:

**a)**  the displacement

**b)**  the block coefficient

**c)**  the prismatic coefficient

**d)**  the midship section area coefficient.

**Q. E1.61:** A conical buoy, 2 m diameter and 3 m high, made of a homogeneous material of density 800 kg/m³, floats apex down in sea water of density 1025 kg/m³. Given that the second moment of area of a circular plane about its diameter is $\pi d^4 \div 64$ m⁴, calculate:

a) the draught at which the buoy will float

b) the metacentric height (*GM*) of the buoy.

**Q. E1.62:** A ship's propeller has a diameter of 5.5 m and a pitch ratio of 1.0. At a speed of 20 knots, the delivered power is 7300 kW at a shaft speed of 2.2 rev/s. The transmission efficiency is 97%, propeller efficiency is 61%, the propulsive coefficient based on shaft power is 0.53 and the real slip ratio is 35%.

Determine:

a) the thrust power

b) the Taylor wake fraction

c) the ship resistance.

**Q. E1.63:**

a) An expression used for the calculation of ship frictional resistance is:

$$R_f = fSV^n$$

i) Explain the terms *S, V* and *n*

ii) What factors influence the value of the coefficient '*f*'?

b) A ship of 12000 tonne displacement travels at a speed of 14.5 knots when developing 3500 kW shaft power. Calculate:

i) the value of the Admiralty Coefficient

ii) the percentage increase in shaft power required to increase the speed by 1.5 knots

iii) the percentage increase in the speed of the ship if the shaft power is increased to 4000 kW.

# SOLUTIONS TO SECTION 1 EXAMINATION QUESTIONS

## A. E1.1:

$$TCP = A_w \times 1.016 \div 100$$
$$= 15 \times 6 \times 0.01016$$
$$\text{Bodily sinkage} = 150 \div 0.914$$
$$= 164 \text{ cm}$$
$$\text{New draught} = 3 + 1.64 = 4.64 \text{ m}$$
$$\text{Load on side} = \rho g A H$$
$$= 1.016 \times 9.81 \times 15 \times 4.64 \times 2.32$$
$$= 1609 \text{ kN}$$
$$\text{Load on end} = 1.016 \times 9.81 \times 6 \times 4.64 \times 2.32$$
$$= 643.8 \text{ kN}$$
$$\text{Load on bottom} = 1.016 \times 9.81 \times 15 \times 6 \times 4.64$$
$$= 4162 \text{ kN}$$

## A. E1.2:

| Mass | Distance | Moment |
|------|----------|--------|
| 6000 | 5.30 | 31800 |
| −1000 | 2.10 | − 2100 |
| (300) | − 2.40 | − 720 |
| 180 | 3.00 | + 540 |
| 460 | 2.20 | + 1012 |
| 5640 | | 30532 |

*Note:* The 300 tonne remains on board and therefore does not alter the final displacement.

$$\text{Centre of gravity above keel} = 30532 \div 5640$$
$$= 5.413 \text{ m}$$

## A. E1.3:

Change in mean draught

$$= (\Delta \times 100 \times 1.025) \div (TPC \times 100) \times (1.025 - 1.000) \div (1.000 \times 1.025)$$
$$= 8000 \times 0.025 \div 14$$
$$= 14.29 \text{ cm increase}$$

TPC in fresh water $= 14 \times 1.000 \div 1.025$

Bodily sinkage $= 300 \div TPC$
$$= 300 \times 1.025 \div 14$$
$$= 21.96 \text{ cm}$$

Total increase $= 14.29 + 21.96$
$$= 36.25 \text{ cm}$$

## A. E1.4:

$$S = \text{wetted surface area of small ship}$$
$$\Delta = \text{displacement of small ship}$$

Then

$$S \div 0.4 = \text{wetted surface area of large ship}$$
$$\Delta + 4750 = \text{displacement of large ship}$$

Now

$\Delta$ is proportional to $S^{3/2}$

$$\Delta \div \Delta_1 = (S \div S_1)^{3/2}$$
$$\Delta \div (\Delta + 4750) = (S \times 0.4 \div S)^{3/2}$$
$$\Delta \div (\Delta + 4750) = 0.2523$$
$$\Delta = 0.2523 \times (\Delta + 4750)$$
$$\Delta(1 - 0.2523) = 0.2523 \times 4750$$
$$\Delta = 0.2523 \times 4750 \div 0.7477$$
$$\Delta = 1603 \text{ tonne}$$

## A. E1.5:

Change in mean draught $= 9000 \times 100 \div 1650 \times (1.025 - 1.000) \div (1.000 \times 1.025)$
$$= 13.30 \text{ cm}$$

Thus new waterline $= 13.30 + 5.0$
$$= 18.30 \text{ cm below SW line}$$

$$TPC = 1650 \times 0.01025$$
$$= 16.91$$

Mass of cargo $= 18.3 \times 16.91$
$$= 309.5 \text{ tonne}$$

**A. E1.6:**

$$At\ 12\ knots:ep = 1400\ kW$$
$$sp = 1400 \div 0.65$$
$$= 2154\ kW$$
$$cons/h = 0.3 \times 2154$$
$$= 646.2\ kg$$

Cons/h is proportional to speed[3]

$$Therefore:\ at\ 10\ knots:cons/h = 646.2 \times (10 \div 12)^3$$
$$= 374\ kg$$
$$Time\ on\ voyage = 10000 \div 10$$
$$= 1000\ h$$
$$Therefore:\ voyage\ consumption = 374 \times 1000$$
$$= 374\ tonne$$

**A. E1.7:**

$$Displacement\ \Delta = L \times B \times d \times C_b \times 1.025$$
$$Area\ of\ waterplane\ A_W = L \times B \times C_W$$
$$\Delta \div A_W = 1.025d \times C_b \div C_W$$
$$Change\ in\ draught = \Delta \times 100 \div A_W \times (\rho_S - \rho_R) \div (\rho_R \times \rho_S)$$
$$1.025 \times 5 \times 0.7 \times 100 \div 0.8 \times (1.025 - 1.000) \div (1.000 \times 1.025)$$
$$= 10.94\ cm$$
$$New\ draught = 5.0 + 0.109$$
$$= 5.109\ m$$

**A. E1.8:**

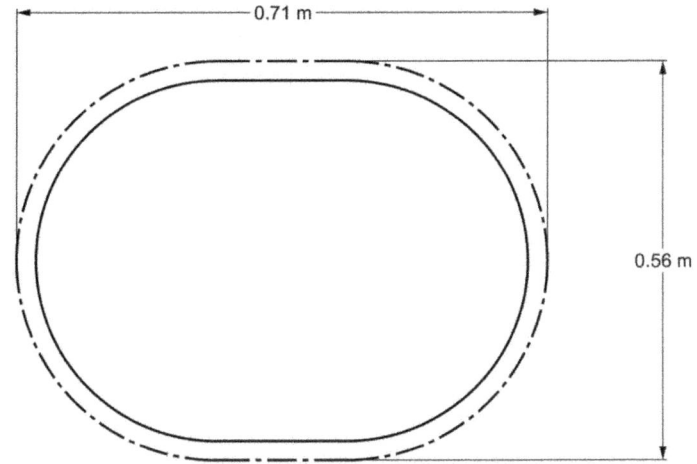

0.71 m

0.56 m

▲ Figure A. E1.1

$$\text{Perimeter of stud line} = \pi \times 0.56 + 2 \times 0.15$$
$$= 2.059\,\text{m}$$
$$\text{Number of studs} = 2.059 \div 0.10$$
$$= 21$$
$$\text{Cross-sectional area of studs} = 21 \times 350$$
$$= 7350\,\text{mm}^2$$
$$= 7.35 \times 10^{-3}\,\text{m}^2$$
$$\text{Effective area of door} = \pi \div 4 \times 0.56^2 + 0.15 \times 0.56$$
$$= 0.2463 + 0.0840$$
$$= 0.3303\,\text{m}^2$$
$$\text{Load on door} = \rho g A\,H$$
$$= 1.025 \times 9.81 \times 0.3303 \times 7.5$$
$$= 24.91\,\text{kN}$$
$$\text{Stress} = \text{load} \div \text{area}$$
$$= 24.91 \div 7.35 \times 10^{-3}$$
$$= 3.389 \times 10^3\,\text{kN/m}^2$$
$$= 3.389\,\text{MN/m}^2$$

**A. E1.9:**

$$KM = 1.98 + 0.45$$
$$= 2.43\,\text{m}$$
$$BM = 2.43 - 0.90$$
$$= 1.53\,\text{m}$$
$$= I \div \nabla$$
$$I = 1.53 \times 150 \div 1.025$$
$$= 223.9\,\text{m}^4$$
$$KG = (150 \times 1.98 + 20 \times 3.6) \div (150 + 20)$$
$$= 297.0 + 72 \div 170$$
$$= 2.17\,\text{m}$$
$$\text{Increase in draught} = 20 \div 2$$
$$= 10\,\text{cm}$$
$$KB_1 = (150 \times 0.9 + 20(1.65 + 0.05)) \div 170$$
$$= (135 + 34) \div 170$$
$$= 0.994\,\text{m}$$
$$BM_1 = 223.9 \div 170 \times 1.025$$
$$= 1.35\,\text{m}$$
$$KM_1 = 0.994 + 1.350$$
$$= 2.344\,\text{m}$$
$$GM_1 = 2.344 - 2.170$$
$$= 0.174\,\text{m}$$

**A. E1.10:**

Fuel cons/day is proportional to $\Delta^{\frac{2}{3}} V^{3.8}$

$$\text{Therefore} : C \div 41 = (13750 \div 10000)^{\frac{2}{3}} \times (17 \div 16^{3.8})$$
$$C = 41 \times (13750 \div 10000)^{\frac{2}{3}} \times (17 \div 16^{3.8})$$
$$\text{Fuel cons/day} = 63.81 \text{ tonne}$$

**A. E1.11:**

| ½ ordinate | SM | Product |
|---|---|---|
| 0 | 1 | – |
| 9 | 4 | 36 |
| 16 | 2 | 32 |
| 23 | 4 | 92 |
| 25 | 2 | 50 |
| 25 | 4 | 100 |
| 22 | 2 | 44 |
| 18 | 4 | 72 |
| 0 | 1 | – |
| | | 426 |

$$h = 320 \div 8$$
$$= 40 \text{ m}$$

a)  $\text{Waterplane area} = \frac{2}{3} \times 40 \times 426$
$$= 11360 \text{ m}^2$$

b)  $\text{TPC} = A \times 0.01025$
$$= 116.44$$

c)  $\text{Waterplane area coefficient} = 11360 \div (320 \times 50)$
$$= 0.710$$

**A. E1.12:**

$$\text{Bottom pressure} = 1.2 \times 10^5 \text{ N/m}^2$$
$$= \rho g h$$
$$h = 1.2 \times 10^5 \div (1025 \times 9.81)$$
$$\text{Height of air pipe above outer bottom} = 11.93 \text{ m}$$
$$\text{Top pressure} = 1.05 \times 10^5 \text{ N/m}^2$$
$$h_1 = 1.05 \times 10^5 \div (1025 \times 9.81)$$
$$\text{Height of air pipe above inner bottom} = 10.44 \text{ m}$$
$$\text{Depth of tank} = 11.93 - 10.44$$
$$= 1.49 \text{ m}$$

**A. E1.13:**

a)
$$C_p = C_b \div C_m$$
$$= 0.759 \div 0.98$$
$$= 0.7745$$

b)
$$\text{Waterplane area} = 125 \times 17.5 \times 0.83$$
$$= 1815.6 \text{ m}^2$$
$$\text{TPC} = 1815.6 \times 0.01025$$
$$= 18.61$$

c)
$$\text{Displacement} = 125 \times 17.5 \times 8 \times 0.759 \times 1.025$$
$$= 13615 \text{ tonne}$$
$$\text{Change in draught} = (13615 \times 100 \div 1815.6) \times (1.025 - 1.016) \div (1.016 \times 1.025)$$
$$= 6.48 \text{ cm}$$

**A. E1.14:**

$$\text{Theoretical speed } V_t = 5 \times 90 \times 60 \div 1852$$
$$= 14.58 \text{ knots}$$
$$\text{Apparent slip } 0.15 = (14.58 - V) \div 14.58$$
$$V = 14.58 \times (1 - 0.15)$$
$$= 12.39 \text{ knots}$$
$$\text{Wake fraction } 0.10 = (12.39 - V_a) \div 12.39$$
$$V_a = 12.39 \times (1 - 0.10)$$
$$= 11.15 \text{ knots}$$
$$\text{Real slip} = (14.58 - 11.15) \div 14.58$$
$$= 0.2352$$
$$\text{or } 23.52\%$$

**A. E1.15:**

$$\text{Mass of mud in hopper} = 2 \times 1.025 \times 15 \times 5 \times 2$$
$$\text{Mass of buoyancy in hopper} = 1.025 \times 15 \times 5 \times 2$$

When the doors are opened, the mud drops out, but at the same time the buoyancy of the hopper is lost. Since the reduction in displacement exceeds the reduction in buoyancy, there will be a reduction in draught.

$$\text{Net loss of displacement} = 1.025 \times 15 \times 5 \times 2$$
$$\text{Net volume of lost displacement} = 15 \times 5 \times 2 \, \text{m}^3$$
$$\text{Area of intact waterplane} = 50 \times 10 - 15 \times 5$$
$$= 425 \, \text{m}^2$$
$$\text{Reduction in draught} = 15 \times 5 \times 2 \div 425$$
$$= 0.353 \, \text{m}$$
$$\text{New draught} = 2.0 - 0.353$$
$$= 1.647 \, \text{m}$$

**A. E1.16:**

$$\text{Let } C = \text{original daily consumption}$$
$$V = \text{original speed}$$
$$K = \text{original voyage consumption}$$
$$C \text{ is proportional to } V^3$$
$$K \text{ is proportional to } V^2$$
$$K \div ((1 - 0.23) \times K) = (V \div (V - 2.2))^2$$
$$1 \div 0.77 = (V \div (V - 2.2))^2$$
$$1 \div \sqrt{0.77} = V \div (V - 2.2)$$
$$V - 2.2 = \sqrt{0.77} \times V$$
$$V - 2.2 = 0.8775V$$
$$V(1 - 0.8775) = 2.2$$
$$\text{Original speed } V = 17.97 \text{ knots}$$
$$C \div (C - 43) = (V \div (V - 2.2))^3$$
$$C \div (C - 43) = (17.97 \div 15.77)^3$$
$$C \div (C - 43) = 1.480$$
$$C = 1.480 \times (C - 43)$$
$$0.48C = 1.480 \times 43$$
$$C = 1.480 \times 43 \div 0.48$$
$$\text{Original consumption} = 132.6 \text{ tonne/day}$$

**A. E1.17:**

$$\text{Total change in draught} = 6.5 - 6.0$$
$$= 0.5\,\text{m}$$
$$\text{Total change in displacement} = 50 \times 21$$
$$= 1050\,\text{tonne}$$
$$\text{Let mass of additional cargo} = m$$
$$\text{Then } 1050 = +300 - 130 + m$$
$$m = 1050 - 300 + 130$$
$$= 880\,\text{tonne}$$

Let mass of cargo in one hold $= x$

Then mass of cargo in other hold $= 880 - x$

Taking moments about the keel:

| Mass | Kg | Moment |
|------|------|------|
| 7000 | 6.0 | 42000 |
| + 300 | 1.0 | + 300 |
| − 130 | 5.0 | − 650 |
| + x | 5.0 | + 5x |
| + (880 − x) | 7.0 | + (6160 − 7x) |
| 8050 | | 47810 − 2x |

Final *KG*

$$5.8 = (47810 - 2x) \div 8050$$
$$5.8 \times 8050 = (47810 - 2x)$$
$$2x = 47810 - 46690$$
$$x = 560\,\text{tonne}$$

Therefore: mass of cargo in holds *Kg* 5 m and *Kg* 7 m respectively are 560 tonne and 320 tonne.

**A. E1.18:**

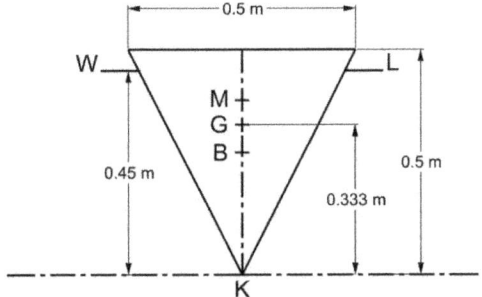

$$\text{Width at waterline} = 0.45 \text{ m}$$
$$KB = \tfrac{2}{3}d$$
$$= 0.30 \text{ m}$$
$$KG = \tfrac{2}{3}D$$
$$= 0.333 \text{ m}$$
$$BM = b^2 \div (6d)$$
$$= 0.45^2 \div (6 \times 0.45)$$
$$= 0.075 \text{ m}$$
$$KM = KB + BM$$
$$= 0.30 + 0.075$$
$$= 0.375 \text{ m}$$
$$GM = KM - KG$$
$$= 0.375 - 0.333$$
$$\text{Metacentric height} = 0.042 \text{ m}$$

**A. E1.19:**

| Waterplane area | SM | Product |
|---|---|---|
| 1670 | 1 | 1670 |
| 1600 | 4 | 6400 |
| 1540 | 2 | 3080 |
| 1420 | 4 | 5680 |
| 1270 | 2 | 2540 |
| 1080 | 4 | 4320 |

| Waterplane area | SM | Product |
|:---|:---:|:---:|
| 690 | 1 | 690 |
| | | 24380 |

$$h = 1.4\,m$$
$$\text{Displacement in fresh water} = (1.4 \div 3) \times 24380 \times 1.000$$
$$= 11377\ \text{tonne}$$
$$\text{Reduction in draught} = (11377 \times 100 \div 1670) \times (1.025 - 1.000) \div (1.000 \times 1.025)$$
$$\text{Reduction in draught} = 16.62\ \text{cm}$$
$$\text{Draught in sea water} = 8.40 - 0.166$$
$$= 8.234\ m$$

## A. E1.20:

The dock must rise 0.5 m before the ship touches. Thus if the ship rises 1.2 m, the dock must rise 1.7 m.

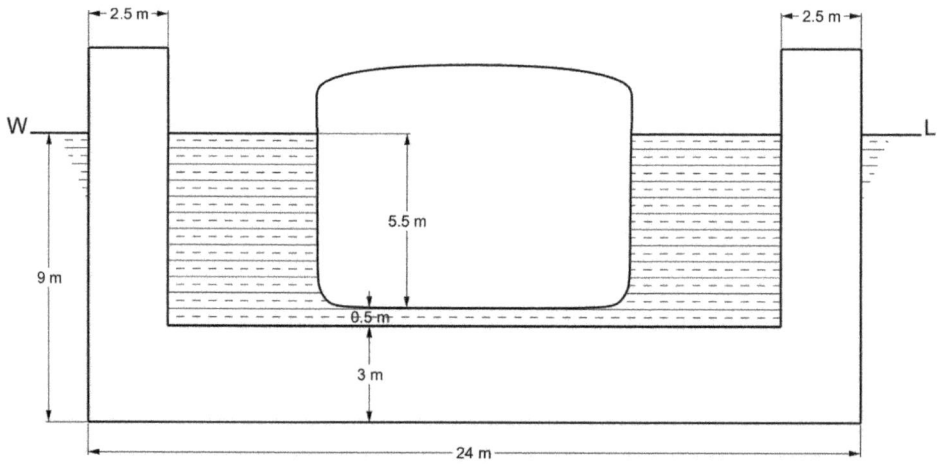

▲ Figure A. E1.3

$$\text{Mass removed to raise dock } 1.7\,m = 150 \times (2.5 + 2.5) \times 1.70 \times 1.025$$
$$= 1307\ \text{tonne}$$
$$\text{Mass removed to raise ship } 1.2m = 4000 - 1307$$
$$= 2693\ \text{tonne}$$
$$\text{Therefore}: \ 20 \times TPC = 2693$$
$$\text{Mean TPC} = 2693 \div 120$$
$$= 22.44$$

**A. E1.21:**

| Mass | Kg | Moment |
|------|-----|--------|
| 7500 | 6.5 | 48750 |
| 300 | 4.8 | 1440 |
| 1000 | 0.7 | 700 |
| 8800 | | 50890 |

a)  New $KG = 50890 \div 8800$

$= 5.783\,m$

b)  After burning oil:

New $KG = (50890 - 500 \times 0.7) \div (8800 - 500)$

$= 6.089\,m$

**A. E1.22:**

With displacement of 15000 tonne:

Reduction in draught $= (15000 \times 100 \div 2150) \times (1.025 - 1.000) \div (1.000 \times 1.025)$

$= 17.02\,cm$

Increase in draught required $= 17.02\,cm$

$TCP = 2150 \times 0.01025$

$= 22.04$

Mass of oil required $= 17.02 \times 22.04$

$= 375.1\,tonne$

**A. E1.23:**

Load on bulkhead $= \rho g AH$

$= 1.025 \times 9.81 \times 12 \times 9 \times 4.5$

$= 4887\,kN$

Shearing force at bottom $= \tfrac{2}{3} \times 4887$

$= 3258\,kN$

Number of rivets $= 12 \div 0.080$

$= 150$

Total cross-sectional area of rivets $= 150 \times (\pi \div 4) \times 20^2 \times 10^{-6}$

$= 47.13 \times 10^{-3}\,m^2$

Stress in rivets $=$ load $\div$ area

$= 3258 \div (47.13 \times 10^{-3})$

$= 69.13 \times 10^3\,KN/m^2$

$= 69.13\,MN/m^2$

**A. E1.24:**

| ½ ordinate | SM | Product |
|---|---|---|
| 1.2 | 1 | 1.2 |
| 3.9 | 4 | 15.6 |
| 5.4 | 2 | 10.8 |
| 6.0 | 4 | 24.0 |
| 6.3 | 2 | 12.6 |
| 6.3 | 4 | 25.2 |
| 6.3 | 2 | 12.6 |
| 5.7 | 4 | 22.8 |
| 4.5 | 2 | 9.0 |
| 2.7 | 4 | 10.8 |
| 0 | 1 | — |
| | | 144.6 |

$$h = 9.6\,\text{m}$$
$$\text{Waterplane area} = \tfrac{2}{3} \times 9.6 \times 144.6$$
$$= 925.44\,\text{m}^2$$
$$\text{TPC} = 925.44 \times 0.01025$$
$$= 9.486$$
$$\text{Mass of oil in tank} = (7.2 \times 6.0 \times 1.2) \div 1.15$$
$$= 45.08\,\text{tonne}$$
$$\text{Therefore: mass removed} = 22.54\,\text{tonne}$$
$$\text{Reduction in draught} = 22.54 \div 9.486$$
$$= 2.38\,\text{cm}$$
$$\text{Final draught} = 4.50 - 0.024$$
$$= 4.476\,\text{m}$$

**A. E1.25:**

Let $\Delta$ = original displacement

Change in draught due to removal of mass

$$a = 0.06\Delta \div \text{TPC}\ \text{cm}$$
$$a = 0.06\Delta \div 19$$
$$= 0.003158\,\Delta\ \text{cm}\ \text{reduction}$$

Change in draught due to change in density

$$b = (0.94\Delta \times 1.023) \div (TPC \times 100) \times (1.023 - 1.006) \div (1.006 \times 1.023)$$
$$b = (0.94\Delta \times 0.017) \div (19 \times 1.006)$$
$$= 0.000836\, \Delta\, cm\, increase$$

Assuming $a$ to be greater than $b$

$$20 = a - b$$
$$20 = 0.003158\, \Delta - 0.000836\, \Delta$$
$$20 = 0.002322\, \Delta$$
$$\text{Original displacement } \Delta = 20 \div 0.002322$$
$$= 8613\, tonne$$

The draught will be reduced by 20 cm.

If the above assumption were wrong, the displacement would work out as a negative value.

**A. E1.26:**

$$\text{Let } V = normal\, speed\, in\, knots$$
$$9200 \div 5710 = ((V + 1.5) \div (V - 1.5))^{2.95}$$
$$(V + 1.5) \div (V - 1.5) = (9200 \div 5710)^{1/2.95}$$
$$(V + 1.5) \div (V - 1.5) = 1.176$$
$$V + 1.5 = 1.176V - 1.176 \times 1.5$$
$$0.176\, V = (1 + 1.176) \times 1.5$$
$$V = 2.176 \times 1.5 \div 0.176$$
$$\text{Normal speed } V = 18.55\, knots$$

**A. E1.27:**

$$L = 7.6 \times B$$
$$B = 2.85 \times d$$
$$\nabla = L \times B \times d \times C_b$$
$$\nabla = 7.6B \times B \times (B \div 2.85) \times 0.69$$
$$\nabla = 7.6 \times 0.69B^3 \div 2.85$$
$$S = 1.7Ld + \nabla \div d$$
$$S = 1.7 \times 7.6B \times (B \div 2.85) + (7.6 \times 0.69 \div 2.85)B^3 \times (2.85 \div B)$$
$$7000 = 4.533B^2 + 5.244B^2$$
$$7000 = 9.777B^2$$
$$B^2 = 7000 \div 9.777$$
$$B = 26.76\, m$$
$$L = 7.6 \times 26.76$$
$$= 203.38m$$
$$d = 26.76 \div 2.85$$
$$= 9.389\, m$$

a)  Displacement $\Delta = 203.38 \times 26.76 \times 9.389 \times 0.69 \times 1.025$
$$= 36140 \text{ tonne}$$

b) Area of immersed midship section

$$A_m = B \times d \times C_b \div C_p$$
$$= 26.76 \times 9.389 \times 0.69 \div 0.735$$
$$= 235.9 \text{ m}^2$$

c) Waterplane area

$$A_w = L \times B \times C_w$$
$$= 203.38 \times 26.76 \times 0.81$$
$$= 4408 \text{ m}^2$$

**A. E1.28:**

$$\text{Original displacement} = 120 \times 17 \times 7.2 \times 0.76 \times 1.025$$
$$= 11442 \text{ tonne}$$
$$\text{Additional displacement} = 6 \times 17 \times 7.2 \times 0.96 \times 1.025$$
$$= 723 \text{ tonne}$$
$$\text{New displacement} = 11442 + 723$$
$$= 12165 \text{ tonne}$$
$$\text{New length} = 126 \text{ m}$$
$$\text{New block coefficient} = 12165 \div (126 \times 17 \times 7.2 \times 1.025)$$
$$= 0.770$$

**A. E1.29:**

$$sp = (\Delta^{\frac{2}{3}} V^3) \div C$$
$$\Delta^{\frac{2}{3}} = 7200 \times 355 \div 15^3$$
$$\text{Displacement } \Delta = 20840 \text{ tonne}$$

sp is proportional to $V^3$

$$\text{Therefore: } sp_1 = 7200 \times (0.84V \div V)^3$$
$$= 4267 \text{ kW}$$

**A. E1.30:**

$$\Delta = L \times B \times d \times C_b \times \rho$$
$$C_b = C_p \times C_m$$
$$8100 = 120 \times 16 \times d \times 0.70 \times 0.98 \times 1.025$$
$$d = 8100 \div (120 \times 16 \times 0.70 \times 0.98 \times 1.025)$$
$$= 6m$$

Let $l$ = length of compartment

Volume of lost buoyancy = $l \times 16 \times 6 \times 0.98$
Area of intact waterplane = $120 \times 16 \times 0.82 - l \times 16$
Increase in draught = $7.5 - 6$
$$= 1.5 \text{ m}$$
$$1.5 = (l \times 16 \times 6 \times 0.98) \div (120 \times 16 \times 0.85 - l \times 16)$$
$$2361.6 - 24\,l = 94.08\,l$$
$$94.08\,l + 24\,l = 2361.6$$
$$118.08\,l = 2361.6$$
$$l = 2361.6 \div 118.08$$
Length of compartment = 20 m

**A. E1.31:**

**a)** Change in draught due to density

$$= 22000 \times 100 \div 3200 \times (1.026 - 1.008) \div (1.008 \times 1.026)$$
$$= 11.97 \text{ cm reduction}$$
Change in displacement = $11.97 \times 3200 \times 1.026 \times 10^{-2}$
$$= 393 \text{ tonne}$$
% difference = $393 \div 22000 \times 100$
$$= 1.79\% \text{ increase}$$

**b)** New draught = $9 - 0.1197$
$$= 8.8803 \text{ m}$$
Final draught = 8.55 m
Change in draught = 0.3303 m
Change in displacement = $0.3303 \times 3200 \times 1.026$
$$= 1084 \text{ tonne}$$
% difference = $1084 \times 100 \div 22000$
$$= 4.93\% \text{ reduction}$$

**A. E1.32:**

Fuel cons/day is proportional to $\Delta^{\frac{2}{3}} V^3$

$$\text{Let } V = \text{new ship speed in knots}$$
$$sp = (\Delta^{\frac{2}{3}} V^3) \div C$$
$$\text{Then new cons/day} = (14500 \div 15500)^{\frac{2}{3}} \times (V \div 14)^3 \times 60$$
$$= 0.02092\, V^3 \text{ tonne}$$
$$\text{Number of days} = 640 \div (24\, V)$$
$$\text{Therefore: voyage cons} = 640 \div (24\, V) \times 0.02092\, V^3$$
$$175 = 640 \div 24 \times 0.02092\, V^2$$
$$V^2 = 313.7$$
$$\text{Ship speed } V = 17.71 \text{knots}$$

**A. E1.33:**

| TPC | SM | Product |
|------|-----|---------|
| 19.0 | 1 | 19.0 |
| 18.4 | 4 | 73.6 |
| 17.4 | 2 | 34.8 |
| 16.0 | 4 | 64.0 |
| 13.8 | 2 | 27.6 |
| 11.0 | 4 | 44.0 |
| 6.6 | 1 | 6.6 |
| | | 269.6 |

$$h = 1.2\,\text{m}$$
$$\text{Displacement} = 1.2 \div 3 \times 269.6 \times 100$$
$$= 10784 \text{ tonne}$$
$$\text{Load draught} = 1.2 \times 6$$
$$= 7.2\,\text{m}$$

**A. E1.34:**

$$\text{Change in draught due to density} = (12000 \times 100) \div (17.7 \times 100) \times$$
$$1.025 \times (1.025 - 1.012) \div$$
$$(1.012 \times 1.025)$$
$$= 8.71\,\text{cm reduction}$$

$$\text{Change in draught due to removal of 130 tonne} = 130 \div 17.7$$
$$= 7.345\,\text{cm reduction}$$

$$\text{Total change in draught} = 8.71 + 7.344$$
$$= 16.055\,\text{cm}$$

ie Summer Load Line would have been submerged 16.05 cm

**A. E1.35:**

The effective area of the bulkhead will depend upon whether the water level is above or below the top. Assume that the water is at the top.

Load on bulkhead $= \rho g A H$

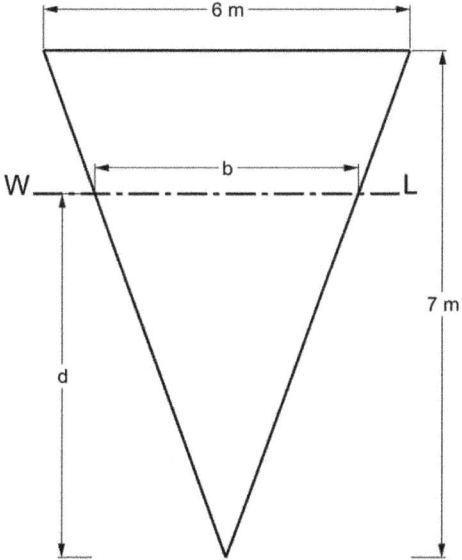

6 m

W ————— b ————— L

7 m

d

▲ Figure A. E1.4

$$\text{Load on bulkhead} = 1.025 \times 9.81 \times (6 \times 7 \div 2) \times (7 \div 3)$$
$$= 492.7\,\text{kN}$$

Thus the water must lie below the top.

$$\text{Let } d = \text{depth of water}$$
$$b = \text{width of bulkhead at water level}$$

$$b = 6 \div 7 \times d$$
$$\text{Load on bulkhead} = 1.025 \times 9.81 \times (b \times d \div 2) \times (d \div 3)$$
$$195 = 1.025 \times 9.81 \times (6 \div 7 \times d) \times (d \div 2) \times (d \div 3)$$
$$d^3 = 195 \times 7 \times 2 \times 3 \div (1.025 \times 9.81 \times 6)$$
$$\text{Depth of water } d = 5.14\,\text{m}$$

**A. E1.36:**

$$GM = KB + BM - KG$$
$$BM = KG + GM - KB$$
$$= 6.6 + 0.46 - 3.6$$
$$= 3.46\,\text{m}$$

But

$$BM = I \div \nabla$$
$$I = BM \times \nabla$$
$$= 3.46 \times 8000 \div 1.025$$
$$\text{Second moment of area} = 27005\,\text{m}^4$$

**A. E1.37:**

When 120 tonne is transferred:

$$\text{Shift in CG} = m \times d \div \Delta$$
$$d = \Delta \times GG_j$$
$$d = 9800 \times 0.75 \div 120$$
$$\text{Distance between tanks} = 61.25\,\text{m}$$

When 420 tonne is used:

Let x = distance from original CG to CG of forward tank. Taking moments about the original CG:

$$(9800 - 420)(-0.45) = 9800 \times 0.75 - 420 \times x$$
$$-4221 = 7350 - 420x$$
$$420x = 7350 + 4221$$
$$x = 11571 \div 420$$
$$= 27.55\,\text{m}$$
$$d - x = 61.25 - 27.55$$
$$= 33.7\,\text{m}$$

Thus the forward tank is 27.55 m from the vessel's original CG and the after tank is 33.7 m from the vessel's original CG.

**A. E1.38:**

$$\text{Fuel cons day} = 18000^{\frac{2}{3}} \times 12^3 \div 47250$$
$$= 25.12 \text{ tonne}$$
$$\text{Number of days} = 500 \div 25.12$$
$$= 19.90$$

a)    $\text{Range of operation} = 19.90 \times 12 \times 243$
$$= 5731 \text{ nautical miles}$$

b)    $\text{New speed} = 12 \times 1.05$
$$= 12.6 \text{ knots}$$
$$\text{New cons/day} = 25.12 \times (12.6 \div 12)^3$$
$$= 29.08 \text{ tonne}$$
$$\text{Number of days} = 500 \div 29.08$$
$$= 17.19$$
$$\text{Range of operation} = 17.19 \times 12.6 \times 24$$
$$= 5198 \text{ nautical miles}$$
$$\text{Percentage difference} = (5734 - 5198) \div 5734 \times 100$$
$$= 9.35\% \text{ reduction}$$

**A. E1.39:**

$$\text{TPC in seawater} = 1625 \times 0.01024$$
$$= 16.64$$
$$\text{Bodily sinkage} = 1650 \div 16.64$$
$$= 99.16 \text{ cm}$$

Thus at 12000 tonne displacement:

$$\text{Mean draught in sea water} = 9.08 - 0.992$$
$$= 8.088 \text{ m}$$
$$\text{Thus change in draught due to density} = 8.16 - 8.088$$
$$= 0.072 \text{ m}$$
$$\textit{Let } \rho_R = \text{density of harbour water in t/m}^3$$

Then:

$$7.2 = 12000 \times 100 \div 1650 \times (1.024 - \rho_R) \div (1.024\rho_R)$$
$$1.024\,\rho_R \times 7.2 = 727.3(1.024 - \rho_R)$$
$$\rho_R(7.37 + 727.3) = 727.3 \times 1.024$$
$$\rho_R = 727.3 \times 1.024 \div 734.67$$

Density of harbour water $= 1.014\,t/m^3$

**A. E1.40:**

$$\text{Mean deflection} = \tfrac{1}{4}(73 + 80 + 78 + 75)$$
$$= 76.5\,mm$$
$$GM = m \times d \div (\Delta \tan\theta)$$
$$GM = 5 \times 12 \times 6 \div (8000 \times 0.0765)$$
$$= 0.588\,m$$
$$KG = KM - GM$$
$$= 5.10 - 0.588$$
$$= 4.512\,m$$

**A. E1.41:**

a)  Load on bulkhead $= \rho gAH$
$$= 1.025 \times 9.81 \times 17 \times 6 \times (3 + 2.5)$$
$$= 5641\,kN$$

b)  Pressure at top of bulkhead $= \rho gh$
$$= 1.025 \times 9.81 \times 2.5$$
$$= 25.14\,kN/m^2$$
Pressure at bottom of bulkhead $= 1.025 \times 9.81 \times (6 + 2.5)$
$$= 85.47\,kN/m^3$$

**A. E1.42:**

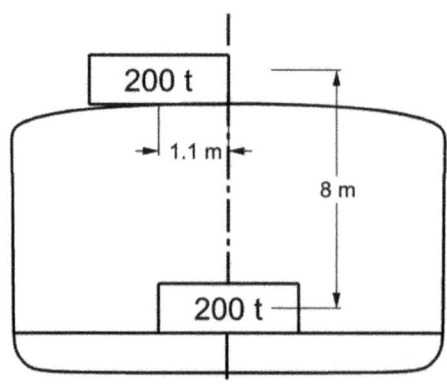

▲ Figure A. E1.5

$$\text{Rise in G} = 200 \times 8 \div 4600$$
$$= 0.348\,\text{m}$$
$$\text{New } GM = 0.77 - 0.348$$
$$= 0.422\,\text{m}$$
$$\tan\theta = (200 \times 1.1) \div (4600 \times 0.422)$$
$$= 0.1133$$
$$\text{Angle of heel } \theta = 6.46°$$

## A. E1.43:

$TPC = A_w \times 0.01025$

| Draught | $A_w$ | TPC | SM$_1$ | Product | SM$_2$ | Product |
|---------|-------|-------|--------|---------|--------|---------|
| 7.5 | 2100 | 21.52 | 1 | 21.52 | | |
| 6.0 | 1930 | 19.78 | 4 | 79.12 | | |
| 4.5 | 1720 | 17.63 | 2 | 35.26 | 1 | 17.63 |
| 3.0 | 1428 | 14.64 | 4 | 58.56 | 4 | 58.56 |
| 1.5 | 966 | 9.90 | 1 | 9.90 | 1 | 9.90 |
| | | | | 204.36 | | 86.09 |

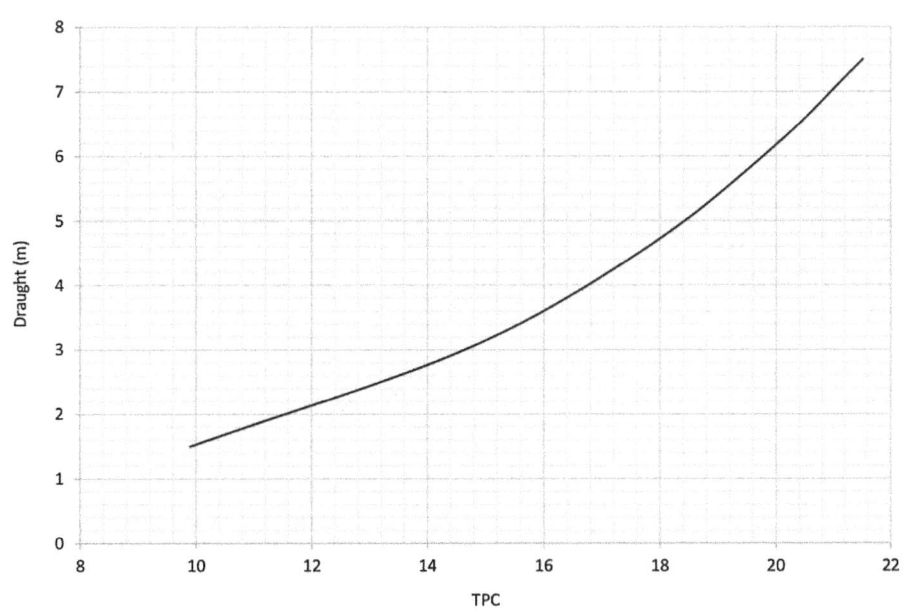

▲ Figure A. E1.6

**a)** Displacement at 7.5m draught $= 711 + (1.5 \div 3) \times 100 \times 204.36$
$$= 711 + 10218$$
$$= 10929 \text{ tonne}$$

**b)** Displacement at 4.5 m draught $= 711 + (1.5 \div 3) \times 100 \times 86.09$
$$= 711 + 4305$$
$$= 5016 \text{ tonne}$$

## A. E1.44:

The centre of buoyancy is the centroid of the immersed part of a ship or floating object, and is the point at which the buoyancy force experienced by the object can be assumed to act.

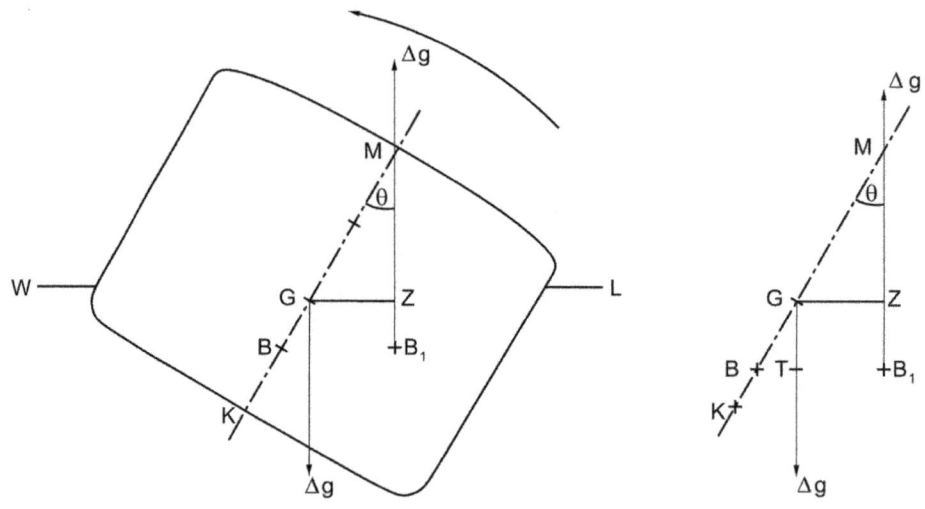

▲ **Figure A. E1.7**

$$\text{Righting moment} = \Delta \times GZ$$
$$570 = 8000 \times GZ$$
$$GZ = 570 \div 8000$$
$$= 0.0713 \text{ m}$$
$$BT = BG \sin \theta$$
$$= 1.0 \times 0.1219$$
$$= 0.1219 \text{ m}$$
$$\text{Horizontal movement of } B = BB_1$$
$$= BT + TB_1$$
$$= BT + GZ$$
$$= 0.1219 + 0.0713$$
$$= 0.1932 \text{ m}$$

**A. E1.45:**

$$\text{Speed of advance} = 12 \times 0.5144$$
$$= 6.173 \text{ m/s}$$
$$\text{Real slip} = (V_t - V_a) \div V_t$$
$$0.30 = (V_t - 6.173) \div V_t$$
$$V_t = 6.173 \div (1 - 0.3)$$
$$= 8.819 \text{ m/s}$$
$$= P \times n$$

a) $\qquad P = 8.819 \div 2$
$$= 4.410 \text{ m}$$

b) $\qquad$ Thrust power $= 300 \times 6.173$
$$= 1852 \text{ kW}$$

c) $\qquad$ Delivered power $= 250 \times 2\pi \times 2$
$$= 3142 \text{ kW}$$

**A. E1.46:**

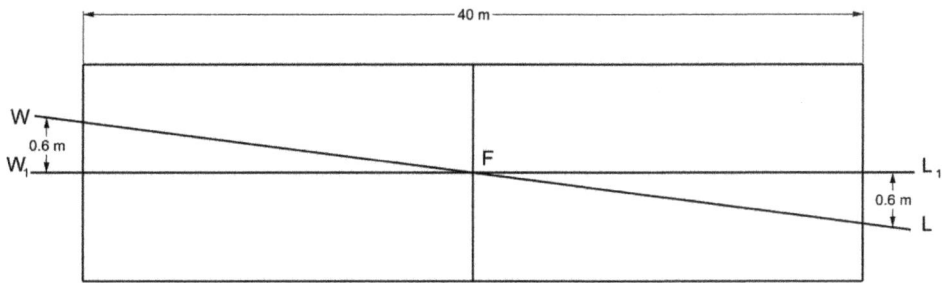

▲ Figure A. E1.8

If it is assumed that the mass is first added amidships, there will be a bodily increase in draught without change of trim.

To obtain a level keel draught, the wedge of buoyancy $WFW_1$ must be transferred to $L_1FL$.

$$\text{Mass of wedge} = 20 \times 7.5 \times 0.6 \times \tfrac{1}{2} \times 1.025$$
$$= 46.125 \text{ tonne}$$
$$\text{Distance moved} = \tfrac{2}{3} \times 40$$
$$\text{Let } x = \text{distance moved forward by 90 tonne, then}$$
$$90x = 46.125 \times \tfrac{2}{3} \times 40$$
$$x = 46.125 \times 2 \times 40 \div (90 \times 3)$$
$$= 13.67 \text{ m}$$

**A. E1.47:**

$$\text{TPC} = L \times B \times 1.025 \div 100$$
$$L = 17 \times 100 \div (16 \times 1.025)$$
$$= 103.7 \text{ m}$$
$$\text{Volume of lost buoyancy} = 20 \times 16 \times 6 \times 0.8$$
$$\text{Area of intact waterplane} = 103.6 \times 16 - 20 \times 16 \times 0.8$$
$$\text{Increase in draught} = (20 \times 16 \times 6 \times 0.8) \div (103.6 \times 16 - 20 \times 16 \times 0.8)1536 / 1401.6$$
$$= 1.096 \text{ m}$$

a)     $$\text{New draught} = 6 + 1.096$$
$$= 7.096 \text{ m}$$

b)     $$KB = 7.096 \div 2$$
$$= 3.548 \text{ m}$$
$$BM = 75000 \div (103.6 \times 16 \times 6)$$
$$= 7.541 \text{m}$$
$$KM = 3.548 + 7.541$$
$$= 11.089 \text{ m}$$

**A. E1.48:**

$$KB = d$$
$$BM = B^2 \div (12d)$$
$$BM = 7.2^2 \div (12d)$$
$$BM = 4.32 \div d$$

| Draught | KB | BM | KM |
|---------|-----|------|------|
| 0 | 0 | ∞ | ∞ |
| 1 | 0.5 | 4.32 | 4.82 |
| 2 | 1.0 | 2.16 | 3.16 |
| 3 | 1.5 | 1.44 | 2.94 |
| 4 | 2.0 | 1.08 | 3.08 |
| 5 | 2.5 | 0.86 | 3.36 |
| 6 | 3.0 | 0.72 | 3.72 |

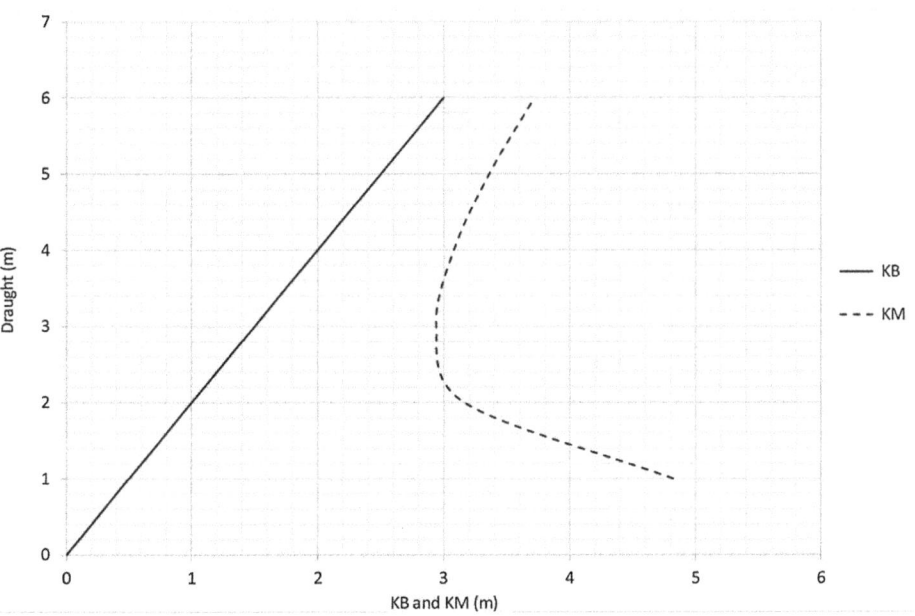

▲ Figure A. E1.9

**A. E1.49:**

$$\text{Propeller pitch} = 1.1 \times 4.28$$
$$= 4.708 \text{ m}$$
$$V_t = 4.708 \times 2$$
$$= 9.416 \text{ m/s}$$
$$\text{Apparent slip } 0.7 = (9.416 - V) \div 9.416 \times 100$$
$$V = 9.416(1 - 0.007)$$
$$= 9.35 \text{ m/s}$$
$$\text{Real slip } 12 = (9.416 - V_a) \div 9.416 \times 100$$
$$V_a = 9.416(1 - 0.12)$$
$$= 8.286 \text{ m/s}$$
$$\text{Wake speed} = 9.35 - 8.286$$
$$= 1.064 \text{ m/s}$$
$$= 1.064 \div 0.5144$$
$$= 2.068 \text{ knots}$$

**A. E1.50:**

$$\text{At 12 knots : cons/day} = (0.55 \times 1710 \times 24) \div 1000$$
$$= 22.572 \text{ tonne}$$

a)    $$\text{At 10 knots : cons/day} = 22.575 \times (10 \div 12)^3$$
$$= 13.062 \text{ tonne}$$
$$\text{Number of days} = 7500 \div (10 \times 24)$$
$$= 31.25$$
$$\text{Fuel required} = 13.062 \times 31.25$$
$$= 408.2 \text{ tonne}$$

b)    $$\text{Fuel coefficient} = 6000^{2/3} \times 12^3 \div 22.572$$
$$= 25278$$

**A. E1.51:**

$$KG = KB + BM - GM$$
$$KB = 4.5 \text{ m}$$
$$BM = I \div \nabla$$
$$= 42.5 \times 10^3 \times 1.025 \div 12000$$
$$= 3.63 \text{ m}$$
$$GM = 0.6 \text{ m}$$
$$KG = 4.50 + 3.63 - 0.60$$
$$= 7.53 \text{ m}$$

**A. E1.52:**

$$\text{Original displacement} = 37 \times 6.4 \times 2.5 \times 1.025$$
$$= 606.8 \text{ tonne}$$
$$\text{New mean draught} = (2.4 + 3.8) \div 2$$
$$= 3.1 \text{m}$$
$$\text{New displacement} = 37 \times 6.4 \times 3.1 \times 100$$
$$= 734.08 \text{ tonne}$$
$$\text{Mass added} = 734.08 - 606.8$$
$$= 127.28 \text{ tonne}$$

**A. E1.53:**

$$\text{Mass of fuel/m depth} = 495 \div 9.9$$
$$= 50 \text{ tonne}$$

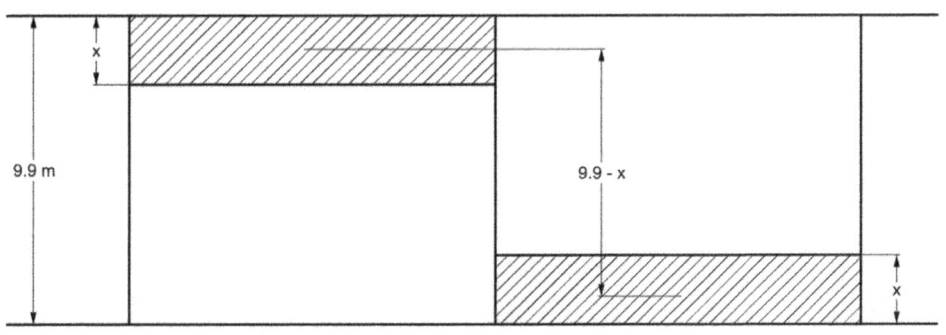

▲ Figure A. E1.10

$$\text{Let } x = \text{depth of fuel that must be transferred}$$
$$\text{Then } 50x = \text{mass of fuel transferred}$$
$$9.9 - x = \text{distance that fuel is transferred}$$
$$\text{Shift in } CG = (m \times d) \div \Delta$$
$$0.12 = 50x \times (9.9 - x) \div 7200$$
$$0.12 = x \times (9.9 - x) \div 144$$
$$0.12 \times 144 = 9.9x - x^2$$
$$x^2 - 9.9x + 17.28 = 0$$
$$x = (9.96 \sqrt{(9.9^2 - 4 \times 17.28)}) \div 2$$
$$x = (9.965 \times 5.375) \div 2$$
$$x = 7.637 \text{ m} \text{ or } 2.263 \text{ m}$$
$$\text{Mass of oil transferred} = 50 \times 7.638$$
$$= 381.9 \text{ tonne}$$
$$\text{or } 50 \times 2.262$$
$$= 113.1 \text{ tonne}$$

**A. E1.54:**

Final KG = 7.30 − 0.5

= 6.8 m

Let $x$ = distance of original CG above the keel

| Mass | $Kg$ | Moment |
|------|------|--------|
| 7000 | $X$ | $7000x$ |
| 150 | $x + 3$ | $150x + 450$ |
| 60 | $x - 5.5$ | $60x - 330$ |
| −76 | 0.6 | −45.6 |
| 7134 | | $7210x + 74.4$ |

$$7134 \times 6.8 = 7210x + 74.4$$
$$7210x = 48511 - 74$$
$$x = 48437 \div 7210$$

Original KG = 6.72 m

**A. E1.55:**

| Station | ½ breadth | SM | Product for area |
|---------|-----------|-----|------------------|
| 0 | 0.1 | 1 | 0.1 |
| 1 | 2.4 | 4 | 9.6 |
| 2 | 5.1 | 2 | 10.2 |
| 3 | 7.4 | 4 | 29.6 |
| 4 | 8.4 | 2 | 16.8 |
| 5 | 8.4 | 4 | 33.6 |
| 6 | 8.4 | 2 | 16.8 |
| 7 | 7.4 | 4 | 29.6 |
| 8 | 5.1 | 2 | 10.2 |
| 9 | 2.4 | 4 | 9.6 |
| 10 | 0.1 | 1 | 0.1 |
| | | | 166.2 |

$$Common\ interval = 120 \div 10\ m$$
$$= 12\ m$$
$$Waterplane\ area = \tfrac{2}{3} \times 12 \times 166.2$$
$$= 1329.6\ m^2$$
$$Intact\ waterplane\ area = 1329.6 - 0.5(24 \times 8.4 \times 2)$$
$$= 1128\ m^2$$
$$Volume\ of\ lost\ buoyancy = 0.5(24 \times 8.4 \times 2 \times 3.5)$$
$$= 705.6\ m^3$$

Increase in draught = volume of lost buoyancy ÷ area of intact waterplane

$$= 705.6 \div 1128$$
$$= 0.625m$$
$$New\ draught = 3.5 + 0.625$$
$$= 4.125m$$

**A. E1.56:**

a) See notes in Chapter 4 for full description, but could include slack mooring lines, still water, pressed tanks, no unnecessary equipment on board, etc.

b)
$$GM = (m \times d) \div (\Delta \times tan\ \theta)$$
$$= 15 \times 12 \times 8 \div (5500 \times 0.25)$$
$$= 1.047\ m$$
$$KM = 7.5\ m$$
$$KG = KM - GM$$
$$= 6.453\ m$$

| Mass | Kg | Moment |
|------|------|--------|
| 5500 | 6.453 | 35491.5 |
| −150 | 9.75 | −1462.5 |
| +220 | 9.00 | +1980 |
| 5570 | | 36009 |

$$New\ KG = 36009 \div 5570$$
$$= 6.465\ m$$

**A. E1.57:**

$$\text{Block coefficient } C_b = 15500 \div (138 \times 18.5 \times 8.5 \times 1.025)$$
$$= 0.697$$
$$W_t = 0.5C_b - 0.048$$
$$= 0.5 \times 0.696 - 0.048$$
$$= 0.3$$
$$\text{Ship speed} = 15.5 \text{ knots}$$
$$= 15.5 \times 0.5144$$
$$= 7.973 \text{ m/s}$$
$$0.3 = 1 - (V_a \div 7.973)$$
$$V_a = 0.7 \times 7.973$$
$$= 5.581 \text{ m/s}$$
$$\text{Real slip} = (V_t - V_a) \div V_t$$
$$V_t(1 - 0.35) = 5.581$$
$$V_t = 8.586 \text{ m/s}$$
$$P \times n = V_t$$
$$\text{Pitch } P = 8.586 \div 1.92$$
$$= 4.472 \text{ m}$$
$$\text{Pitch ratio } p = P \div D$$
$$\text{Diameter } D = 4.472 \div 0.83$$
$$= 5.388 \text{ m}$$

**A. E1.58:**

$$\text{a)} \qquad B \div D = b \div d$$
$$\text{Breadth at waterline } b = (14 \div 7) \times 5$$
$$= 10 \text{ m}$$
$$\text{Displacement } \Delta = 40 \times 10 \times 5 \times \tfrac{1}{2} \times 1.025$$
$$= 1025 \text{ tonne}$$

b) In normal bilging questions, the waterplane area is regarded as constant over the increased draught. In a triangular cross-sectional vessel, this is quite inaccurate. The best method to use in this question is to consider a vessel of the same displacement but of reduced length:

$$\text{Let } d_1 = \text{new draught}$$
$$b_1 = \text{new breadth at waterline}$$
$$= 2d_1$$
$$L_1 = \text{new length}$$
$$= 40 - 6$$
$$= 34 \text{ m}$$
$$\text{Displacement } \Delta = 34 \times 2d_1 \times 1.025 \times \tfrac{1}{2}$$
$$d_1^2 = 1025 \times 2 \div (34 \times 2 \times 1.025)$$
$$\text{New draught } d_1 = 5.423 \text{ m}$$

**A. E1.59:**

$$V_t = P \times n$$
$$= 5.4 \times 0.875 \times 1.87$$
$$= 8.836 \text{ m/s}$$
$$V_a = 8.836(1 - 0.28)$$
$$= 6.362 \text{ m/s}$$

**a)**    Thrust power tp $= T \times V_a$
$$= 860 \times 6.362$$
$$= 5471 \text{ kW}$$

**b)**  Delivered power dp = tp ÷ propeller efficiency

$$= 5471 \div 0.68$$
$$= 8046 \text{ kW}$$
Delivered power $= 2\pi n \times$ torque

**c)**    Torque $= 8046 \div (2\pi \times 1.87)$
$$= 684.8 \text{ kNm}$$

**d)**    Blade area $= (\pi \div 4) \times 5.4^2 \times 0.46$
$$= 10.535 \text{ m}^2$$
Pressure = thrust ÷ area
$$= 860 \div 10.535$$
$$= 81.63 \text{ kN/m}^2$$

**A. E1.60:**

| Station | ½ CSA | SM | Product for volume |
|---------|-------|-----|--------------------|
| 0 | 10 | 1 | 10 |
| 1 | 26 | 4 | 104 |
| 2 | 40 | 2 | 80 |
| 3 | 59 | 4 | 236 |
| 4 | 60 | 2 | 120 |
| 5 | 60 | 4 | 240 |

| Station | ½ CSA | SM | Product for volume |
|---------|-------|-----|--------------------|
| 6 | 60 | 2 | 120 |
| 7 | 56 | 4 | 224 |
| 8 | 38 | 2 | 76 |
| 9 | 14 | 4 | 56 |
| 10 | 4 | 1 | 4 |
| | | | 1270 |

$$h = 150 \div 10$$
$$\text{Volume of displacement} = 15 \times \tfrac{2}{3} \times 1270$$
$$= 12700\,\text{m}^3$$

a) $$\text{Displacement} = 12700 \times 1.025$$
$$= 13017\,\text{tonne}$$

b) $$C_b = 12700 \div (150 \times 18 \times 7)$$
$$= 0.672$$

c) $$C_p = 12700 \div (150 \times 2 \times 60)$$
$$= 0.706$$

d) $$C_m = (2 \times 60) \div (18 \times 7)$$
$$= 0.952$$

**A. E1.61:**

a) $$\text{Volume of cone} = \pi \times (\pi \div 4) \times 2^2 \times 3$$
$$= 3.142\,\text{m}^3$$
$$\text{Displacement} = 3.142 \times 0.8$$
$$= 2.514\,\text{tonne}$$
$$\text{Let } d = \text{draught}$$
$$b = \text{diameter at waterline}$$
$$b = \tfrac{2}{3} \times d$$
$$\text{Then displacement} = \tfrac{1}{3} \times (\pi \div 4) \times (\tfrac{2}{3}d)^2 \times d \times 1.025$$
$$d^3 = 2.514 \times (3 \div 1.025) \times (4 \div \pi) \times (9 \div 4)$$
$$\text{Draught } d = 2.762\,\text{m}$$

**b)** $\qquad$ Diameter at waterline $= \frac{2}{3} \times 2.762$

$$= 1.841 \, \text{m}$$

$$KG = \frac{3}{4} \times 3$$

$$= 2.25 \, \text{m}$$

$$KB = \frac{3}{4} \times 2.762$$

$$= 2.071$$

$$I = (\pi b^4) \div 64$$

$$\nabla = \frac{1}{3} \times (\pi \div 4) \times b^2 \times d$$

$$BM = I \div \nabla$$

$$= (12b^2) \div (64d)$$

$$= 12 \times 1.841^2 \div (64 \times 2.762)$$

$$= 0.23 \, \text{m}$$

Metacentric height $GM = KB + BM - KG$

$$= 2.071 + 0.230 - 2.25$$

$$= 0.051 \, \text{m}$$

**A. E1.62:**

$$\text{Ship speed} = 20 \, \text{knots}$$

$$= 20 \times 0.5144$$

$$= 10.29 \, \text{m/s}$$

$$\text{Shaft power} = 7300 \div 0.97$$

$$= 7526 \, \text{kW}$$

**a)** $\qquad$ Thrust power $= 7526 \times 0.53$

$$= 3989 \, \text{kW}$$

$$V_t = P \times n$$

$$= 5.5 \times 1.0 \times 2.2$$

$$= 12.1 \, \text{m/s}$$

$$\text{Real slip } 0.35 = (12.1 - V_a) \div 12.1$$

$$\text{Speed of advance } V_a = 12.1(1 - 0.35)$$

$$= 7.865 \, \text{m/s}$$

**b)** $\qquad$ Taylor wake fraction $W_t = (V - V_a) \div V$

$$= (10.29 - 7.865) \div 10.29$$

$$= 0.236$$

c)      Effective power $ep = tp \times \eta$

$$= 3989 \times 0.61$$
$$= 2433 \text{ kW}$$
$$= R_t \times V$$

Ship resistance $R_t = 2433 \div 10.29$
$$= 236.5 \text{ kN}$$

**A. E1.63:**

**a)** See notes in Chapter 7 for full description:

i)   $S$ is the wetted surface area in m$^2$

$V$ is the ship speed in knots

$n$ is an index of about 1.825

ii)  $f$ is influenced by the length of the ship, surface roughness and water density.

**b)**

i)      Admiralty Coefficient $= (\Delta^{\frac{2}{3}} V^3) \div sp$

$$= (12000^{\frac{2}{3}} 14.5^3) \div 3500$$
$$= 456.6$$

ii)  $sp$ is proportional to $V^3$

$$sp_1 = 3500 \times (16 \div 14.5)^3$$
$$= 4702 \text{ kW}$$

Percentage increase $= (4702 \div 3500) \times 100 - 100$
$$= 34.34\%$$

or $sp_1 \div sp = (16 \div 14.5)^3$
$$= 1.3435$$

ie percentage increase $= 34.35$

iii)     $(V_2 \div V_1) = (4000 \div 3500)^{\frac{1}{3}}$
$$= 1.0455$$

ie speed increase $= 4.55\%$

# SECTION 2

**Q. E2.1:** A ship 120 m long has draughts of 6.6 m forward and 6.9 m aft. The TPC is 20, MCT 1 cm 101 tonne m and the centre of flotation 3.5 m aft of midships. Calculate the maximum position aft at which 240 tonne may be added so that the after draught does not exceed 7.2 m.

**Q. E2.2:** A vessel when floating at a draught of 3.6 m has a displacement of 8172 tonne, *KB* 1.91 m and LCB 0.15 m aft of midships. From the following information, calculate the displacement, *KB* and position of the LCB for the vessel when floating at a draught of 1.2 m:

| Draught (m) | TPC | LCF from midships |
|---|---|---|
| 1.2 | 23.0 | 1.37 F |
| 2.4 | 24.2 | 0.76 A |
| 3.6 | 25.0 | 0.92 A |

**Q. E2.3:** An oil tanker has LBP 142 m, beam 18.8 m and draught 8 m. It displaces 17000 tonne in sea water of 1.025 t/m$^3$. The face pitch ratio of the propeller is 0.673 and the diameter 4.8 m. The results of the speed trial show that the true slip may be regarded as constant over a range of speeds of 9 to 12 knots and is 35%. The wake fraction may be calculated from the equation:

$$w = 0.5C_b - 0.05$$

If the vessel uses 20 tonne of fuel per day at 12 knots, and the consumption varies as (speed)$^3$, find the consumption per day at 100 rev/min.

**Q. E2.4:** A ship of 5000 tonne displacement has a *KM* of 6.4 m. When 5 tonne are moved 15 m across the ship, a pendulum 6 m long has a deflection of 12 cm. A double bottom tank 7.5 m long, 9 m wide and 1.2 m deep is half-full of sea water. Calculate the *KG* of the light ship.

**Q. E2.5:** A propeller has a pitch ratio of 0.95. When turning at 120 rev/min, the real slip is 30%, the wake fraction 0.28 and the ship speed 16 knots. The thrust is found to be 400 kN, the torque 270 kN m and the QPC 0.67.
Calculate:
**a)** the propeller diameter
**b)** the shaft power
**c)** the propeller efficiency
**d)** the thrust deduction factor.

**Q. E2.6:** A pontoon has a constant cross section as shown below. The metacentric height is 2.5 m. Find the height of the centre of gravity above the keel.

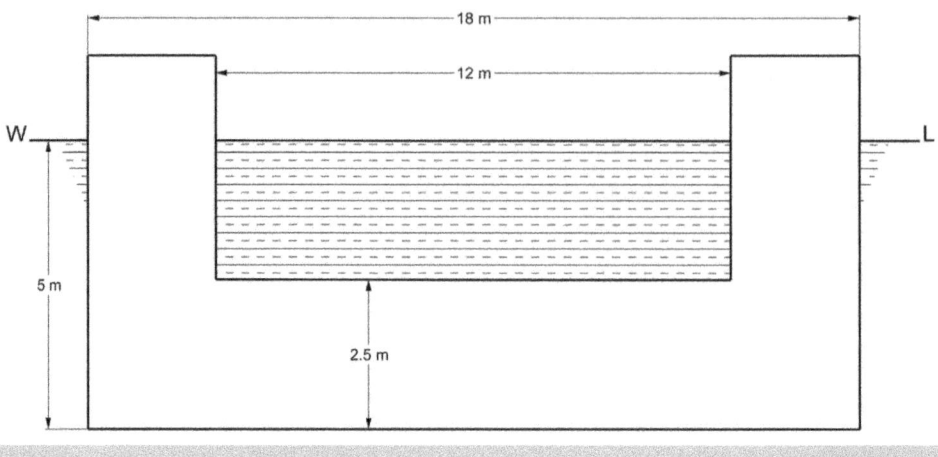

▲ Figure Q. E2.1

**Q. E2.7:**

a) Derive the Admiralty Coefficient formula and show how this may be modified to suit a fast ship.

b) A ship of 14000 tonne displacement requires 23000 kW shaft power to drive it at 24 knots. Using the modified Admiralty Coefficient formula, calculate the shaft power required for a similar ship of 12000 tonne displacement at 21 knots.

**Q. E2.8:** A ship 80 m long has equally spaced immersed cross-sectional areas of 0, 11.5, 27, 38.5, 44, 45, 44.5, 39, 26.5, 14.5 and 0 m² respectively.
Calculate:
a) displacement
b) distance of centre of buoyancy from midships
c) prismatic coefficient.

**Q. E2.9:** The following data refer to two similar ships

|        | L   | S    | V  | $ep_n$ | $f_{sw}$ |
|--------|-----|------|----|------|-------|
| Ship A | 160 | 4000 | 18 | 6400 | 0.420 |
| Ship B | 140 |      |    |      | 0.425 |

Calculate $ep_n$ for ship B at the corresponding speed.

**Q. E2.10:** A ship of 11200 tonne displacement has a double bottom tank containing oil, whose centre of gravity is 16.5 m forward and 6.6 m below the centre of gravity of the ship. When the oil is used, the ship's centre of gravity moves 380 mm.

Calculate:

**a)** the mass of oil used

**b)** the angle that the centre of gravity moves relative to the horizontal.

**Q. E2.11:** A watertight door is 1.2 m high and 0.75 m wide, with a 0.6 m sill. The bulkhead is flooded with sea water to a depth of 3 m on one side and 1.5 m on the other side. Draw the load diagram and from it determine the resultant load and position of the centre of pressure on the door.

**Q. E2.12:** A ship of 10000 tonne displacement has KM 8 m and GM 0.6 m. A rectangular double bottom tank is 1.5 m deep, 18 m long and 15 m wide. Assuming that KM remains constant, determine the new GM when the tank is now:

**a)** filled with sea water

**b)** half-filled with sea water.

**Q. E2.13:** A propeller 6 m diameter has a pitch ratio of 0.9, BAR 0.48 and, when turning at 110 rev/min, has a real slip of 25% and wake fraction 0.30. If the propeller delivers a thrust of 300 kN and the propeller efficiency is 0.65, calculate:

**a)** blade area

**b)** ship speed

**c)** thrust power

**d)** shaft power

**e)** torque.

**Q. E2.14:** When a ship is 800 nautical miles from port, its speed is reduced by 20%, thereby reducing the daily fuel consumption by 42 tonne and arriving in port with 50 tonne on board. If the fuel consumption in t/h is given by the expression $(0.136 + 0.001 \, V^3)$ where $V$ is the speed in knots, estimate:

**a)** the reduced consumption per day

**b)** the amount of fuel on board when the speed was reduced

**c)** the percentage decrease in consumption for the latter part of the voyage

**d)** the percentage increase in time for this latter period.

**Q. E2.15:** An oil tanker 160 m long and 22 m beam floats at a draught of 9 m in sea water. $C_w$ is 0.865. The midship section is in the form of a rectangle with 1.2 m radius at the bilges. A midship tank 10.5 m long has twin longitudinal bulkheads and contains oil of 1.4 m³/t to a depth of 11.5 m. The tank is holed to the sea for the whole of its transverse section. Find the new draught.

**Q. E2.16:** A ship 160 m long and of 8700 tonne displacement floats at a waterline with ½ ordinates of: 0, 2.4, 5.0, 7.3, 7.9, 8.0, 8.0, 7.7, 5.5, 2.8 and 0 m respectively. While floating at this waterline, the ship develops a list of 10° due to instability. Calculate the negative metacentric height when the vessel is upright in this condition.

**Q. E2.17:** The speed of a ship is increased to 18% above normal for 7.5 hours, then reduced to 9% below normal for 10 hours. The speed is then reduced for the remainder of the day so that the consumption for the day is the normal amount. Find the percentage difference between the distance travelled in that day and the normal distance travelled per day.

**Q. E2.18:** A double bottom tank containing sea water is 6 m long, 12 m wide and 1 m deep. The inlet pipe from the pump has its centre 75 mm above the outer bottom. The pump has a pressure of 70 kN/m² and is left running indefinitely. Calculate the load on the tank top:

**a)** if there is no outlet

**b)** if the overflow pipe extends 5 m above the tank top.

**Q. E2.19:** A ship 120 m long displaces 12000 tonne. The following data are available from trial results:

| V (knots) | 10 | 11 | 12 | 13 | 14 | 15 |
|---|---|---|---|---|---|---|
| sp (kW) | 880 | 1155 | 1520 | 2010 | 2670 | 3600 |

**a)** Draw the curve of Admiralty Coefficients against speed.

**b)** Estimate the shaft power required for a similar ship 140 m long at 14 knots.

**Q. E2.20:** A box barge 60 m long and 10 m wide floats at a level keel draught of 3 m. Its centre of gravity is 2.5 m above the keel. Determine the end draughts if an empty, fore end compartment 9 m long is laid open to the sea.

**Q. E2.21:** A vessel of constant triangular cross section floats apex down at a draught of 4 m, the width of the waterplane being 8 m, when its keel just touches a layer of mud having relative density twice that of the water. The tide now falls 2 m. Calculate the depth to which the vessel sinks in the mud.

**Q. E2.22:** The following information relates to a model propeller of 400 mm pitch:

| Rev/min | | 400 | 450 | 500 | 550 | 600 |
|---|---|---|---|---|---|---|
| Thrust | N | 175 | 260 | 365 | 480 | 610 |
| Torque | Nm | 16.8 | 22.4 | 28.2 | 34.3 | 40.5 |

**a)** Plot curves of thrust and torque against rev/min.

**b)** When the speed of advance of the model is 150 m/min and slip 0.20, calculate the efficiency.

**Q. E2.23:** A ship of 8100 tonne displacement floats upright in sea water. $KG = 7.5$ m and $GM = 0.45$ m. A tank, whose centre of gravity is 0.5 m above the keel and 4 m from the centreline, contains 100 tonne of water ballast. Neglecting free surface effect, calculate the angle of heel when the ballast is pumped out.

**Q. E2.24:** The ½ ordinates of a waterplane 120 m long are 0.7, 3.3, 5.5, 7.2, 7.5, 7.5, 7.5, 6.8, 4.6, 2.2 and 0 m respectively. The ship displaces 11000 tonne. Calculate the transverse $BM$.

**Q. E2.25:** A box barge 85 m long, 18 m beam and 6 m draught floats in sea water of 1.025 t/m³. A midship compartment 18 m long contains cargo stowing at 1.8 m³/t and having a density of 1.600 t/m³. There is a watertight flat 6 m above the keel. Calculate the new draught if this compartment is bilged below the flat.

**Q. E2.26:** The ½ ordinates of a waterplane 90 m long are as follows:

| Station | AP | ½ | 1 | 2 | 3 | 4 | 5 | 6 | 7 | 7½ | FP |
|---|---|---|---|---|---|---|---|---|---|---|---|
| ½ ordinate | 0.6 | 2.7 | 4.6 | 6.0 | 6.3 | 6.3 | 6.3 | 5.7 | 4.8 | 2.0 | 0 m |

Calculate the area of the waterplane and the distance of the centre of flotation from midships.

**Q. E2.27:** A ship of 6600 tonne displacement has $KG$ 3.6 m and $KM$ 4.3 m. A mass of 50 tonne is now lifted from the quay by one of the ship's derricks, whose head is 18 m above the keel. The ship heels to a maximum of 9.5° while the mass is being transferred. Calculate the outreach of the derrick from the ship's centreline.

**Q. E2.28:** A ship 120 m long displaces 10500 tonne and has a wetted surface area of 3000 m². At 15 knots the shaft power is 4100 kW, propulsive coefficient 0.6 and 55% of the thrust is available to overcome frictional resistance. Calculate the shaft power required for a similar ship 140 m long at the corresponding speed. $f = 0.42$ and $n = 1.825$.

**Q. E2.29:** A box-shaped vessel 30 m long and 9 m wide floats in water of 1.025 t/m³ at a draught of 0.75 m when empty. The vessel moves from water of 1.000 t/m³ to water of 1.025 t/m³ in a partially laden state and, on reaching the sea water, it is found that the mean draught is reduced by 3.2 cm. Calculate the mass of cargo on board.

**Q. E2.30:** A ship of 5000 tonne displacement has a double bottom tank 12 m long. The ½ breadths of the top of the tank are 5, 4 and 2 m respectively. The tank has a watertight centreline division. Calculate the free surface effect if the tank is partially full of fresh water *on one side only.*

**Q. E2.31:** The following data apply to a ship operating at a speed of 15 knots:

$$Shaft\ power = 3050\ kW$$
$$Propeller\ speed = 1.58\ rev/s$$
$$Propeller\ thrust = 360\ kN$$
$$Apparent\ slip = 0$$

Calculate the propeller pitch, real slip and the propulsive coefficient if the Taylor wake fraction and thrust deduction factor are 0.31 and 0.20 respectively.

**Q. E2.32:** The force acting normal to the plane of a rudder at angle a is given by:

$$F_n = 577AV^2 \sin \alpha\ N$$

$$where\ A = area\ of\ rudder = 22\ m^2$$
$$and\ V = water\ speed\ in\ m/s.$$

When the rudder is turned to 35°, the centre of effort is 1.1 m from the centreline of stock. Allowing 20% for race effect, calculate the diameter of the stock if the maximum ship speed is 15 knots and the maximum allowable stress is 70 MN/m². If the effective diameter is reduced by corrosion and wear to 330 mm, calculate the speed at which the vessel must travel so that the above stress is not exceeded.

**Q. E2.33:** A ship 100 m long and 15 m beam floats at a mean draught of 3.5 m. The semi-ordinates of the waterplane at equal intervals are: 0, 3.0, 5.5, 7.3, 7.5, 7.5, 7.5, 7.05, 6.10, 3.25 and 0 m respectively. The section amidships is constant and parallel for 20 m and the submerged cross-sectional area is 50 m² at this section.

Calculate the new mean draught when a midship compartment 15 m long is opened to the sea. Assume the vessel to be wall-sided in the region of the waterplane.

**Q. E2.34:** A ship 85 m long displaces 8100 tonne when floating in sea water at draughts of 5.25 m forward and 5.55 m aft. TPC 9.0, $GM_L$ 96 m, LCF 2 m aft of midships. It is decided to introduce water ballast to completely submerge the propeller and a draught aft of 5.85 m is required. A ballast tank 33 m aft of midships is available. Find the least amount of water required and the final draught forward.

**Q. E2.35:** A solid block of wood has a square cross section of side $S$ and length $L$ greater than $S$. Calculate the relative density of the wood if it floats with its sides vertical in fresh water.

**Q. E2.36:** A ship travelling at 15.5 knots has a propeller of 5.5 m pitch turning at 95 rev/min. The thrust of the propeller is 380 kN and the delivered power 3540 kW. If the real slip is 20% and the thrust deduction factor 0.198, calculate the QPC and the wake fraction.

**Q. E2.37:**

a) Describe briefly the inclining experiment and explain how the results are used.

b) A ship of 8500 tonne displacement has a double bottom tank 11 m wide extending for the full breadth of the ship, having a free surface of sea water. If the apparent loss in metacentric height due to slack water is 14 cm, find the length of the tank.

**Q. E2.38:**

a) Derive an expression for the change in draught of a vessel moving from sea water into river water.

b) A ship of 8000 tonne displacement has TPC 17 when at a level keel draught of 7 m in sea water of 1.024 t/m³. The vessel then moves into water of 1.008 t/m³. The maximum draught at which the vessel may enter dock is 6.85 m. Calculate how much ballast must be discharged.

**Q. E2.39:**

a) What is meant by the Admiralty Coefficient and the Fuel Coefficient?

b) A ship of 14900 tonne displacement has a shaft power of 4460 kW at 14.55 knots. The shaft power is reduced to 4120 kW and the fuel consumption at the same displacement is 541 kg/h. Calculate the fuel coefficient for the ship.

**Q. E2.40:** A ship of 12000 tonne displacement has a rudder 15 m² in area, whose centre is 5 m below the waterline. The metacentric height of the ship is 0.3 m and the centre of buoyancy is 3.3 m below the waterline. When travelling at 20 knots, the rudder is turned through 30°. Find the initial angle of heel if the force $F_n$, perpendicular to the plane of the rudder, is given by:

$$F_n = 577AV^2 \sin\alpha \text{ N}$$

Allow 20% for the race effect.

**Q. E2.41:** A ship 120 m long has a light displacement of 4000 tonne and LCG in this condition 2.5 m aft of midships.

The following items are then added:

| Cargo  | 10000 | tonne LCG | 3 m  | forward of | midships |
|--------|-------|-----------|------|------------|----------|
| Fuel   | 1500  | tonne LCG | 2 m  | aft of     | midships |
| Water  | 400   | tonne LCG | 8 m  | aft of     | midships |
| Stores | 100   | tonne LCG | 10 m | forward of | midships |

Using the following hydrostatic data, calculate the final draughts:

| Draught (m) | Displacement (t) | MCT 1 cm (t m) | LCB from midships (m) | LCF from midships (m) |
|-------------|------------------|----------------|-----------------------|-----------------------|
| 8.5 | 16650 | 183 | 1.94$_F$ | 1.20$_A$ |
| 8   | 15350 | 175 | 2.10$_F$ | 0.06$_F$ |

**Q. E2.42:** A box barge 30 m long and 9 m beam floats at a draught of 3 m. The centre of gravity lies on the centreline and $KG$ is 3.5 m. A mass of 10 tonne, which is already on board, is now moved 6 m across the ship.

**a)** Estimate the angle to which the vessel will heel, using the formula:

$$GZ = \sin\theta\left(GM + \frac{1}{2} BM \tan^2\theta\right)$$

**b)** Compare the above result with the angle of heel obtained by the metacentric formula.

**Q. E2.43:** The fuel consumption of a ship at 17 knots is 47 tonne/day. The speed is reduced and the consumption is reduced to 22 tonne/day. At the lower speed, however, the consumption per unit power is 13.2% greater than at 17 knots. Find the reduced speed and the percentage saving on a voyage of 3000 nautical miles.

**Q. E2.44:** A ship of 14000 tonne displacement is 135 m long and floats at draughts of 7.3 m forward and 8.05 m aft. $GM_L$ is 127 m, TPC 18 and LCF 3 m aft of midships. Calculate the new draughts when 180 tonne of cargo are added 40 m forward of midships.

**Q. E2.45:** A propeller has a pitch of 5.5 m. When turning at 93 rev/min, the apparent slip is found to be −S% and the real slip +S%, the wake speed being 10% of the ship's speed. Calculate the speed of the ship, the apparent slip and the real slip.

**Q. E2.46:** The ½ ordinates of a waterplane at 15 m intervals, commencing from aft, are 1, 7, 10.5, 11, 11, 10.5, 8, 4 and 0 m. Calculate:

**a)** TPC

**b)** distance of the centre of flotation from midships

c)   second moment of area of the waterplane about a transverse axis through the centre of flotation.

*Note*: Second moment of area about any axis y–y, which is parallel to an axis N–A through the centroid and distance x from it, is given by:

$$I_{yy} = I_{NA} + Ax^2$$

**Q. E2.47:**  A ship travelling at 12 knots has a metacentric height of 0.25 m. The distance between the centre of gravity and the centre of lateral resistance is 2.7 m. If the vessel turns in a circle of 600 m radius, calculate the angle to which it will heel.

**Q. E2.48:**  The following data are available for a twin-screw vessel:

| V | (knots) | 15 | 16 | 17 | 18 |
|---|---------|------|------|------|------|
| $ep_n$ | (kW) | 3000 | 3750 | 4700 | 5650 |
| QPC | | 0.73 | 0.73 | 0.72 | 0.71 |

Calculate the service speed if the brake power for each engine is 3500 kW. The transmission losses are 3% and the allowances for weather and appendages 30%.

**Q. E2.49:**  A ship 120 m long displaces 8000 tonne, $GM_L$ is 102 m, TPC 17.5 and LCF 2 m aft of midships. It arrives in port with draughts of 6.3 m forward and 6.6 m aft. During the voyage, the following changes in loading have taken place:

| Fuel used | 200 tonne | 18 m | forward | of midships |
|-----------|-----------|------|---------|-------------|
| Water used | 100 tonne | 3 m | aft | of midships |
| Stores used | 10 tonne | 9 m | aft | of midships |
| Ballast added | 300 tonne | 24 m | forward | of midships |

Calculate the *original* draughts.

**Q. E2.50:**  A propeller has a pitch of 5.5 m. When turning at 80 rev/min, the ship speed is 13.2 knots, speed of advance 11 knots, propeller efficiency 70% and delivered power 3000 kW. Calculate:

a)   real slip
b)   wake fraction
c)   propeller thrust.

**Q. E2.51:**  A watertight bulkhead 6 m deep is supported by vertical inverted angle stiffeners 255 mm × 100 mm × 12.5 mm, spaced 0.6 m apart. The ends of the stiffeners in contact with the tank top are welded all round, and the thickness of weld at its throat is 5 mm.

Calculate the shear stress in the weld metal at the tank top when the bulkhead is covered on one side, by water of density 1025 kg/m³, to a depth of 4.85 m.

**Q. E2.52:** A ship of 5000 tonne displacement has three rectangular double bottom tanks: A 12 m long and 16 m wide; B 14 m long and 15 m wide; C 14 m long and 16 m wide. Calculate the free surface effect for any one tank and state in which order the tanks should be filled when making use of them for stability correction.

**Q. E2.53:** A box barge 75 m long and 8.5 m beam floats at draughts of 2.13 m forward and 3.05 m aft. An empty compartment is now flooded and the vessel finally lies at a draught of 3 m level keel. Calculate the length and LCG of the flooded compartment.

**Q. E2.54:** A vessel has a maximum allowable draught of 8.5 m in fresh water and 8.25 m in sea water of 1.026 t/m³, the TPC in the sea water being 27.5. The vessel is loaded in river water of 1.012 t/m³ to a draught of 8.44 m. If it now moves into sea water, is it necessary to pump out any ballast, and if so, how much?

**Q. E2.55:** A bulkhead is in the form of a trapezoid 13 m wide at the deck, 10 m wide at the tank top and 7.5 m deep.

Calculate the load on the bulkhead and the position of the centre of pressure if it is flooded to a depth of 5 m with sea water on one side only.

**Q. E2.56:** A double bottom tank is 23 m long. The half breadths of the top of the tank are 5.5, 4.6, 4.3, 3.7 and 3.0 m respectively. When the ship displaces 5350 tonne, the loss in metacentric height due to free surface is 0.2 m. Calculate the density of the liquid in the tank.

**Q. E2.57:** A vessel of constant rectangular cross section is 100 m long and floats at a draught of 5 m. It has a mid-length compartment 10 m long extending right across the vessel, but subdivided by a horizontal watertight flat 3 m above the keel. GM is 0.8 m.

Calculate the new draught and metacentric height if the compartment is bilged below the flat.

**Q. E2.58:**

a) If resistance is proportional to $SV^2$ and S is proportional to $\Delta^{\frac{2}{3}}$, derive the Admiralty Coefficient formula.

b) A ship 160 m long, 22 m beam and 9.2 m draught has a block coefficient of 0.765. The pitch of the propeller is 4 m and when it turns at 96 rev/min the true slip is 33%, the wake fraction 0.335 and shaft power 2900 kW. Calculate the Admiralty Coefficient and the shaft power at 15 knots.

**Q. E2.59:** A ship model 6 m long has a total resistance of 40 N when towed at 3.6 knots in fresh water. The ship itself is 180 m long and displaces 20400 tonne. The wetted surface area may be calculated from the formula:

$$S = 2.57 \sqrt{(\Delta L)}$$

Calculate $ep_n$, for the ship at its corresponding speed in sea water. $f$ $(\text{model})_{Fw} = 0.492; f\,(\text{ship})_{SW} = 0.421; n = 1.825$.

**Q. E2.60:**

a) Why is an inclining experiment carried out? Write a short account of the method adopted.

b) An inclining experiment was carried out on a ship of 8000 tonne displacement. The inclining ballast was moved transversely through 12 m and the deflections of a pendulum 5.5 m long, measured from the centreline, were as follows:

| | |
|---|---|
| 3 tonne port to starboard | 64 mm S |
| 3 tonne port to starboard | 116 mm S |
| Ballast restored | 3 mm S |
| 3 tonne starboard to port | 54 mm P |
| 3 tonne starboard to port | 113 mm P |

Calculate the metacentric height of the vessel.

**Q. E2.61:** A ship of 8000 tonne displacement, 110 m long, floats in sea water of 1.024 t/m³ at draughts of 6 m forward and 6.3 m aft. The TPC is 16, LCB 0.6 m aft of midships, LCF 3 m aft of midships and MCT 1 cm 65 tonne m. The vessel now moves into fresh water of 1.000 t/m³. Calculate the distance a mass of 50 tonne must be moved to bring the vessel to an even keel and determine the final draught.

**Q. E2.62:** A rectangular watertight bulkhead 9 m high and 14.5 m wide has sea water on both sides, the height of water on one side being four times that on the other side. The resultant centre of pressure is 7 m from the top of the bulkhead. Calculate:

a) the depths of water

b) the resultant load on the bulkhead.

**Q. E2.63:** The following values refer to a vessel 143 m in length, which is to have a service speed of 14 knots:

| Service speed (knots) | 13.0 | 14.1 | 15.2 | 16.3 |
|---|---|---|---|---|
| Effective power naked ep$_n$ (kW) | 1690 | 2060 | 2670 | 3400 |

If allowances for the above ep$_n$ for trial and service conditions are 13% and 33% respectively, and the ratio of service indicated power to maximum available indicated power is 0.9, calculate using the data below:
a) the indicated power (ip) of the engine to be fitted
b) the service and trial speed of the vessel if the total available ep were used.

The vessel has the following data:

Quasi-propulsive coefficient (QPC) = 0.72

Shaft losses = 3.5%

Mechanical efficiency of the engine to be fitted = 87%.

**Q. E2.64:** A ship of 15000 tonne displacement has righting levers of 0, 0.38, 1.0, 1.41 and 1.2 m at angles of heel of 0°, 15°, 30°, 45° and 60° respectively and an assumed KG of 7 m. The vessel is loaded to this displacement but the KG is found to be 6.8 m and GM 1.5 m.
a) Draw the amended stability curve.
b) Estimate the dynamical stability at 60°.

**Q. E2.65:** On increasing the speed of a vessel by 1.5 knots, it is found that the daily consumption of fuel is increased by 25 tonne and the percentage increase in fuel consumption for a voyage of 2250 nautical miles is 20. Estimate:
a) the original daily fuel consumption
b) the original speed of the ship.

**Q. E2.66:** The end bulkhead of the wing tank of an oil tanker has the following widths at 3 m intervals, commencing at the deck: 6.0, 6.0, 5.3, 3.6 and 0.6 m. Calculate the load on the bulkhead and the position of the centre of pressure if the tank is full of oil rd 0.8.

**Q. E2.67:** The following data for a ship has been produced from propulsion experiments on a model:

| Ship speed (knot) | 12.50 | 13.25 | 14.00 |
|---|---|---|---|
| Effective power (kW) | 1440 | 1800 | 2230 |
| QPC | 0.705 | 0.713 | 0.708 |
| Propeller efficiency | 0.565 | 0.584 | 0.585 |
| Taylor wake fraction | 0.391 | 0.362 | 0.356 |

Determine the speed of the ship and propeller thrust when the delivered power is 2385 kW.

**Q. E2.68:** A box barge 45 m long and 15 m wide floats at a level keel draught of 2 m in sea water, the load being uniformly distributed over the full length. Two masses, each of 30 tonne, are added at 10 m from each end and 50 tonne is evenly distributed between them. Sketch the shear force diagram and give the maximum shear force.

**Q. E2.69:** The power delivered to a propeller is 3540 kW at a ship speed of 15.5 knots. The propeller rotates at 1.58 rev/s, develops a thrust of 378 kN and has a pitch of 4.87 m.
If the thrust deduction fraction is 0.24, real slip 30% and transmission losses are 3%, calculate:
a) the effective power
b) the Taylor wake fraction
c) the propulsive coefficient
d) the quasi-propulsive coefficient, assuming the appendage and weather allowance is 15%.

**Q. E2.70:** A ship of 14000 tonne displacement is 125 m long and floats at draughts of 7.9 m forward and 8.5 m aft. The TPC is 19, $GM_L$ 120 m and LCF 3 m forward of midships. It is required to bring the vessel to an even keel draught of 8.5 m. Calculate the mass that should be added and the distance of the centre of the mass from midships.

**Q. E2.71:** A ship of 4000 tonne displacement has a mass of 50 tonne on board, on the centreline of the tank top. A derrick, whose head is 18 m above the CG of the mass, is used to lift it. Find the shift in the ship's centre of gravity from its original position when the mass is:
a) lifted just clear of the tank top
b) raised to the derrick head
c) placed on the deck 12 m above the tank top
d) swung outboard 14 m.

**Q. E2.72:** A rectangular bulkhead 8 m wide has water of density 1000 kg/m³ to a depth of 7 m on one side and on the other side oil of density 850 kg/m³ to a depth of 4 m. Calculate:
a) the resultant pressure on the bulkhead
b) the position of the resultant centre of pressure.

**Q. E2.73:** A ship of 91.5 m length between perpendiculars contains ballast water in a forward compartment and has the following equidistant half areas of immersed sections commencing at the after perpendicular (AP).

| Section | O(AP) | 1 | 2 | 3 | 4 | 5 | 6 | 7 | 8 | 9 | 10(FP) |
|---|---|---|---|---|---|---|---|---|---|---|---|
| Half-area of immersed sections (m²) | 0.4 | | 7.6 | 21.4 | 33.5 | 40.8 | 45.5 | 48.4 | 52.0 | 51.1 | 34.4 0 |

If, prior to ballasting, the ship's displacement was 5750 tonne and the position of the longitudinal centre of buoyancy (LCB) was 4.6 m forward of midships, calculate:
a) the mass of water of density 1025 kg/m³ added as ballast
b) the distance of the centre of gravity of the ballast water contained in the forward compartment from midships.

**Q. E2.74:** A ship of 7500 tonne displacement has a double bottom tank 14 m long, 12 m wide and 1.2 m deep full of sea water. The centre of gravity is 6.7 m above the keel and the metacentric height is 0.45 m.
Calculate the new *GM* if half of the water is pumped out of the tank. Assume that *KM* remains constant.

**Q. E2.75:** A ship 120 m long displaces 9100 tonne. It loads in fresh water of 1.000 t/m³ to a level keel draught of 6.7 m. It then moves into sea water of 1.024 t/m³. TPC in sea water 16.8, MCT 1 cm 122 tonne m, LCF 0.6 aft of midships, LCB 2.25 m forward of midships. Calculate the end draughts in the sea water.

**Q. E2.76:** A box-shaped vessel is 20 m long and 10 m wide. The weight of the vessel is uniformly distributed throughout the length and the draught is 2.5 m. The vessel contains ten evenly spaced double bottom tanks, each having a depth of 1 m.
Draw the shear force diagrams:
a) with No. 1 and No. 10 tanks filled
b) with No. 3 and No. 8 tanks filled
c) with No. 5 and No. 6 tanks filled.

Which ballast condition is to be preferred from the strength point of view?

**Q. E2.77:** For a box-shaped barge of 216 tonne displacement, 32 m in length, 5.5 m breadth and floating in water of density 1025 kg/m³, the *KG* is 1.8 m. An item of machinery of mass 81 tonne is loaded amidships and, to maintain a positive metacentric height, 54 tonne of solid ballast is taken aboard and evenly distributed over the bottom of the barge so that the average *Kg* of the ballast is 0.15 m. If in the final condition the *GM* is 0.13 m, calculate the *Kg* of the machinery.

**Q. E2.78:** The maximum allowable draught of a ship in fresh water of 1.000 t/m³ is 9.5 m and in sea water of 1.025 t/m³ is 9.27 m. The vessel is loaded to a draught of 9.5 m in a river, but when it proceeds to sea it is found that 202 tonne of water ballast must be pumped out to prevent the maximum draught being exceeded. If the TPC in the sea water is 23, calculate the density of the river water.

**Q. E2.79:** The following data are recorded from tests carried out on a model propeller 0.3 m diameter rotating at 8 rev/s in water of density 1000 kg/m³.

| Speed of advance $V_a$ (m/s) | 1.22 | 1.46 | 1.70 | 1.94 |
|---|---|---|---|---|
| Thrust (N) | | 93.7 | 72.3 | 49.7 | 24.3 |
| Torque (Nm) | | 3.90 | 3.23 | 2.50 | 1.61 |

Draw graphs of thrust and delivered power against speed of advance $V_a$.

A geometrically similar propeller 4.8 m diameter operates in water of 1025 kg/m³. If the propeller absorbs 3000 kW delivered power and satisfies the law of comparison, determine for the propeller:

**a)** the thrust power

**b)** the efficiency.

*Note:* For geometrically similar propellers, the thrust power and delivered power vary directly as (diameter)$^{3.5}$.

**Q. E2.80:** A ship 128 m in length and 16.75 m in breadth has the following hydrostatic data:

| Draught (m) | 1.22 | 2.44 | 3.66 | 4.88 | 6.10 |
|---|---|---|---|---|---|
| Waterplane area coefficient | 0.78 | 0.82 | 0.85 | 0.88 | 0.90 |
| Position of longitudinal centre of flotation (LCF) from midships (m) | 1.30 for'd | 1.21 for'd | 0.93 for'd | 0.50 for'd | 0.06 aft |

Calculate:

a) the displacement in water of density 1025 kg/m³ of a layer of ship body between the waterplanes at 1.22 m and 6.1 m draught

b) for the layer:

   i) the position of the longitudinal centre of buoyancy

   ii) the position of the vertical centre of buoyancy.

**Q. E2.81:** The following particulars apply to a ship of 140 m length when floating in sea water of 1025 kg/m³ at a level keel draught of 7 m.

| | |
|---|---|
| Displacement ($\Delta$) | 14000 tonne |
| Centre of gravity above keel (KG) | 8.54 m |
| Centre of gravity from midships (LCG) | 0.88 m aft |
| Centre of buoyancy above keel (KB) | 4.27 m |
| Waterplane area (A$_w$) | 2110 m² |
| Second moment of area of waterplane about transverse axis at midships (I) | 2326048 m⁴ |
| Centre of flotation from midships (LCF) | 4.6 m aft |

Fuel oil is transferred from a forward storage tank through a distance of 112 m to an aft settling tank, after which the draught aft is found to be 7.45 m. Using these data, calculate:

a) the moment to change trim 1 cm (MCT 1 cm)

b) the new draught forward

c) the new longitudinal position of the centre of gravity

d) the amount of fuel transferred.

**Q. E2.82:** The wetted surface area of a container ship is 5946 m². When travelling at its service speed, the effective power required is 11250 kW with frictional resistance 74% of the total resistance and specific fuel consumption of 0.22 kg/kWh.

To conserve fuel, the ship speed is reduced by 10%. The daily fuel consumption is then found to be 83 tonne.

Frictional coefficient in sea water is 1.432.

Speed in m/s with index *(n)* 1.825.

Propulsive coefficient may be assumed constant at 0.6.

Determine:
a) the service speed of the ship
b) the percentage increase in specific fuel consumption when running at reduced speed.

**Q. E2.83:** An empty box-shaped vessel of length 60 m and breadth 10.5 m displaces 300 tonne and has a *KG* of 2.6 m. When floating in sea water of density 1025 kg/m³, the following loads are added as indicated:

| Load | Mass (tonne) | *Kg* (m) |
|------|--------------|----------|
| Lower hold cargo | 1000 | 4.7 |
| 'Tween deck cargo | 500 | 6.1 |
| Deep tank cargo | 200 | 3.4 |

a) Determine for the new condition:
   i) the initial metacentric height
   ii) the angle to which the vessel will loll
b) A metacentric height of 0.15 m is required. Calculate the amount of cargo to be transferred from 'tween deck to deep tank at the *Kgs* stated above.

*Note:* For wall-sided vessels $GZ = \sin \theta \, (GM + \frac{1}{2}BM \tan^2\theta)$ m

**Q. E2.84:** A box-shaped barge of uniform construction is 80 m long, 12 m beam and has a light displacement of 888 tonne.
The barge is loaded to a draught of 7 m in sea water of density 1025 kg/m³ with cargo evenly distributed over two end compartments of equal length. The empty midship compartment extending to the full width and depth of the barge is bilged and the draught increases to 10 m. Determine:
a) the length of the midship compartment
b) the longitudinal still water bending moment at midships:
   i) in the loaded intact condition

ii) in the new bilged condition.

**Q. E2.85:** A ship of 355190 tonne displacement is 325 m long, 56 m wide and floats in sea water of density 1025 kg/m³ at a draught of 22.4 m. The propeller has a diameter of 7.4 m, a pitch ratio of 0.85, and when rotating at 1.5 rev/s, the real slip is 48.88% and the fuel consumption is 165 tonne per day. The Taylor wake fraction $W_t$ is given by:

$$W_t = 0.5\, C_b - 0.05$$

Calculate:

a) the ship speed in knots

b) the reduced speed at which the ship should travel if the fuel consumption on a voyage is to be halved

c) the length of the voyage if the extra time on passage is six days when travelling at the reduced speed

d) the amount of fuel required on board, before commencing on the voyage at the reduced speed.

**Q. E2.86:** A ship 137 m long displaces 13716 tonne when floating at a draught of 8.23 m in sea water of density 1025 kg/m³. The shaft power *(sp)* required to maintain a speed of 15 knots is 4847 kW, and the propulsive coefficient is 0.67. Given:

$$\text{Wetted surface area } S = 2.58\, \sqrt{(\Delta \times L)} \text{ m}^2$$

speed in m/s with index *(n)* for both ships 1.825 values of Froude friction coefficient are:

| Length of ship (m) | 130 | 140 | 150 | 160 |
|---|---|---|---|---|
| Coefficient (f) | 1.417 | 1.416 | 1.415 | 1.414 |

Calculate the shaft power for a geometrically similar ship that has a displacement of 18288 tonne and which has the same propulsive coefficient as the smaller ship and is run at the corresponding speed.

# SOLUTIONS TO SECTION 2 EXAMINATION QUESTIONS

**A. E2.1:**

$$\text{Bodily sinkage} = 240 \div 20$$
$$= 12\,\text{cm}$$
$$\text{New draught aft} = 6.9 + 0.12$$
$$= 7.02\,\text{m}$$

ie the after draught may be increased by a further 0.18 m and this becomes the change in trim aft.

$$\text{Change in trim aft} = \text{total change in trim} \times 56.5 \div 120$$
$$18 = t \times 56.5 \div 120$$
$$t = 18 \times 120 \div 56.5$$
$$t = 38.23\,\text{cm}$$
$$\text{But } t = m \times d \div \text{MCT1cm}$$
$$d = 38.23 \times 101 \div 240$$
$$= 16.09\,\text{m} \text{ aft of the centre of flotation}$$
$$= 19.59\,\text{m} \text{ aft of midships}$$

**A. E2.2:**

| Draught | TPC | SM | Product for displacement | Lever* | Product for vertical moment |
|---------|-----|----|--------------------------|--------|------------------------------|
| 1.2 | 23.0 | 1 | 23.0 | 1 | 23.0 |
| 2.4 | 24.2 | 4 | 96.8 | 2 | 193.6 |
| 3.6 | 25.0 | 1 | 25.0 | 3 | 75.0 |
|  |  |  | 144.8 |  | 291.6 |

*Using a lever of 1 at 1.2 m draught produces a vertical moment about the *keel*.

$$\text{Displacement 1.2 m to 3.6 m} = 1.2 \div 3 \times 144.8 \times 100$$
$$= 5792 \text{ tonne}$$
$$\text{Displacement 0 m to 3.6 m} = 8172 \text{ tonne}$$
$$\text{Therefore: displacement 0 m to 1.2 m} = 2380 \text{ tonne}$$
$$\text{Vertical moment 1.2 m to 3.6 m} = 1.2 \div 3 \times 1.2 \times 291.6 \times 100$$
$$= 14000 \text{ tonne m}$$
$$\text{Vertical moment 0 m to 3.6 m} = 8172 \times 1.91$$
$$= 15609 \text{ tonne m}$$
$$\text{Vertical moment 0 m to 1.2 m} = 1610 \text{ tonne m}$$
$$KB \text{ at 1.2 m draught} = 2380$$
$$= 0.676 \text{ m}$$

| Draught | TPC | LCF* | TPC × LCF | SM | Product for longitudinal moment |
|---------|-----|------|-----------|----|---------------------------------|
| 1.2 | 23.0 | −1.37 | −31.51 | 1 | −31.51 |
| 2.4 | 24.2 | +0.76 | +18.39 | 4 | +73.56 |
| 3.6 | 25.0 | +0.92 | +23.0 | 1 | +23.0 |
| | | | | | +65.07 |

Taking forward as negative and aft as positive:

$$\text{Longitudinal moment 1.2 m to 3.6 m} = 1.2 \div 3 \times 65.07 \times 100$$
$$= +2602.8 \text{ tonne m}$$
$$\text{Longitudinal moment 0 m to 3.6 m} = 8172 \times 0.15$$
$$= +1225.8 \text{ tonne m}$$
$$\text{Longitudinal moment 0 m to 1.2 m} = -1376.6 \text{ tonne m}$$
$$LCB \text{ at 1.2 m draught} = -1376.6 \div 2380$$
$$= 0.578 \text{ m forward of midships}$$

ie at 1.2 m draught, the displacement is 2380 tonne, $KB$ is 0.676 m and the LCB is 0.578 m forward of midships.

**A. E2.3:**

$$\text{Block coefficient } C_b = 17000 \div (1.025 \times 142 \times 18.8 \times 8)$$
$$= 0.777$$
$$W_t = 0.5 \times 0.777 - 0.05$$
$$= 0.338$$
$$\text{Pitch} = 4.8 \times 0.673$$
$$= 3.23\,\text{m}$$
$$\text{Theoretical speed} = (3.23 \times 100 \times 60) \div 1852$$
$$= 10.46\,\text{knots}$$
$$\text{Speed of advance} = 10.46 \times 0.65$$
$$= 6.8\,\text{knots}$$
$$\text{Ship speed} = 6.8 \div (1 - 0.338)$$
$$= 10.27\,\text{knots}$$
$$\text{New fuel consumption} = 20 \times (10.27 \div 12)^3$$
$$= 12.54\,\text{tonne per day}$$

**A. E2.4:**

$$\text{Effective } GM = (5 \times 15 \times 6) \div (5000 \times 0.12)$$
$$= 0.75\,\text{m}$$
$$\text{Free surface effect} = (1.025 \div 1.025) \times (7.5 \times 9^3 \times 1.025) \div (12 \times 5000)$$
$$= 0.093\,\text{m}$$
$$\text{Hence, actual } GM = 0.75 + 0.093$$
$$= 0.843\,\text{m}$$
$$KM = 6.4\,\text{m}$$
$$KG = 5.557\,\text{m}$$
$$\text{Mass of water in tank} = 1.025 \times 7.5 \times 9 \times 0.6$$
$$= 41.51\,\text{tonne}$$

Taking moments about the keel:

$$\text{Light ship KG} = (5000 \times 5.557 - 41.51 \times 0.3) \div (5000 - 41.51)$$
$$= 5.601\,\text{m}$$

**A. E2.5:**

$$w = (V - V_a) \div V$$
$$\text{Speed of advance } V_a = 16(1 - 0.28)$$
$$= 11.52 \text{ knots}$$
$$\text{Real slip} = (V_t - V_a) \div V_t$$
$$\text{Theoretical speed } V_t = 11.52 \div (1 - 0.30)$$
$$= 16.46 \text{ knots}$$
$$\text{But } V_t = P \times N \times 60 \div 1852$$
$$\text{Pitch } P = 16.46 \times 1852 \div (120 \times 60)$$
$$= 4.23 \text{ m}$$
$$\text{Pitch ratio } p = P \div D$$

a)     $\text{Diameter } D = 4.23 \div 0.95$
$$= 4.45 \text{ m}$$

b)     $\text{Shaft power sp} = 2\pi \times (120 \div 60) \times 270$
$$= 3393 \text{ kW}$$

(*Note*: This is the power at the after end of the shaft and hence is strictly the delivered power.)

c)     $\text{Thrust power tp} = 400 \times 11.52 \times 0.5144$
$$= 2370 \text{ kW}$$
$$\text{Propeller efficiency} = 2370 \div 3393 \times 100$$
$$= 69.85\%$$

d) ·     $ep = 3393 \times 0.67$
$$= 2273.3$$
$$\text{But } ep = R_t \times V$$
$$R_t = 2273.3 \div (16 \times 0.5144)$$
$$R_t = 276.2 \text{ kN}$$
$$R_t = T(1 - t)$$
$$(1 - t) = 276.2 \div 400$$
$$= 0.69$$
$$\text{Thrust deduction factor } t = 0.31$$

**A. E2.6:**

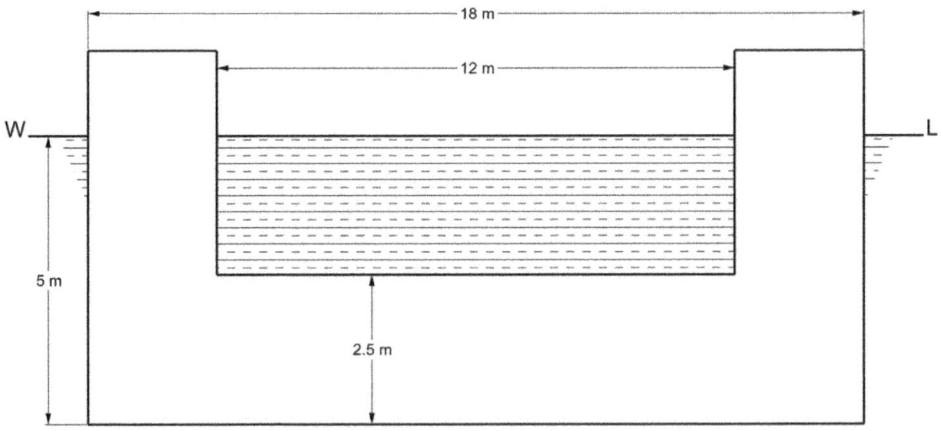

▲ Figure A. E2.1

$$KB = (12 \times 2.5 \times 1.25 + 2 \times 3 \times 5 \times 2.5) \div (12 \times 2.5 + 2 \times 3 \times 5)$$
$$= 1.875\,\text{m}$$
$$I = (1 \div 12)L(18^3 - 12^3)$$
$$= (1 \div 12) \times 4104L$$
$$\nabla = L(12 \times 2.5 + 2 \times 3 \times 5)$$
$$= 60L$$
$$BM = (4014L) \div (12 \times 60L)$$
$$= 5.7\,\text{m}$$
$$KM = 1.875 + 5.7$$
$$= 7.575\,\text{m}$$
$$GM = 2.5\,\text{m}$$
$$KG = 5.075\,\text{m}$$

**A. E2.7:**

a) For derivation, see Chapter 7, section 7.3, but the following formula should be derived:

$$sp = (\Delta^{\frac{2}{3}} V^4) \div C$$

The speed index is 4, as this is the modification for fast craft. From this:

$$sp_1 \div sp_2 = (\Delta_1^{\frac{2}{3}} \div \Delta_2^{\frac{2}{3}})(V_1^4 \div V_2^4)$$

b) $$sp_2 = 23000 \times (12000 \div 14000)^{\frac{2}{3}} \times (21 \div 24)^4$$
$$= 12166\,\text{kW}$$

**A. E2.8:**

| CSA | SM | Product of CSA | level | Product for Moment |
|-----|----|----|-------|-----|
| 0 | 1 | — | +5 | — |
| 11.5 | 4 | 46.0 | +4 | +184.0 |
| 27.0 | 2 | 54.0 | +3 | +162.0 |
| 38.5 | 4 | 154.0 | +2 | +308.0 |
| 44.0 | 2 | 88.0 | +1 | +88.0 |
| 45.0 | 4 | 180.0 | 0 | +742.0 |
| 44.5 | 2 | 89.0 | −1 | −89 |
| 39.0 | 4 | 156.0 | −2 | −312 |
| 26.5 | 2 | 53.0 | −3 | −159 |
| 14.5 | 4 | 58.0 | −4 | −232 |
| 0 | 1 | — | −5 | — |
| | | 878.0 | | −729.0 |

$h = 8$ m

a)     Displacement $= (8 \div 3) \times 878 \times 1.025$
$= 2400$ tonne

b)     LCB from midships $= 8 \times (742 - 792) \div 878$
$= -0.456$ m
$= 0.456$ m forward

c)     Prismatic coefficient $= (8 \times 878) \div (3 \times 80 \times 45)$
$= 0.650$

**A. E2.9:**

Ship A:

$$ep_n = R_t \times V$$
$$R_t = 6400 \div (18 \times 0.5144)$$
$$R_t = 691.1 \text{kN}$$
$$R_f = f \, SV^n$$
$$R_f = 0.42 \times 4000 \times 18^{1.825}$$
$$R_f = 328.2 \text{ kN}$$
$$R_r = 691.1 - 328.2$$
$$R_r = 362.9 \text{ kN}$$

Ship B:

$R_r$ is proportional to $L^3$

$$R_r = 362.9 \times (140 \div 160)^3$$
$$= 243.1 \text{ kN}$$

$S$ is proportional to $L^2$

$$S = 4000 \times (140 \div 160)^2$$
$$S = 3062 \text{ m}^2$$

$V$ is proportional to $\sqrt{L}$

$$V = 18 \times \sqrt{(140 \div 160)}$$
$$V = 16.84 \text{ knots}$$
$$R_f = 0.425 \times 3025 \times 16.84^{1.825}$$
$$R_f = 225.1 \text{ kN}$$
$$R_t = 225.1 + 243.1$$
$$R_t = 468.2 \text{ kN}$$
$$ep_n = 468.2 \times 16.84 \times 0.5144$$
$$= 4055 \text{ kN}$$

## A. E2.10:

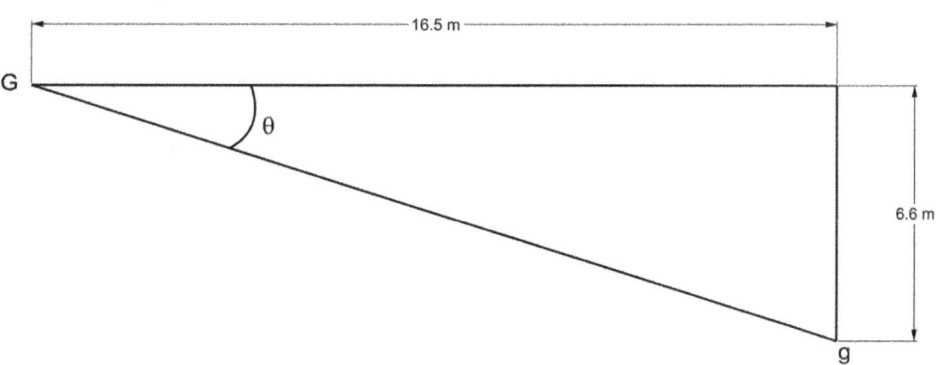

▲ **Figure A. E2.2**

$$Gg = \sqrt{(16.5^2 + 6.6^2)}$$
$$= \sqrt{315.81}$$
$$= 17.77 \text{ m}$$

This is the distance from the centre of gravity of the tank to the original centre of gravity of the ship.

**a)** Let $m$ = mass of oil used

Then shift in centre of gravity = $m \times Gg \div$ final displacement

$$0.38 = (m \times 17.77) \div (11200 - m)$$
$$m = (11200 \times 0.38) \div (17.77 + 0.38)$$
$$= 234.5 \text{ tonne}$$

**b)** $\tan \theta = 6.6 \div 16.5$
$$= 0.4$$

Angle of shift $\theta = 21.8°$

**A. E2.11:**

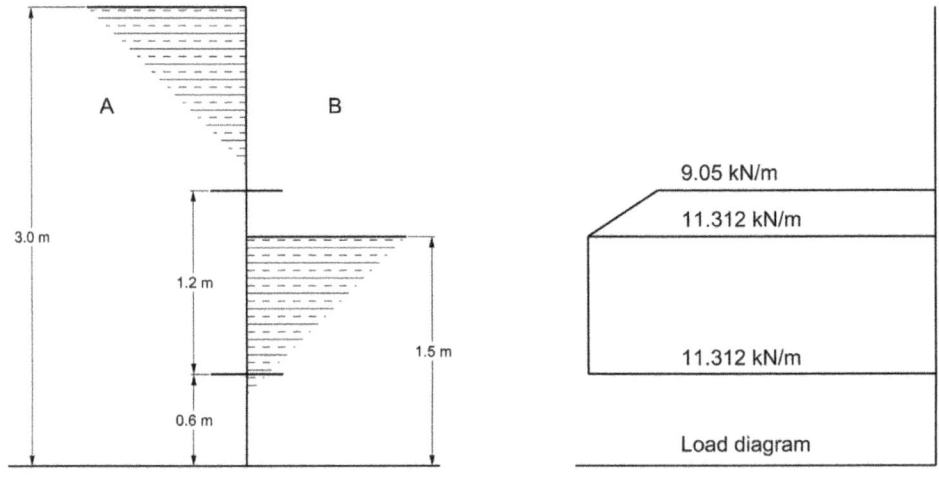

▲ Figure A. E2.3

Load/m at top of door from side A = $1.025 \times 9.81 \times 1.2 \times 0.75$
$$= 9.05 \text{ kN}$$

Load/m 0.3m from top of door side A = $1.025 \times 9.81 \times 1.5 \times 0.75$
$$= 11.312 \text{ kN}$$

Load/m at bottom of door from side A = $1.025 \times 9.81 \times 2.4 \times 0.75$
$$= 18.099 \text{ kN}$$

Load/m at bottom of door from side B = $1.025 \times 9.81 \times 0.9 \times 0.75$
$$= 6.787 \text{ kN}$$

Therefore : net load/m at bottom of door = $18.099 - 6.787$
$$= 11.312 \text{ kN}$$

Thus the load diagram is in the form shown above. The area of this diagram represents the load, while the centroid represents the position of the centre of pressure.

Taking moments about the top of the door:

Centre of pressure from top

$$
\begin{aligned}
&= (11.312 \times 12 \times 0.6 - (11.312 - 9.05) \times 0.3 \times \tfrac{1}{2} \times 0.10) \div \\
&\quad (11.312 \times 1.2 - (11.312 - 9.05) \times 0.3 \times \tfrac{1}{2}) \\
&= (8.145 - 0.0339) \div (13.574 - 0.339) \\
&= 8.111 \div 13.235 \\
&= 0.613\,\text{m}
\end{aligned}
$$

$$\text{Resultant load} = 13.235\,\text{kN}$$

**A. E2.12:**

a)  Mass of water in tank $= 18 \times 15 \times 1.5 \times 1.025$

$$
\begin{aligned}
&= 415.1\,\text{tonne} \\
\text{New KG} &= (10000 \times 7.4 + 415.1 \times 0.75) \div (10000 + 415.1) \\
&= (74000 + 311) \div 10415.1 \\
&= 7.135\,\text{m} \\
\text{New KG} &= 8.00 - 7.135 \\
&= 0.865\,\text{m}
\end{aligned}
$$

b)  Mass of water in tank $= 415.1 \div 2$

$$
\begin{aligned}
&= 207.55\,\text{tonne} \\
\text{New } KG &= (74000 + 207.55 \times 0.375) \div (10000 + 207.55) \\
&= 74077.83 \div 10207.55 \\
&= 7.257\,\text{m} \\
\text{Free surface effect} &= (1.025 \times 18 \times 15^3 \times 1.025) \div (1.025 \times 12 \times 10207.55) \\
&= 63825.47 \div 125552.87 \\
&= 0.508\,\text{m} \\
\text{New } GM &= 8.00 - 7.257 - 0.508 \\
&= 0.235\,\text{m}
\end{aligned}
$$

**A. E2.13:**

$$a) \quad \text{Blade area} = 0.48 \times (\pi \div 4) \times 6^2$$
$$= 13.57 \, \text{m}^2$$

$$b) \quad \text{Theoretical speed } V_t = 6 \times 0.9 \times 110 \div 60$$
$$= 9.9 \, \text{m/s}$$
$$\text{Real slip } 0.25 = (9.9 - V_a) \div 9.9$$
$$\text{Speed of advance } V_a = 9.9 \times (1 - 0.25)$$
$$= 7.425 \, \text{m/s}$$
$$\text{Wake fraction } 0.30 = (V - 7.425) \div V$$
$$\text{Ship speed } V = 7.425 \div (1 - 0.30)$$
$$= 10.61 \, \text{m/s}$$
$$= 10.61 \div 0.5144$$
$$= 20.62 \, \text{knots}$$

$$c) \quad \text{Thrust power tp} = 300 \times 7.425$$
$$= 2227.5 \, \text{kW}$$

$$d) \quad \text{Shaft power sp} = 2227.5 \div 0.65$$
$$= 3427 \, \text{kW}$$

$$e) \quad \text{sp} = 2\pi n Q$$
$$\text{Torque } Q = (3427 \times 60) \div (2\pi \times 110)$$
$$= 297 \, \text{kNm}$$

**A. E2.14:**

$$\text{Let } C = \text{normal cons/h at } V \text{ knots}$$
$$C_1 = \text{cons/h at reduced speed of 0.8 } V \text{ knots}$$
$$\text{Then } C_1 = C - (42 \div 24) \text{ tonne/h}$$
$$\text{Now } C = 0.136 + 0.001V^3$$
$$C - (42 \div 24) = 0.136 + 0.001(0.8V)^3$$
$$\text{Subtracting} : (42 \div 24) = 0.001V^3 - 0.001(0.512 \, V^3)$$
$$= 0.001V^3 (1 - 0.512)$$
$$V^3 = 42 \div (24 \times 0.001 \times 0.488)$$
$$V = 15.31 \, \text{knots}$$
$$\text{Reduced speed} = 0.8 \times 15.31$$
$$= 12.25 \, \text{knots}$$
$$C = 0.136 + 0.001 \times 15.31^3$$
$$= 3.725 \, \text{tonne/h}$$
$$\text{Nominal cons/day} = 3.725 \times 24$$
$$= 89.39 \, \text{tonne}$$

$$a) \quad \text{Reduced cons/day} = 89.39 - 42$$
$$= 47.39 \, \text{tonne}$$

**b)** Time taken to travel 800 nautical miles at normal speed $= 800 \div 15.31$

$= 52.25\,\text{h}$

Time taken at reduced speed $= 800 \div 12.25$

$= 65.31\,\text{h}$

Fuel consumption for 800 nautical miles at reduced speed $= 47.39 \times 65.31 \div 24$

$= 129.0\,\text{tonne}$

Fuel on board when speed reduced $= 129 + 50$

$= 179\,\text{tonne}$

**(c)** Normal cons for 800 Nm $= 89.39 \times 52.25 \div 24$

$= 194.6\,\text{tonne}$

% reduction in consumption $= (194.6 - 129) \div 194.6 \times 100$

$= 65.6 \div 194.6 \times 100$

$= 33.71\%$

**d)** % increase in time $= (65.31 - 52.25) \div 52.25 \times 100$

$= 13.06 \div 52.25 \times 100$

$= 25\%$

**A. E2.15:**

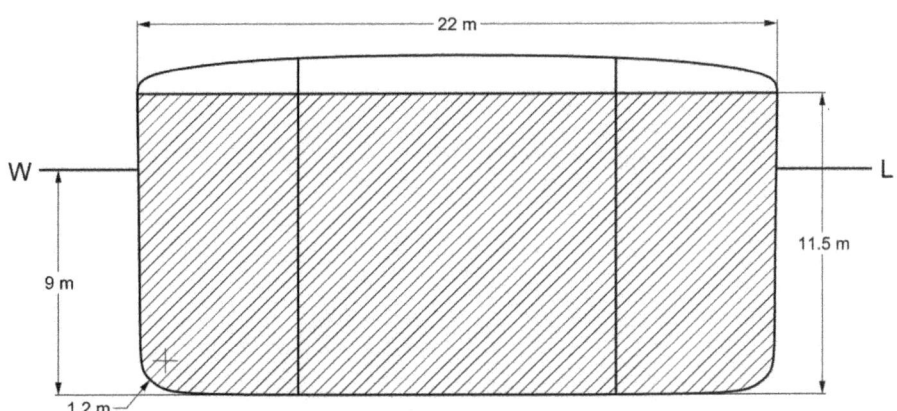

▲ Figure A. E2.4

Complete waterplane area $= 160 \times 22 \times 0.865$

$= 3044.8\,\text{m}^2$

Intact waterplane area $= 3045.8 - 10.5 \times 22$

$= 2814.8\,\text{m}^2$

It may be assumed that the whole of the mass of the oil is taken from the ship and that all the buoyancy of the compartment is lost.

$$\text{Cross-sectional area of oil} = \frac{1}{2} \times 1.2^2 + (22 - 2.4) \times 1.2 + 22 \times (11.5 - 1.2)$$
$$= 2.26 + 23.52 + 226.6$$
$$= 252.38 \text{ m}^2$$
$$\text{Immersed cross-sectional area} = 252.38 - 22 \times 2.5$$
$$= 197.38 \text{ m}^2$$
$$\text{Mass of oil in compartment} = 252.38 \times 10.5 \div 1.4$$
$$= 1892.85 \text{ tonne}$$
$$\text{Mass of buoyancy lost} = 197.38 \times 10.5 \times 1.025$$
$$= 2124.30 \text{ tonne}$$
$$\text{Net loss in buoyancy} = 2124.30 - 1892.85$$
$$= 231.45 \text{ tonne}$$
$$\text{Equivalent volume} = 231.55 \div 1.025$$
$$= 225.9 \text{ m}^3$$
$$\text{Increase in draught} = \text{net volume of lost buoyancy} \div \text{area of intact waterplane}$$
$$= 225.9 \div 2814.8$$
$$\text{Increase in draught} = 0.0802 \text{ m}$$
$$\text{New draught} = 9.08 \text{ m}$$

**A. E2.16:**

| ½ ord | ½ ord³ | SM | Product |
|---|---|---|---|
| 0 | 0 | 1 | — |
| 2.4 | 13.82 | 4 | 55.28 |
| 5.0 | 125.00 | 2 | 250.00 |
| 7.3 | 389.02 | 4 | 1556.08 |
| 7.9 | 493.04 | 2 | 986.08 |
| 8.0 | 512.00 | 4 | 2048.00 |
| 8.0 | 512.00 | 2 | 1024.00 |
| 7.7 | 456.53 | 4 | 1826.12 |
| 5.5 | 166.38 | 2 | 332.76 |
| 2.8 | 21.95 | 4 | 87.80 |
| 0 | 0 | 1 | — |
| | | | 8166.12 |

$$h = 16\,m$$
$$I = (2 \div 9) \times 16 \times 8166.12$$
$$= 29035\,m^4$$
$$BM = 29035 \div 8700 \times 1.025$$
$$= 3.421\,m$$

At angle of loll $\tan \theta = \sqrt{(-2GM \div BM)}$

$$GM = -\tfrac{1}{2}BM \tan^2 \theta$$
$$= -\tfrac{1}{2} \times 3.421 \times 0.1763^2$$
$$= -0.053\,m$$

**A. E2.17:**

Let $V$ = normal speed
$C$ = normal consumption per hour
Then $24\,C$ = normal consumption per day.

For first 7.5 hours:

$$Cons/h = C \times (V_1 \div V)^3$$
$$= C \times (1.18V \div V)^3$$
$$= 1.643C$$
$$\text{Cons for 7.5 hours} = 7.5 \times 1.643C$$
$$= 12.32C$$

For next 10 hours:

$$Cons/h = C \times (0.91V \div V)^3$$
$$= 0.7536C$$
$$\text{Cons for 10 hours} = 7.536C$$
$$\text{ie cons for 17.5 hours} = 12.32C + 7.536C$$
$$= 19.856C$$
$$\text{Cons for remaining 6.5 hours} = 24C - 19.856C$$
$$= 4.144C$$
$$Cons/h = 4.144C \div 6.5$$
$$= 0.637C$$
$$\text{Reduced speed } V_3 = V \times (0.637C \div C)^{\frac{1}{3}}$$
$$= 0.86\,V$$
$$\text{Normal distance travelled/day} = 24\,V$$
$$\text{New distance travelled/day} = 1.18V \times 7.5 + 0.91V \times 10 + 0.86V \times 6.5$$
$$= 23.54\,V$$
$$\text{\% reduction in distance/day} = (24V - 23.54V) \div 24V \times 100$$
$$= 1.92\%$$

**A. E2.18:**

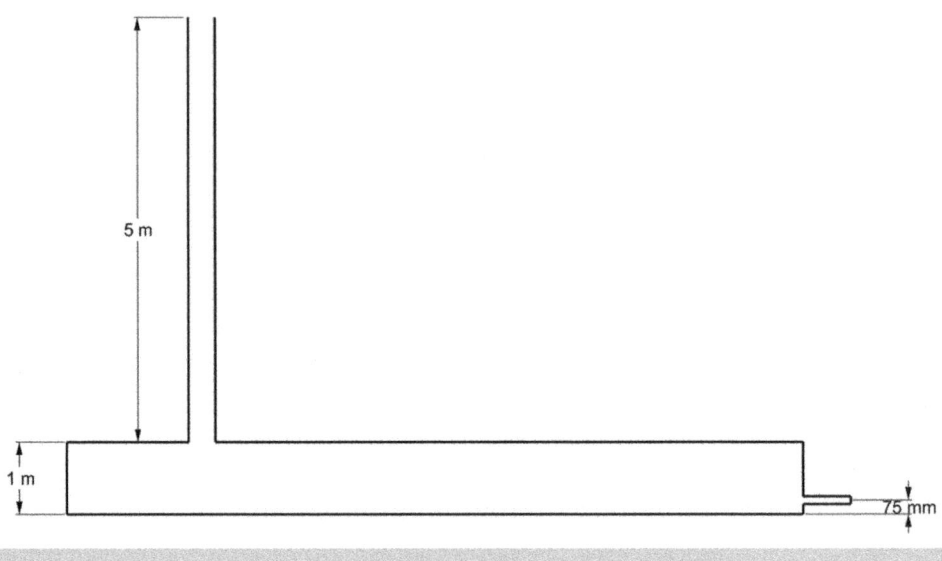

▲ Figure A. E2.5

a)  Pressure at tank top = pressure exerted by pump − pressure due to head of water

$$= 70 - 1.025 \times 9.81 \times (1 - 0.075)$$
$$= 60.70 \text{ kN/m}^2$$
$$\text{Load on tank top} = 60.70 \times 6 \times 12$$
$$= 4370 \text{ kN}$$
$$= 4.37 \text{ MN}$$

b)  With 70 kN/m² pressure:

$$\text{maximum head above inlet} = 70 \div (1.025 \times 9.81)$$
$$= 6.962 \text{ m}$$
$$\text{Maximum head above tank top} = 6.968 - 0.925$$
$$= 6.043 \text{ m}$$

Hence the water will overflow and the maximum head above the tank top is therefore 5 m.

$$\text{Load on tank top} = 1.025 \times 9.81 \times 6 \times 12 \times 5$$
$$= 3619.89 \text{ kN}$$
$$= 3.62 \text{ MN}$$

**A. E2.19:**

a)

| V | V³ | Δ²ᐟ³ | SP | Ad Coeff |
|----|------|-------|------|----------|
| 10 | 1000 | 524.1 | 880 | 595.6 |
| 11 | 1331 | 524.1 | 1155 | 604.0 |
| 12 | 1728 | 524.1 | 1520 | 595.8 |
| 13 | 2197 | 524.1 | 2010 | 572.9 |
| 14 | 2744 | 524.1 | 2670 | 538.6 |
| 15 | 3375 | 524.1 | 3600 | 491.3 |

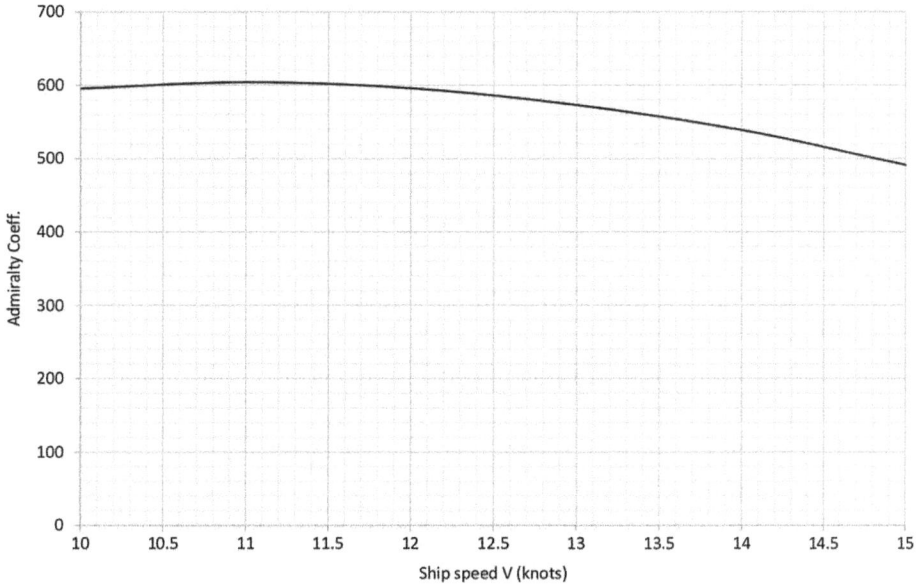

▲ Figure A. E2.6

b) Corresponding speed of 120 m ship to 14 knots for 140 m ship.

$$= 14\sqrt{(120 \div 140)}$$
$$= 12.96 \text{ knots}$$

From graph at 12.96 knots, the Admiralty Coefficient is 574.6 $\Delta$ is proportional to $L^3$

$$\text{Therefore: } \Delta = 12000 \times (140 \div 120)^3$$
$$= 19056 \text{ tonne}$$
$$\text{Hence shaft power} = 19056^{2/3} \times 12.96^3 \div 574.6$$
$$= 2703 \text{ kW}$$

**A. E2.20:**

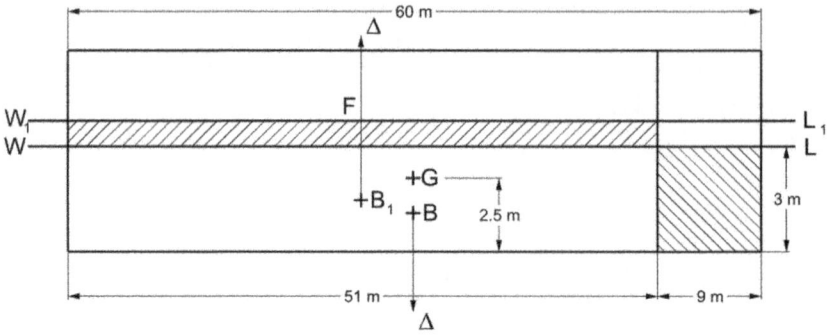

▲ Figure A. E2.7

$$\text{Increase in mean draught} = (9 \times 10 \times 3) \div (51 \times 10)$$
$$= 0.529\,\text{m}$$
$$\text{New mean draught } d_1 = 3 + 0.529$$
$$= 3.529\,\text{m}$$
$$KB_1 = 3.529 \div 2$$
$$= 1.765\,\text{m}$$
$$I_F = (1 \div 12) \times 51^3 \times 10$$
$$\nabla = 60 \times 10 \times 3$$
$$BM_L = (51^3 \times 10) \div (12 \times 60 \times 10 \times 3)$$
$$= 61.41\text{m}$$
$$GM_L = 1.765 + 61.41 - 2.50$$
$$= 60.675\,\text{m}$$
$$BB_1 = 9 \div 2$$
$$= 4.5\,\text{m}$$
$$\text{Change in trim} = (\Delta \times 4.5) \div (\Delta \times 60.675) \times 60$$
$$= 4.45\,\text{m by the head}$$
$$\text{Change forward} = +(4.45 \div 60) \times (60 \div 2 + 4.5)$$
$$= +2.559\,\text{m}$$
$$\text{Change aft} = -(4.45 \div 60) \times (60 \div 2 - 4.5)$$
$$= -1.891\text{m}$$
$$\text{New draught forward} = 3.259 + 2.559$$
$$= 5.818\,\text{m}$$
$$\text{New draught aft} = 3.259 - 1.891$$
$$= 1.368\,\text{m}$$

**A. E2.21:**

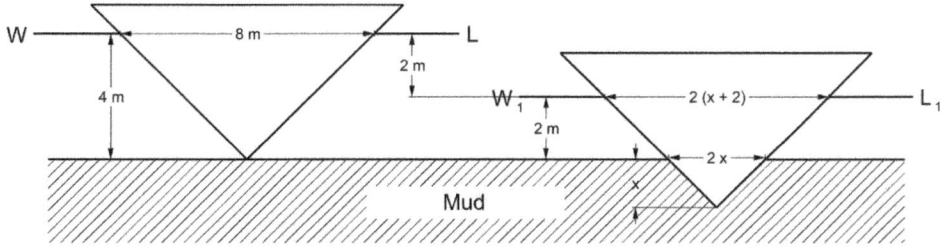

▲ Figure A. E2.8

Let $L$ = length of vessel in m

$\rho$ = density of water in tonne/m³

$2\rho$ = density of mud in tonne/m³

$x$ = depth to which vessel sinks in mud

Original displacement = $\rho \times L \times (8 \times 4) \div 2$

$= 16\rho L$ tonne

This displacement remains constant.

Displacement of part in mud = $2\rho \times L \times (2x \times x) \div 2$

$= 2\rho Lx^2$

Displacement of part in water = $\rho \times L \times 2 \times (2x + 2x + 4) \div 2$

$= \rho L(4x + 4)$

Hence $16\rho L = 2\rho Lx^2 + 4\rho L \times + 4\rho L$

$16 = 2x^2 + 4x + 4$

$2x^2 + 4x - 12 = 0$

from which $x = 1.646$ m

ie vessel sinks 1.646 m into the mud.

**A. E2.22:**

a)

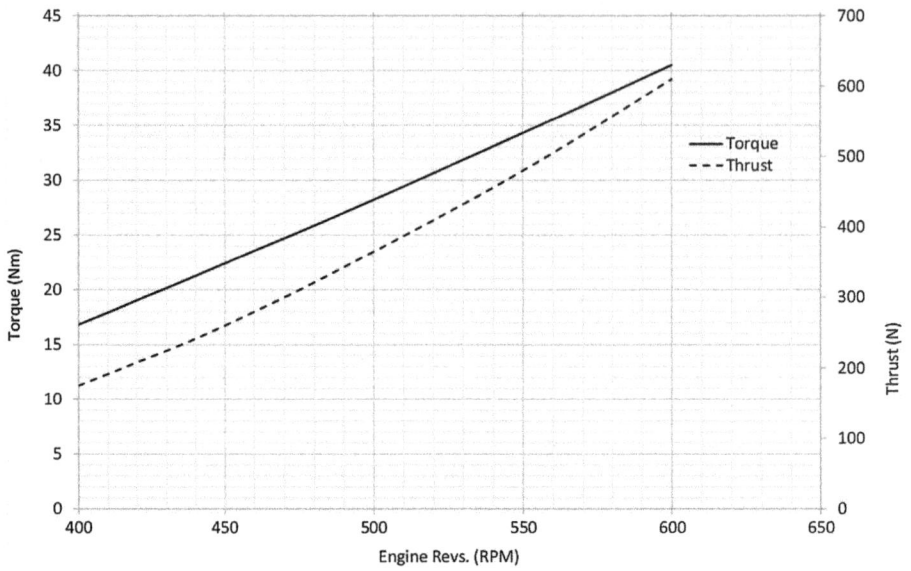

▲ Figure A. E2.9

b) $$V_t = P \times N$$
$$V_a = 0.8 \, V_t$$
$$150 = 0.8 \times 0.4 \times N$$
Rev/min $N = 469$

At 469 rev/min, $T = 298$ N and $Q = 24.6$ Nm

Thrust power tp $= 298 \times 150 \div 60$
$= 745$ W
Delivered power dp $= 24.6 \times 2\pi \times 469 \div 60$
$= 1208$ W
Propeller efficiency $=$ tp $\div$ dp
$= 745 \div 1208$
$= 61.67\%$

**A. E2.23:**

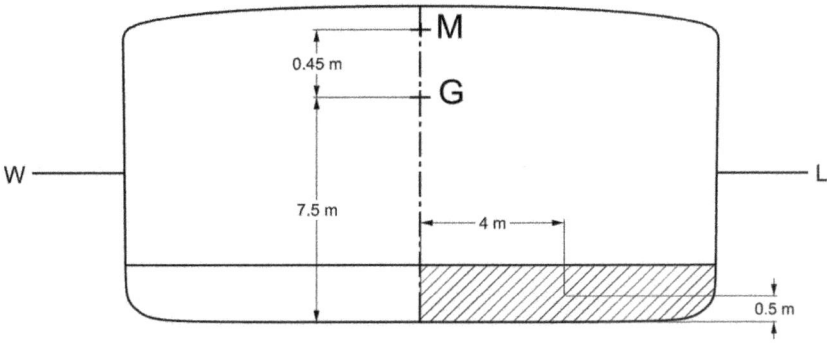

▲ Figure A. E2.10

$$\text{New } KG = (8100 \times 7.5 - 100 \times 0.5) \div (8100 - 100)$$
$$= (60750 - 50) \div 8000$$
$$= 7.588 \text{ m}$$
$$\text{New } GM = 7.5 + 0.45 - 7.588$$
$$= 0.362 \text{ m}$$
$$\text{Heeling moment} = 100 \times 4$$
$$\tan\theta = (100 \times 4) \div (8000 \times 0.362)$$
$$= 0.1381$$
$$\text{Angle of heel } \theta = 7.862°$$

**A. E2.24:**

| ½ ord | ½ ord³ | SM | Product |
|---|---|---|---|
| 0.7 | 0.34 | 1 | 0.34 |
| 3.3 | 35.94 | 4 | 143.76 |
| 5.5 | 166.38 | 2 | 332.76 |
| 7.2 | 373.25 | 4 | 1493.00 |
| 7.5 | 421.88 | 2 | 843.76 |
| 7.5 | 421.88 | 4 | 1687.52 |
| 7.5 | 421.88 | 2 | 843.76 |
| 6.8 | 314.43 | 4 | 1257.72 |
| 4.6 | 97.34 | 2 | 194.68 |
| 2.2 | 10.65 | 4 | 42.60 |
| 0 | — | 1 | — |
| | | | 6839.90 |

$$h = 12\,m$$
$$I = (2 \div 9) \times 12 \times 6839.9$$
$$= 18240\,m^4$$
$$BM = 18240 \times 1.025 \div 11000$$
$$= 1.7\,m$$

## A. E2.25:

1 tonne of stowed cargo occupies 1.8 m³

1 tonne of solid cargo occupies (1 ÷ 1.6) or 0.625 m³

Hence in every 1.8 m³ of space 0.625 m³ is occupied by cargo and the remaining 1.175 m³ is available for water.

$$\text{Permeability} = 1.175 \div 1.8$$
$$= 0.653$$
$$\text{Volume of lost buoyancy} = 18 \times 18 \times 6 \times 0.653$$
$$\text{Area of intact waterplane} = 85 \times 18$$
$$\text{Increase in draught} = (18 \times 18 \times 6 \times 0.653) \div (85 \times 18)$$
$$= 0.83\,m$$
$$\text{New draught} = 6 + 0.83$$
$$= 6.83\,m$$

## A. E2.26:

| Station | ½ ord | SM | Product | Lever | Product |
|---------|-------|-----|---------|-------|---------|
| AP | 0.6 | ½ | 0.3 | +4 | +1.2 |
| ½ | 2.7 | 2 | 5.4 | +3½ | +18.9 |
| 1 | 4.6 | 1½ | 6.9 | +3 | +20.7 |
| 2 | 6.0 | 4 | 24.0 | +2 | +48.0 |
| 3 | 6.3 | 2 | 12.6 | +1 | +12.6 |
| 4 | 6.3 | 4 | 25.2 | 0 | +101.4 |
| 5 | 6.3 | 2 | 12.6 | −1 | −12.6 |
| 6 | 5.7 | 4 | 22.8 | −2 | −45.6 |
| 7 | 4.8 | 1½ | 7.2 | −3 | −21.6 |
| 7½ | 2.0 | 2 | 4.0 | −3½ | −14.0 |
| FP | 0 | ½ | — | −4 | — |
| | | | 121.0 | | −93.8 |

$$\text{Common interval} = 90 \div 8$$
$$\text{Area} = \tfrac{2}{3} \times (90 \div 8) \times 121.0$$
$$= 907.5 \, \text{m}^3$$
$$\text{LCF from midships} = (90 \div 8) \times ((101.4 - 93.8) \div 121.0)$$
$$= 0.707 \, \text{m aft}$$

**A. E2.27:**

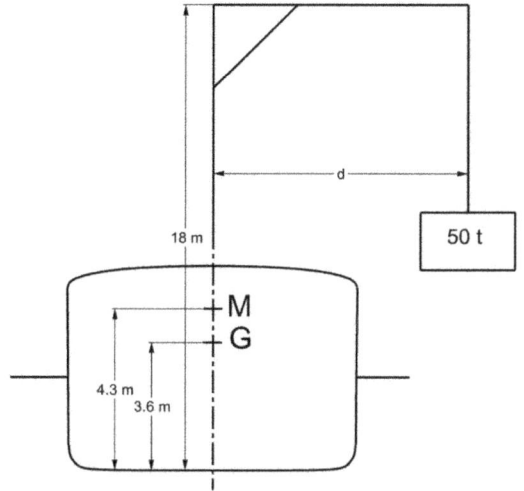

▲ **Figure A. E2.11**

$$\text{New } KG = (6600 \times 3.6 + 50 \times 18) \div (6600 + 50)$$
$$= (23760 + 900) \div 6650$$
$$= 3.708 \, \text{m}$$
$$\text{New } GM = 4.30 - 3.708$$
$$= 0.592 \, \text{m}$$
$$\tan \theta = m \times d \div (\Delta \times GM)$$
$$d = 6650 \times 0.592 \times \tan 9.5° \div 50$$
$$\text{Outreach of derrick} = 13.18 \, \text{m}$$

**A. E2.28:**

120 m ship:

$$\text{Shaft power} = 4100\,\text{kW}$$
$$\text{Effective power} = 4100 \times 0.6$$
$$= 2460\,\text{kW}$$
$$= R_t \times V$$
$$R_t = 2460 \div (15 \times 0.5144)$$
$$= 318.8\,\text{kN}$$
$$R_f = 0.55\,R_t$$
$$R_r = 0.45\,R_t$$
$$R_r = 143.46\,\text{kN}$$

140 m ship: $R_r$ is proportional to $L^3$

$$\text{Therefore: } R_r = 143.46 \times (140 \div 120)^3$$
$$= 227.81\,\text{kN}$$

$S$ is proportional to $L^2$

$$S = 3000 \times (140 \div 120)^2$$
$$= 4083\,\text{m}^2$$

$V$ is proportional to $\sqrt{L}$

$$V = 15 \times \sqrt{(140 \div 120)}$$
$$V = 16.2\,\text{knots}$$
$$R_f = 0.42 \times 4083 \times 16.2^{1.825}$$
$$= 276.44\,\text{kN}$$
$$R_t = 276.44 + 227.81$$
$$= 504.25\,\text{kN}$$
$$\text{ep} = 504.25 \times 16.2 \times 0.5144$$
$$= 4202\,\text{kW}$$
$$\text{sp} = \text{ep} \div \text{pc}$$
$$= 4202 \div 0.6$$
$$= 7003\,\text{kW}$$

**A. E2.29:**

$$\text{Original displacement} = 30 \times 9 \times 0.75 \times 1.025$$
$$= 207.6 \text{ tonne}$$
$$\text{Change in mean draught} = (\Delta \times 100 \div A_w) \times (\rho_s - \rho_r) \div (\rho_s \rho_r) \text{ cm}$$
$$3.2 = (\Delta \times 100 \div (30 \times 9)) \times (1.025 - 1.000) \div (1.025 \times 1.000)$$
$$\Delta = 3.2 \times 30 \times 9 \times 1.025 \div (100 \times 0.025)$$
$$= 354.2 \text{ tonne}$$
$$\text{Cargo added} = 354.2 - 207.5$$
$$= 146.7 \text{ tonne}$$

**A. E2.30:**

| $\frac{1}{2}$ breadth | SM | f of a | $(\frac{1}{2}b)^2$ | SM | f of m | $(\frac{1}{2}b)^3$ | SM | f of i |
|---|---|---|---|---|---|---|---|---|
| 5 | 1 | 5 | 25 | 1 | 25 | 125 | 1 | 125 |
| 4 | 4 | 16 | 16 | 4 | 64 | 64 | 4 | 256 |
| 2 | 1 | 2 | 4 | 1 | 4 | 8 | 1 | 8 |
| | | 23 | | | 93 | | | 389 |

$$\text{Area of tank surface } a = (6 \div 3) \times 23$$
$$= 46 \text{ m}^2$$
$$\text{Centroid from centreline} = 93 \div (2 \times 23)$$
$$= 2.022 \text{ m}$$
$$\text{Second moment of area about centreline} = (6 \div 9) \times 389$$
$$= 259.33 \text{ m}^4$$
$$\text{Second moment of area about centroid } I = 259.33 - 46 \times 2.022^2$$
$$= 71.26 \text{ m}^4$$
$$\text{Free surface effect} = \rho_L I \div (\rho \nabla)$$
$$= 1.000 \times 71.26 \div (1.025 \times 5000)$$
$$= 0.0139 \text{ m}$$

**A. E2.31:**

$$\text{Ship speed} = 15 \times 0.5144$$
$$= 7.716 \text{ m/s}$$
$$\text{Apparent slip} = (V_t - V) \div V_t$$
$$0 = (V_t - V)$$
$$V_t = V$$
$$= 7.716 \text{ m/s}$$
$$\text{Propeller pitch } P = 7.716 \div 1.58$$
$$= 4.884 \text{ m}$$
$$w_t = (V - V_a) \div V$$
$$V_a = 7.716(1 - 0.31)$$
$$V_a = 5.324 \text{ m/s}$$
$$\text{Real slip} = (7.716 - 5.324) \div 7.716$$
$$= 0.31$$
$$R_t = T(1 - t)$$
$$= 360(1 - 0.20)$$
$$= 288 \text{ kN}$$
$$\text{ep} = 288 \times 7.716$$
$$= 2222.2 \text{ kW}$$
$$\text{Propulsive coefficient} = 2222.2 \div 3050$$
$$= 0.729$$

**A. E2.32:**

$$\text{Torque} = \text{force} \times \text{level}$$
$$= F_n \times b$$
$$= 577 \times 22 \times (1.2 \times 15 \times 0.5144)^2 \times \sin 35° \times 1.1$$
$$= 686780 \text{ Nm}$$
$$\text{But } T \div J = q \div r$$
$$\text{And } J = (\pi \div 2)r^4$$
$$686780 \div (\pi \div 2 \times r^4) = 70 \times 10^6 \div r$$
$$r^3 = 686780 \times 2 \div (70 \times 10^6 \pi)$$
$$r = 0.184 \text{ m}$$
$$\text{Diameter of stock} = 368 \text{ mm}$$

If the diameter is reduced to 330 mm:

$$T \div (\pi \div 2 \times 0.165^4) = 70 \times 10^6 \div 0.165$$
$$T = 70 \times 10^6 \times \pi \times 0.165^3 \div 2$$
$$T = 493920\, \text{Nm}$$
$$493920 = 577 \times 22 \text{x} (1.2 \times V \times 0.5144)^2 \times \sin 35° \times 1.1$$
$$V^2 = 493920 \div (577 \times 22 \times 1.2^{2 \times} 0.5144^2 \times 0.5736 \times 1.1)$$
$$V^2 = 493920 \div 3051.861$$
$$V^2 = 161.842$$
$$\text{Ship speed } V = 12.72\, \text{knots}$$

**A. E2.33:**

| ½ ord | SM | Product |
|-------|-----|---------|
| 0 | 1 | — |
| 3.0 | 4 | 12.0 |
| 5.5 | 2 | 11.0 |
| 7.3 | 4 | 29.2 |
| 7.5 | 2 | 15.0 |
| 7.5 | 4 | 30.0 |
| 7.5 | 2 | 15.0 |
| 7.05 | 4 | 28.2 |
| 6.10 | 2 | 12.2 |
| 3.25 | 4 | 13.0 |
| 0 | 1 | — |
| | | 165.6 |

$$h = 10\, \text{m}$$
$$\text{Waterplane area} = \tfrac{2}{3} \times 10 \times 165.6$$
$$= 1104\, \text{m}^2$$
$$\text{Intact waterplane area} = 1104 - (15 \times 15)$$
$$= 879\, \text{m}^2$$
$$\text{Immersed cross-sectional area} = 50\, \text{m}^2$$
$$\text{Volume of lost buoyancy} = 15 \times 50$$
$$= 750\, \text{m}^2$$
$$\text{Increase in draught} = 750 \div 879$$
$$= 0.853\, \text{m}$$
$$\text{New draught} = 4.353\, \text{m}$$

**A. E2.34:**

$$\text{Let } m = \text{mass of ballast required}$$
$$\text{MCT 1 cm} = \Delta \times GM_L \div (100L)$$
$$= (8100 \times 96) \div (100 \times 85)$$
$$= 91.48 \text{ tonne m}$$
$$\text{Trimming moment} = m(33 - 2)$$
$$= 31m$$
$$\text{Change of trim } t = 31m \div 91.48 \text{ cm by the stern}$$
$$\text{change aft} = +t \div 85(85 \div 2 - 2)$$
$$= 0.476t \text{ cm}$$
$$\text{Bodily sinkage} = m \div 9.0 \text{ cm}$$
$$\text{New draught aft} = \text{old draught aft} + (m \div 900) + (0.476t \div 100)$$
$$5.85 = 5.55 + 0.00111m + (0.476 \div 100) + (31m \div 91.48)$$
$$0.30 = 0.002726m$$
$$\text{Ballast required } m = 110 \text{ tonne}$$
$$\text{Bodily sinkage} = 110 \div 9$$
$$= 12.22 \text{ cm}$$
$$\text{Change in trim} = (31 \times 110) \div 91.48$$
$$= 37.28 \text{ cm by the stern}$$
$$\text{Change forward} = -37.28 \div 85 \times (85 \div 2 + 2)$$
$$= -19.52 \text{ cm}$$
$$\text{New draught forward} = 5.25 + 0.122 - 0.195$$
$$= 5.177 \text{ m}$$

**A. E2.35:**

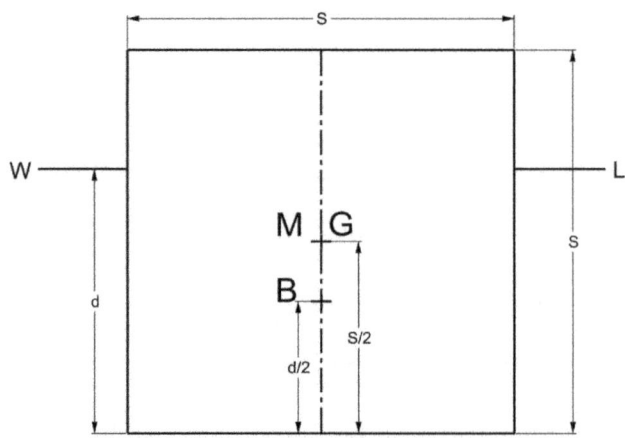

▲ Figure A. E2.12

Let $x$ = relative density of wood

Then draught $d = Sx$

The limit of stability occurs when $G$ and $M$ coincide

$$KG = S \div 2$$
$$KB = d \div 2$$
$$BM = S^2 \div (12d)$$

Since $KG = KM$

$$S \div 2 = d \div 2 + S^2 \div (12d)$$
$$S \div 2 = (S\,x) \div 2 + S^2 \div (12d)$$

Multiplying by 12:

$$6x = 6x^2 + 1$$
$$6x^2 - 6 \times + 1 = 0$$
$$x = (6 \pm \sqrt{(36 - 24)}) \div 12$$

Relative density $x = 0.212$ or $0.788$

It may be seen on referring to the metacentric diagram that the block will be unstable between these limits. Thus the relative density must be below 0.212 or between 0.788 and 1.0.

**A. E2.36:**

$$\text{Theoretical speed } V_t = 5.5 \times 95 \div 60$$
$$= 8.708 \text{ m/s}$$
$$\text{Real slip s} = (V_t - V_a) \div V_t$$
$$V_a = 8.708 \times 0.8$$
$$= 6.966 \text{ m/s}$$
$$\text{Thrust power tp} = T \times V_a$$
$$= 380 \times 6.966$$
$$= 2647 \text{ kW}$$
$$R_t = T(1 - t)$$
$$= 380(1 - 0.198)$$
$$= 304.8 \text{ kN}$$
$$\text{ep} = R_t \times V$$
$$= 304.8 \times 15.5 \times 0.5144$$
$$= 2430 \text{ kW}$$
$$\text{QPC} = \text{ep} \div \text{dp}$$
$$= 2430 \div 3540$$
$$= 0.686$$
$$\text{Wake fraction} = (V - V_a) \div V$$
$$= (15.5 - (6.966 \div 0.5144)) \div 15.5$$
$$= 0.126$$

**A. E2.37:**

a) See notes in Chapter 4 for full description.

b) $\quad$ Free surface effect $= \rho i \div (\rho \nabla)$
$$= lb^3 \div 12\nabla$$
$$0.14 = (l \times 11^3) \div (12 \times 8500) \times 1.025$$
$$l = (0.14 \times 12 \times 8500) \div 1331 \times 1.025$$
$$= 10.997 \text{ m}$$

**A. E2.38:**

a) See derivation in Chapter 5, section 5.4 m, which leads to the following formula:

$$\text{Change in mean draught} = (\Delta \times 100 \div A_w) \times (\rho_s - \rho_r) \div (\rho_s \rho_r) \text{ cm}$$

b) Change in draught due to density

$$= (8000 \times 100) \div (\text{TPC} \times 100) \times 1.024 \times (1.024 - 1.008) \div$$
$$(1.008 \times 1.024)$$
$$= (8000 \div 17) \times 1.024 \times (0.016 \div 1.032)$$
$$= 7.47 \text{ cm increase}$$

$$\text{New mean draught} = 7.075 \text{ m}$$
$$\text{Maximum allowable draught} = 6.85 \text{ m}$$
$$\text{Required reduction in draught} = 0.225 \text{ m}$$
$$\text{Mass of ballast discharged} = 22.5 \times 17 \times 1008 \div 1.024$$
$$= 376.5 \text{ tonne}$$

**A. E2.39:**

a) $\quad$ Admiralty Coefficient $= (\Delta^{\frac{2}{3}} V^3) \div \text{sp}$
$$\text{Fuel coefficient} = \Delta^{\frac{2}{3}} V^3 \div (\text{fuel cons/day})$$

b) With constant displacement and Admiralty Coefficient

sp is proportional to $V^3$

$$4460 \div 4120 = (14.55 \div V_1)^3$$
$$V_1 = 14.55(4120 \div 4460)^{\frac{1}{3}}$$
$$= 14.17 \text{ knots}$$
$$\text{At 14.17 knots, fuel cons} = 541 \text{kg/h}$$
$$= 541 \times 24 \times 10^{-3}$$
$$= 12.98 \text{ tonne/day}$$
$$\text{Fuel coefficient} = (14900^{\frac{2}{3}} \times 14.17^3) \div 12.98$$
$$= 132700$$

**A. E2.40:**

$$\text{Normal rudder force } F_n = 577AV^2 \sin \alpha \text{(in N)}$$
$$\text{Transverse force } F_t = 577AV^2 \sin \alpha \cos \alpha$$
$$= 577 \times 15 \times (1.2 \times 20 \times 0.5144)^2 \times 0.5 \times 0.860$$
$$= 567.23 \text{ kN}$$
$$\text{Heeling moment} = 567.23 \times (5 - 3.3) \cos \theta$$
$$= 964.29 \cos \theta \text{ kNm}$$
$$\text{Righting moment} = \Delta g GZ$$
$$= \Delta g GM \sin \theta$$

Steady heel will be produced when the heeling moment is equal to the righting moment.

$$12000 \times 9.81 \times 0.3 \sin \theta = 964.29 \cos \theta$$
$$\tan \theta = 964.29 \div (12000 \times 9.81 \times 0.3)$$
$$= 0.0273$$
$$\text{Angle of heel } \theta = 1.564°$$

**A. E2.41:**

| Item | Mass | LCG | Moment forward | Moment aft |
|---|---|---|---|---|
| Cargo | 10000 | 3.0F | 30000 | |
| Fuel | 1500 | 2.0A | | 3000 |
| Water | 400 | 8.0A | | 3200 |
| Stores | 100 | 10.0F | 1000 | |
| Light ship | 4000 | 2.5A | | 10000 |
| Displacement 16000 | | | 31000 | 16200 |

$$\text{Excess moment forward} = 31000 - 16200$$
$$= 14800 \text{ tonne m}$$
$$\text{LCG from midships} = 14800 \div 16000$$
$$= 0.925 \text{ m forward}$$

From hydrostatic data at 16000 tonne displacement: $d = 8.25$ m; MCT 1 cm = 179 t m; LCB = 2.02 m fwd; LCF = 0.57 m aft.

$$\text{Trimming lever} = 2.02 - 0.925$$
$$= 1.095\,\text{m aft}$$
$$\text{Trim} = (16000 \times 1.095) \div 179$$
$$= 97.88\,\text{cm by the stern}$$
$$\text{Change forward} = -(97.88 \div 120) \times (120 \div 2 + 0.57)$$
$$= -49.4\,\text{cm}$$
$$\text{Change aft} = (97.88 \div 120) \times (120 \div 2 - 0.57)$$
$$= 48.48\,\text{cm}$$
$$\text{Draught forward} = 8.250 - 0.494$$
$$= 7.756\,\text{m}$$
$$\text{Draught aft} = 8.250 + 0.485$$
$$= 8.735\,\text{m}$$

**A. E2.42:**

$$KB = 3.0 \div 2$$
$$= 1.5\,\text{m}$$
$$BM = 9^2 \div (12 \times 3)$$
$$= 2.25\,\text{m}$$
$$GM = 1.5 + 2.25 - 3.5$$
$$= 0.25\,\text{m}$$
$$\text{Displacement } \Delta = 30 \times 9 \times 3 \times 1.025$$
$$= 830.25\,\text{tonne}$$
$$\text{Righting moment} = \Delta \times GZ$$
$$\text{Heeling moment} = 10 \times 6$$

**a)**
$$10 \times 6 = 830.25 GZ$$
$$60 = 830.25 \sin\theta(GM + \tfrac{1}{2}BM \tan^2\theta)$$
$$60 = 830.25 \sin\theta(0.25 + (2.25 \div 2)\tan^2\theta)$$
$$0.07227 = \sin\theta(0.25 + 1.125\tan^2\theta)$$

This expression may be solved graphically.

| θ | tan θ | tan² θ | 1.125 tan² θ | sin θ | GZ |
|---|---|---|---|---|---|
| 5° | 0.0875 | 0.00766 | 0.00861 | 0.0872 | 0.0226 |
| 10° | 0.1763 | 0.03108 | 0.03497 | 0.1736 | 0.0495 |
| 15° | 0.2680 | 0.07182 | 0.08080 | 0.2588 | 0.0856 |
| 20° | 0.3640 | 0.13250 | 0.14906 | 0.3420 | 0.1365 |

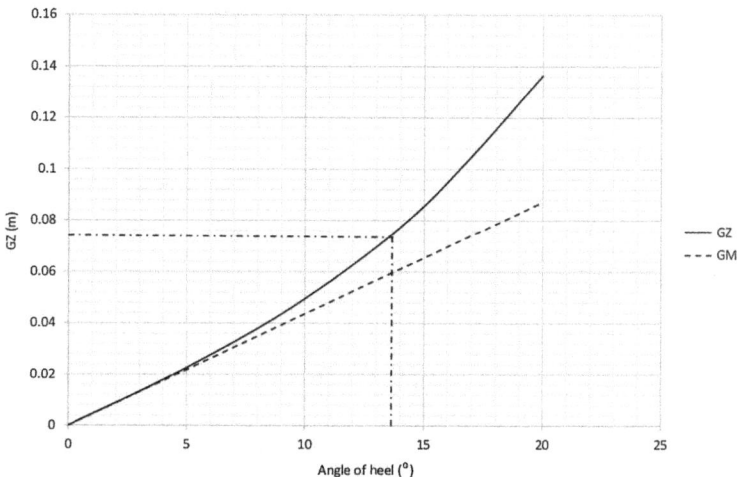

▲ **Figure A. E2.13**

From graph when $GZ = 0.07227$
Angle of heel $= 13.5°$
$GZ = GM \sin\theta$
$0.07227 = 0.25 \sin\theta$
$\sin\theta = 0.07227 \div 0.25$
$= 0.28908$
Angle of heel $= 16.8°$

**A. E2.43:**

Let $V =$ reduced speed in knots

Normally at $V$ knots the consumption per day would be:

$$47 \times (V \div 17)^3 \text{ tonne}$$
$$\text{Actual cons/day } 22 = 1.132 \times 47 \times (V \div 17)^3$$
$$V^3 = 22 \times 17^3 \div (1.132 \times 47)$$
$$\text{Reduced speed } V = 12.66 \text{ knots}$$
$$\text{At 17 knots, time taken} = 3000 \div (17 \times 24)$$
$$= 7.353 \text{ days}$$
$$\text{and voyage consumption} = 7.353 \times 47$$
$$= 345.6 \text{ tonne}$$
$$\text{At 12.66 knots, time taken} = 3000 \div (12.66 \times 24)$$
$$= 9.872 \text{days}$$
$$\text{and voyage consumption} = 9.872 \times 22$$
$$= 217.2 \text{ tonne}$$
$$\text{Difference in consumption} = (345.16 - 217.2) \div 345.6 \times 100$$
$$= 37.15\%$$

**A. E2.44:**

$$\text{Bodily sinkage} = 180 \div 18$$
$$= 10 \text{ cm}$$
$$\text{Trimming moment} = 180 \times (40 + 3)$$
$$\text{MC 1 cm} = (14000 \times 127) \div (100 \times 135)$$
$$\text{Change in trim} = (180 \times 43 \times 100 \times 135) \div (14000 \times 127)$$
$$= 58.76 \text{ cm by the head}$$
$$\text{Change forward} = +(58.76 \div 135) \times (135 \div 2 + 3)$$
$$= +30.69 \text{ cm}$$
$$\text{Change aft} = -(58.76 \div 135) \times (135 \div 2 - 3)$$
$$= -28.07 \text{cm}$$
$$\text{New draught forward} = 7.30 + 0.10 + 0.307$$
$$= 7.707 \text{ m}$$
$$\text{New draught aft} = 8.05 + 0.10 - 0.281$$
$$= 7.869 \text{ m}$$

**A. E2.45:**

$$\text{Theoretical speed } V_t = (5.5 \times 93 \times 60) \div 1852$$
$$= 16.57 \text{ knots}$$
$$\text{Let ship speed} = V$$
$$\text{Then} - S = (V_t - V) \div V_t$$
$$-S V_t = V_t - V \text{(1)}$$

and

$$+S = (V_t - 0.9V) \div V_t$$
$$+S V_t = V_t - 0.9V \text{ (2)}$$

Adding (1) and (2)

$$0 = 2V_t - 1.9V$$
$$V = 2 \times 16.57 \div 1.9$$

Ship speed V = 17.44 knots

Substituting for V:

$$-S = (16.57 - 17.44) \div 16.57$$
$$-S = -0.0525$$
$$\text{ie apparent slip} = -5.25\%$$
$$\text{and real slip} = +5.25\%$$

**A. E2.46:**

| ½ ord | SM | Product | Lever | Product | Lever | Product |
|-------|-----|---------|-------|---------|-------|---------|
| 1 | 1 | 1 | +4 | +4 | +4 | +16 |
| 7 | 4 | 28 | +3 | +84 | +3 | +252 |
| 10.5 | 2 | 21 | +2 | +42 | +2 | +84 |
| 11 | 4 | 44 | +1 | +44 | +1 | +44 |
| 11 | 2 | 22 | 0 | +174 | 0 | 0 |
| 10.5 | 4 | 42 | −1 | −42 | −1 | +42 |
| 8 | 2 | 16 | −2 | −32 | −2 | +64 |
| 4 | 4 | 16 | −3 | −48 | −3 | +144 |
| 0 | 1 | − | −4 | − | −4 | − |
| | | 190 | | −122 | | +646 |

$h = 15$ m

   **a)**    Waterplane area $A = \frac{2}{3} \times 15 \times 190$

$$= 1900 \text{ m}^2$$
$$TPC = 1900 \times 0.01025$$
$$= 19.475$$

   **b)**    LCF from midships $x = 15 \times (174 - 122) \div 190$
$$= 4.11 \text{m aft}$$

   **c)**    Second moment about midships $= \frac{2}{3} \times 15^3 \times 646$
$$= 1453500 \text{ m}^4$$
$$Ax^2 = 1900 \times 4.11^2$$
$$= 32095 \text{ m}^4$$
Second moment about centroid $= 1453500 - 32095$
$$= 1421405 \text{ m}^4$$

**A. E2.47:**

$$\text{Centrifugal force} = \Delta V^2$$
$$= (\Delta \div 600) \times (12 \times 0.5144)^2$$
$$= 0.06352\,\Delta$$
$$\text{Heeling moment} = CF \times GL \cos\theta$$
$$= 0.06352\,\Delta \times 2.7\cos\theta$$
$$\text{Righting moment} = \Delta gGM \sin\theta$$
$$= 0.25 \times 9.81\Delta \sin\theta$$
$$0.25 \times 9.81\Delta \sin\theta = 0.06352 \times 2.7\Delta\cos\theta$$
$$\tan\theta = (0.06352 \times 2.7) \div (0.25 \times 9.81)$$
$$= 0.069933$$
$$\text{Angle of heel } \theta = 4°$$

**A. E2.48:**

| V | $ep_n$ | $ep = ep_n \times 1.3$ | QPC | dp |
|---|---|---|---|---|
| 15 | 3000 | 3900 | 0.73 | 5342 |
| 16 | 3750 | 4875 | 0.73 | 6678 |
| 17 | 4700 | 6110 | 0.72 | 8486 |
| 18 | 5650 | 7345 | 0.71 | 10345 |

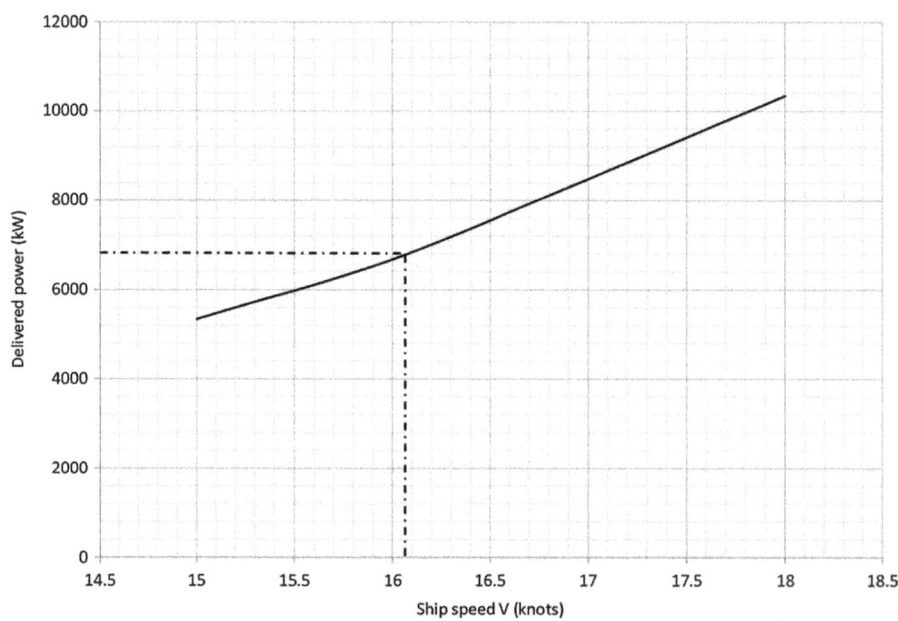

▲ Figure A. E2.14

$$\text{Total brake power} = 2 \times 3500$$
$$= 7000 \text{ kW}$$
$$\text{Total delivered power} = 7000(1 - 0.03)$$
$$= 6790 \text{ kW}$$
$$\text{From graph: service speed} = 16.06 \text{ knots}$$

**A. E2.49:**

$$\text{MCT 1 cm} = (8000 \times 102) \div (100 \times 120)$$
$$= 68 \text{ tonne m}$$

*Note*: The distances must be measured from the LCF.

|  | Mass | Distance from F | Moment forward | Moment aft |
|---|---|---|---|---|
| Fuel used | +200 | 20F | +4000 | |
| Water used | +100 | 1A | | +100 |
| Stores used | +10 | 7A | | +70 |
| Ballast added | − 300 | 26F | − 7800 | |
|  | +10 | | − 3800 | +170 |

$$\text{Bodily sinkage} = 10 \div 17.5$$
$$= 0.57 \text{ cm}$$
$$\text{Nett moment aft} = 170 - (-3800)$$
$$= 3970 \text{ tonne m}$$
$$\text{Change in trim} = 3970 \div 68$$
$$= 58.38 \text{ cm by the stern}$$
$$\text{Change forward} = -(58.38 \div 120) \times (120 \div 2 + 2)$$
$$= -30.16 \text{ cm}$$
$$\text{Change aft} = +(58.38 \div 120) \times (120 \div 2 - 2)$$
$$= -28.22 \text{ cm}$$
$$\text{Original draught forward} = 6.30 + 0.006 - 0.302$$
$$= 6.004 \text{ m}$$
$$\text{Original draught aft} = 6.60 + 0.006 + 0.282$$
$$= 6.888 \text{ m}$$

**A. E2.50:**

$$V_t = 5.5 \times 80 \times 60 \div 1852$$
$$= 14.25 \text{ knots.}$$

**a)**     Real slip $= (14.25 - 11) \div 14.25$
$$= 0.2281$$
or 22.81%.

**b)**     Wake fraction $= (13.2 - 11) \div 13.2$
$$= 0.167$$

Thrust power $= 3000 \times 0.7$
$$= 2100 \text{ kW}$$

**c)**     But thrust power $= T \times V_a$

Thrust $T = 2100 \div (11 \times 0.5144)$
$$= 371.1 \text{ kN}$$

**A. E2.51:**

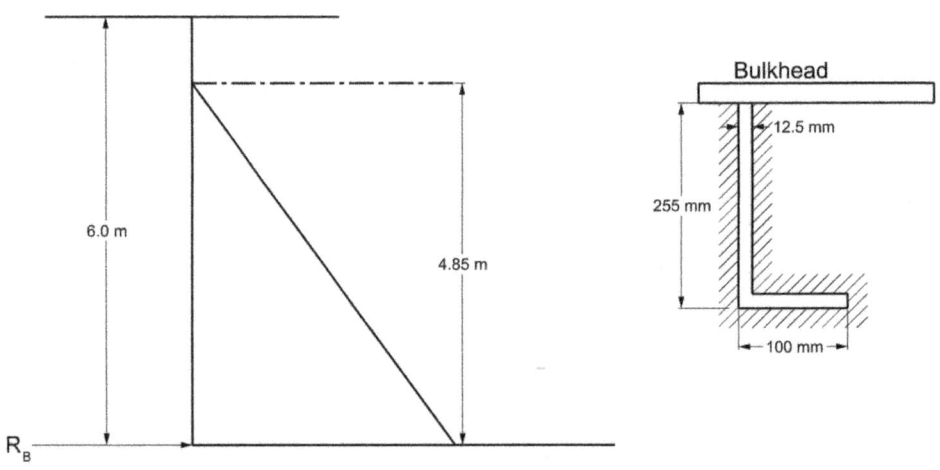

▲ Figure A. E2.15

Load on stiffener $= \rho g A H$
$$= 1.025 \times 9.81 \times 4.85 \times 0.6 \times 4.85 \div 2$$
$$= 70.96 \text{ kN}$$

Centre of pressure from surface $= \tfrac{2}{3} \times 4.85$
$$= 3.233 \text{ m}$$

Centre of pressure from top $= 3.233 + 1.15$
$$= 4.383 \text{ m}$$

Taking moments about the top

$$R_B = 70.96 \times 4.383 \div 6$$
$$= 51.84 \text{ kN}$$

This is also the shear force at the bottom of the stiffener.

$$\text{Length of weld metal} = 255 + 255 + 100 + 100 - 12.5$$
$$= 697.5 \text{ mm}$$
$$\text{Area of weld metal} = 697.5 \times 5 = 3487.5 \text{ mm}^2$$
$$\text{Shear stress in weld} = 51.84 \times 10^3 \div (3487.5 \times 10^{-6})$$
$$= 14.86 \times 10^6 \text{ N/m}^2$$
$$= 14.86 \text{ MN/m}^2$$

**A. E2.52:**

$$\text{Tank A, free surface effect} = (\rho Li) \div (\rho \nabla)$$
$$= i \div \nabla \text{ since } \rho L = \rho$$
$$(12 \times 16^3 \times 1.025) \div (12 \times 5000)$$
$$= 0.840 \text{ m}$$
$$\text{Tank B, free surface effect} = (14 \times 15^3 \times 1.025) \div (12 \times 5000)$$
$$= 0.807 \text{ m}$$
$$\text{Tank C, free surface effect} = (14 \times 16^3 \times 1.025) \div (12 \times 5000)$$
$$= 0.980 \text{ m}$$

The tank with the lowest free surface effect is filled first and thus they should be filled in the order B, A, C.

*Note*: Since the difference in free surface effect depends upon the product ($l \times b^3$), this value could have been calculated for each tank instead of the complete free surface effect.

**A. E2.53:**

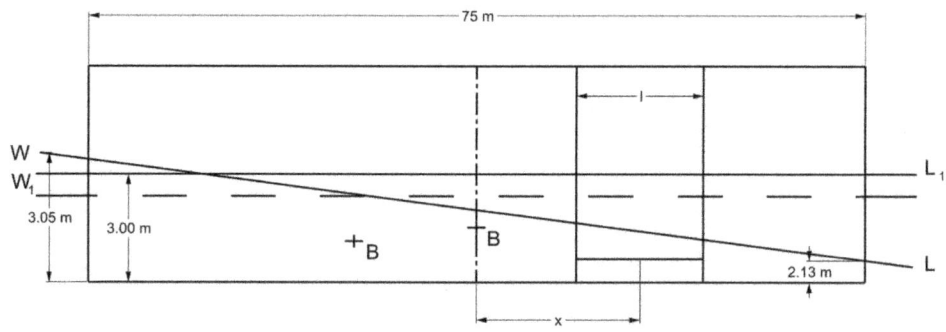

▲ **Figure A. E2.16**

$$\text{Original mean draught} = (3.05 + 2.13) \div 2$$
$$= 2.59 \text{ m}$$
$$\text{Volume of displacement} = 75 \times 8.5 \times 2.59$$
$$= 1651 \text{ m}^3$$
$$\text{LCB from midships} = (37.5 \times (0.46 \div 2) \times (4 \div 3) \times 37.5) \div (75 \times 2.59)$$
$$= 2.22 \text{ m aft}$$
$$\text{Final volume of displacement} = 75 \times 8.5 \times 3.0$$
$$= 1912 \text{ m}^3$$

LCB at midships

$$\text{Increase in volume of displacement} = 1912 - 1651$$
$$= 261 \text{ m}^3$$

The effect of this added volume is to bring the vessel to an even keel.

Let $x$ = distance of LCG of compartment forward of midships

Taking moment about midships:

$$1912 \times 0 = 1651 \times 2.22 - 261 \times x$$
$$x = (1651 \times 2.22) \div 261$$
$$= 14.04 \text{ m}$$
$$\text{Immersed cross-sectional area} = 8.5 \times 3$$
$$= 25.5 \text{ m}^2$$
$$\text{Length of compartment} = 261 \div 25.5$$
$$= 10.24 \text{ m}$$

**A. E2.54:**

$$\text{TPC} = A_w \times 0.01026$$
$$A_w = 27.5 \div 0.01026 \text{ m}^2$$
$$\text{Change in draught} = (\Delta \times 100 \div A_w) \times (\rho_s - \rho_r) \div (\rho_s \rho_r) \text{ cm}$$
$$25 = (\Delta \times 100 \times 0.01026 \div 27.5) \times (1.026 - 1.000) \div (1.026 \times 1.000)$$
$$25 = \Delta \times 0.026 \div 27.5$$
$$\Delta = 25 \times 27.5 \div 0.026$$
$$= 26442 \text{ tonne}$$

If, with this displacement, the vessel moves into the river water, then:

$$\text{Change in draught} = (26442 \times 1.026 \div 27.5) \times (1.026 - 1.012) \div (1.012 \times 1.026)$$
$$= 13.3 \text{ cm}$$

Thus the maximum allowable draught in the river water

$$= 8.25 + 0.133$$
$$= 8.383\,m$$
$$\text{Actual draught} = 8.44\,m$$
$$\text{Excess draught} = 0.057\,m$$
$$\text{TPC in river water} = 27.5 \times 1.012 \div 1.026$$
$$= 27.12$$
$$\text{Therefore: excess mass} = 5.7 \times 27.12$$
$$= 154.6\,tonne$$

**A. E2.55:**

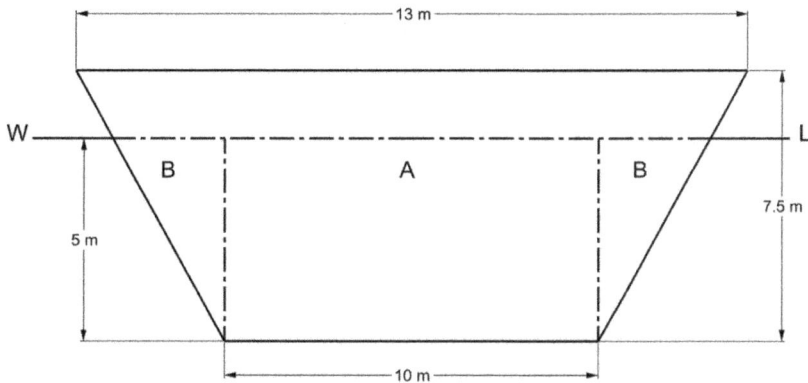

▲ Figure A. E2.17

$$\text{Breadth at water level} = 10 + (3 \div 7.5 \times 5)$$
$$= 12\,m$$

Divide into a rectangle A and two triangles B.

$$\text{Load on A} = \rho g A H$$
$$= 1.025 \times 9.81 \times 10 \times 5 \times 2.5$$
$$= 1256.91\,kN$$
$$\text{Centre of pressure from WL} = \tfrac{2}{3} \times 5$$
$$= 3.33\,m$$
$$\text{Load on B} = 1.025 \times 9.81 \times \tfrac{2}{3} \times 5 \times (5 \div 3) \times 2$$
$$= 83.79\,kN$$
$$\text{Centre of pressure from WL} = \tfrac{1}{2} \times 5$$
$$= 2.5\,m$$
$$\text{Total load} = 1256.91 + 83.79$$
$$= 1340.7\,kN$$

Taking moments about the waterline:

$$\text{Centre of pressure from WL} = (1256.91 \times 3.33 + 83.79 \times 2.5) \div (1256.91 + 83.79)$$
$$= (4185.5 + 209.5) \div 1340.7$$
$$= 3.278 \text{ m}$$

Therefore: centre of pressure is 5.778 m from the top of the bulkhead.

**A. E2.56:**

| ½ ord | ½ ord³ | SM | Product |
|-------|--------|----|---------|
| 5.5 | 166.38 | 1 | 166.38 |
| 4.6 | 97.34 | 4 | 389.36 |
| 4.3 | 79.51 | 2 | 159.02 |
| 3.7 | 50.65 | 4 | 202.60 |
| 3.0 | 27.00 | 1 | 27.00 |
| | | | 944.36 |

$$h = 23 \div 4$$
$$= 5.75 \text{ m}$$

Second moment of area about centerline:

$$I = (2 \div 9) \times 5.75 \times 944.36$$
$$= 1206.7 \text{ m}^4$$
$$\text{Free surface effect} = (\rho_L i) \div (\rho \nabla)$$
$$0.2 = (\rho_L \times 1206.7 \times 1.025) \div (1.025 \times 5350)$$
$$\text{Density of liquid } \rho_L = (0.2 \times 5350) \div 1206.7$$
$$= 0.887 \text{ t/m}^3$$

**A. E2.57:**

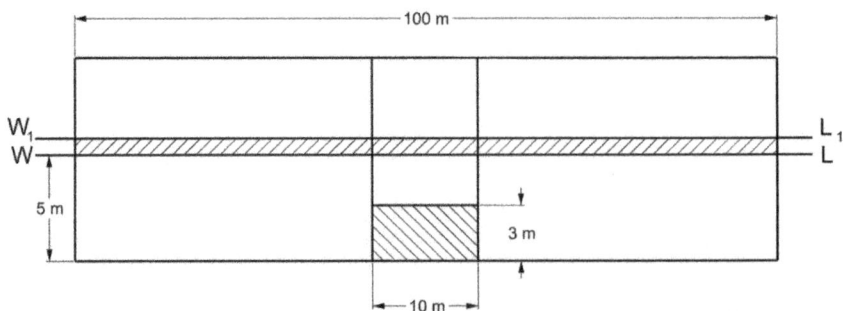

▲ Figure A. E2.18

$$\text{Before bilging}: KB = 5 \div 2$$
$$= 2.5\,\text{m}$$
$$BM = B^2 \div (12 \times 5)$$
$$KG = 2.5 + (B^2 \div 60) - 0.8$$
$$= 1.7 + (B^2 \div 60)$$

After bilging:

$$\text{Increase in draught} = (10 \times B \times 3) \div (10 \times B)$$
$$= 0.3\,\text{m}$$
$$\text{New draught} = 5.3\,\text{m}$$
$$KB_1 = (100 \times 5.3 \times (5.3 \div 2) - 10 \times 3 \times (3 \div 2)) \div (100 \times 5.3 - 10 \times 3)$$
$$= (1404.5 - 45) \div (530 - 30)$$
$$= 2.719\,\text{m}$$
$$B_1M_1 = (100 \times B^3) \div (12 \times 100 \times B \times 5)B_1$$
$$M_1 = B^2 \div 60$$
$$\text{New metacentric height } GM_1 = KB_1 + B_1M_1 - KG$$
$$= 2.719 + (B^2 \div 60) - (1.7 + (B^2 \div 60))$$
$$= 1.019\,\text{m}$$

**A. E2.58:**

a)      See Chapter 7, section 7.3 for the derivation of : $C = (\Delta^{\frac{2}{3}} V^3) \div \text{sp}$

b)      $\text{Displacement} = 160 \times 22 \times 9.2 \times 0.765 \times 1.025$
$$= 25393\,\text{tonne}$$
$$\text{Theoretical speed } V_t = 4.0 \times 96 \times 60 \div 1852$$
$$= 12.44\ \text{knots}$$
$$\text{Real slip } 0.33 = (12.44 - V_a) \div 12.44$$
$$V = 12.44(1 - 0.33)$$
$$= 8.335\ \text{knots}$$
$$\text{Wake fraction } 0.335 = (V - V_a) \div V$$
$$0.665\,V = V_a$$
$$V = 8.335 \div 0.665$$
$$= 12.53\,\text{knots}$$
$$\text{Admiralty Coefficient} = 25393^{\frac{2}{3}} \times 12.53^3 \div 2900$$
$$= 585.9$$
$$\text{At 15 knots: shaft power} = 2900 \times (15 \div 12.53)^3$$
$$= 4976\,\text{kW}$$

**A. E2.59:**

$$\text{Wetted surface area of ship} = 2.57\sqrt{(24000\times180)}$$
$$\text{Wetted surface area of model} = 5341\text{ m}^2$$
$$= 5341\times(6\div180)^2$$
$$= 5.934\text{ m}^2$$
$$\text{Model}: R_t = 40\text{ N in FW}$$
$$R_f = 0.492\times5.934\times3.6^{1.825}$$
$$= 30.24\text{ N in FW}$$
$$R_r = 40-30.24$$
$$= 9.76\text{N in FW}$$
$$= 9.76\times1.025$$
$$= 10.004\text{ N in SW}$$
$$\text{Ship}: R_r \text{ is proportional to } L^3$$
$$\text{Therefore}: R_r = 10.004\times(180\div6)^3$$
$$= 270108\text{N}$$

*V* is proportional to √L

$$V = 3.6\sqrt{(180\div6)}$$
$$= 19.72\text{ knots}$$
$$R_f = 0.421\times5341\times19.72^{1.825}$$
$$R_f = 518930\text{ N}$$
$$Rt = 518930+270110$$
$$= 789040\text{ N}$$
$$ep_n = 789040\times19.72\times0.5144$$
$$= 8004\text{ kW}$$

**A. E2.60:**

a)  See Chapter 4 for inclining experiment description.

b)  From the data given, create the following table using the equation:

$$GM = (m\times d)\div(\Delta\times\tan\theta)$$

Calculate the metacentric height of the vessel.

| Mass | Moment | Deflection | Tan (θ) From pendulum | GM (m) |
|---|---|---|---|---|
| 3 tonne (S) | 36 tonne m | 64 mm S | 0.0116 | 0.387 |
| 6 tonne (S) | 72 tonne m | 116 mm S | 0.0211 | 0.427 |
| 3 tonne (P) | 36 tonne m | 54 mm P | 0.0098 | 0.458 |
| 6 tonne (P) | 72 tonne m | 113 mm P | 0.0205 | 0.438 |
| | | | Mean | 0.427 |

$GM = 0.427$ m

## A. E2.61:

Change in mean draught $= (\Delta \times 100 \div A_w) \times (\rho_s - \rho_r) \div (\rho_s \rho_r)$ cm

$$= (8000 \times 100 \times 1.024) \div (16 \times 100) \times (1.024 - 1.000) \div (1.024 \times 1.000)$$

$$= 12 \text{ cm increase}$$

Shift in centre of buoyancy

$$= (\rho_s - \rho_r) \div (\rho_s) \times FB$$
$$= (1.024 - 1.000) \div 1.024 \times (3.0 - 0.6)$$
$$BB_1 = 0.05625 \text{ m aft}$$
$$\text{Change in trim} = (8000 \times 0.05625) \div 65$$
$$= 6.92 \text{ cm by the head}$$
$$\text{New trim} = 30 - 6.92$$
$$= 23.08 \text{ cm}$$
$$\text{Moment required} = 23.08 \times 65$$
$$\text{Distance moved by mass} = 23.08 \times 65 \div 50$$
$$= 30 \text{ m}$$
$$\text{Total change in trim} = 30 \text{ cm by the head}$$
$$\text{Change forward} = +(30 \div 110) \times (110 \div 2 + 3)$$
$$= +15.8 \text{ cm}$$
$$\text{Final level keel draught} = 6 + 0.12 + 0.158$$
$$= 6.278 \text{ m}$$

**A. E2.62:**

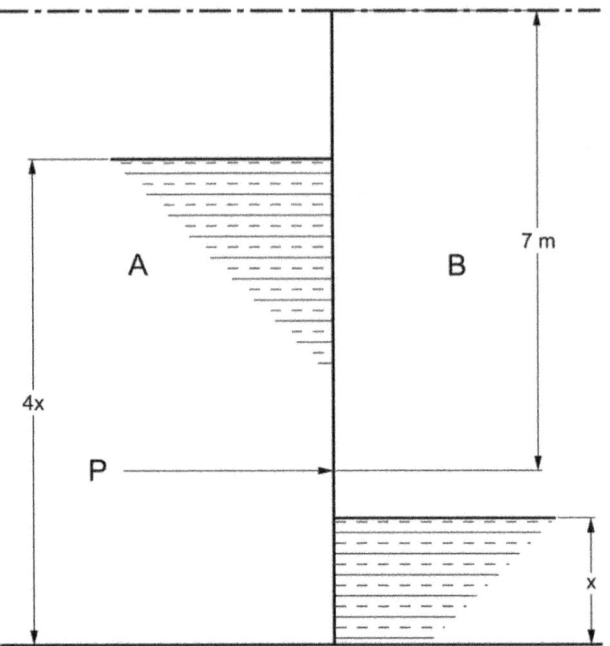

▲ Figure A. E2.19

**a)** Let $x$ = height of water on side B

$4x$ = height of water on side A

Load on side A = $\rho g A H$

$= 1.025 \times 9.81 \times 4x \times 14.5 \times 2x$

$= 145.8 \times 8x^2$

Centre of pressure on side A = $\frac{2}{3} \times 4x$ from surface

$= \frac{1}{3} \times 4x$ from bottom

Load on side B = $1.025 \times 9.81 \times x \times 14.5 \times 0.5x$

$= 145.8 \times 0.5x^2$

Centre of pressure on side B = $\frac{1}{3} \times x$ from bottom

Taking moments about the bottom of the bulkhead:

$$2 \times (145.8 \times 8x^2 - 145.8 \times 0.5x^2) = 145.8 \times 8x^2 \times (4 \div 3)x - 145.8 \times 0.5x^2 \times (x \div 3)$$

Dividing by $148.5x^2$

$$2\times(8-0.5)=(32\div3)\times-(0.5\div3)x$$
$$x=(2\times7.5\times3)\div31.5$$
$$=1.429\,m$$

Height on side B $= 1.429\,m$

Height on side A $= 5.716\,m$

b) Resultant load $= 145.8x^2(8-0.5)$
$$=2233\,kN$$
$$=2.233\,MN$$

### A. E2.63:

There are several methods of approach with this question. Probably the most straightforward is to plot curves of ep (trial) and ep (service).

| $V$ | 13.0 | 14.1 | 15.2 | 16.3 |
|-----|------|------|------|------|
| $ep_n$ | 1690 | 2060 | 2670 | 3400 |
| $ep_t$ | 1909.7 | 2327.8 | 3017.1 | 3842 |
| $ep_s$ | 2247.7 | 2739.8 | 3551.1 | 4522 |

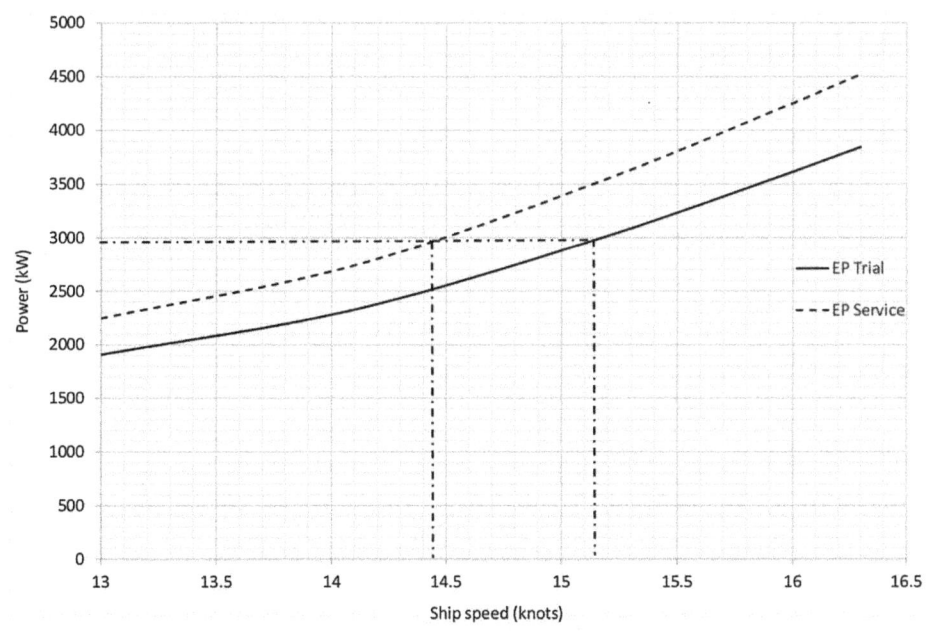

▲ **Figure A. E2.20**

$$\text{At 14 knots, ep(service)} = 2675\,\text{kW}$$
$$\text{ep(max)} = 2675 \div 0.9$$
$$= 2972\,\text{kW}$$

**a)** $\qquad$ Required ip $= 2970 \div (0.72 \times 0.965 \times 0.87)$
$$= 4913\,\text{kW}$$

**b)** At ep 2970 kW, from graph:

$$\text{Service speed} = 14.46\,\text{knots}$$
$$\text{Trial speed} = 15.14\,\text{knots}$$

## A. E2.64:

a)

| $GG_1 = 0.20$ m | | | | | | |
|---|---|---|---|---|---|---|
| $\theta$ | $\sin \theta$ | $GG_1 \sin \theta$ | $GZ$ | $G_1Z$ | SM | Product |
| 0 | 0 | 0 | 0 | 0 | 1 | 0 |
| 15° | 0.259 | 0.0518 | 0.38 | 0.43 | 4 | 1.72 |
| 30° | 0.500 | 0.100 | 1.00 | 1.10 | 2 | 2.20 |
| 45° | 0.707 | 0.1414 | 1.41 | 1.55 | 4 | 6.20 |
| 60° | 0.866 | 0.1732 | 1.20 | 1.37 | 1 | 1.37 |
| | | | | | | 11.49 |

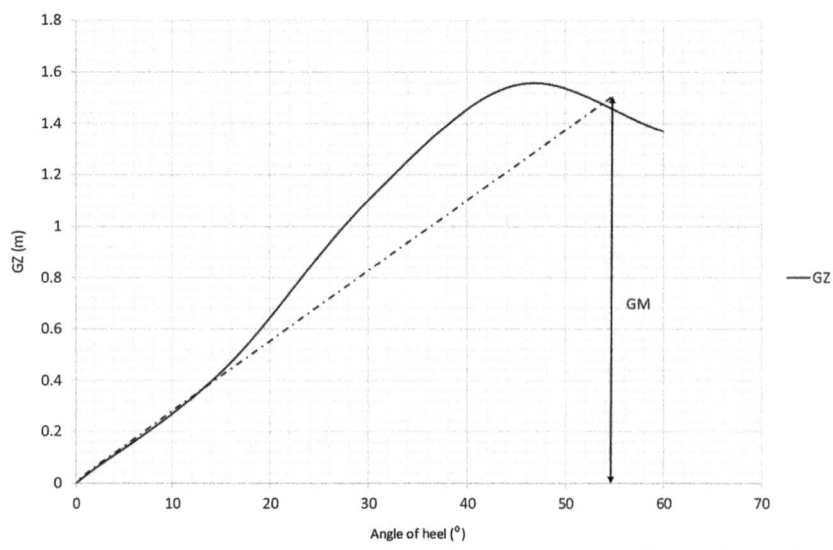

▲ Figure A. E2.21

**b)**  Dynamical stability $= \frac{1}{3} \times (15 \div 57.3) \times 11.49 \times 9.81 \times 15000$

$= 147535 \, \text{kN m}$

$= 147.5 \, \text{MJ}$

## A. E2.65:

Let $V$ = original speed

$C$ = original cons/day

$K$ = original cons for the voyage

Cons/day is proportional to speed$^3$

Therefore:

$$C \div (C + 25) = (V \div (V + 1.5))^3$$

Voyage cons is proportional to speed$^2$

$$K \div 1.2K = (V \div (V + 1.5))^2$$
$$\sqrt{(1 \div 1.2)} = V \div (V + 1.5)$$
$$V + 1.5 = \sqrt{1.2} \times V$$
$$V + 1.5 = 1.095V$$
$$V = 1.5 \div 0.095$$

Therefore: original speed $V = 15.79 \, \text{knots}$

$$C \div (C + 25) = (15.79 \div (15.79 + 1.5))^3$$
$$C \div (C + 25) = 0.7617$$
$$C = 0.7617(C + 25)$$
$$C(1 - 0.7617) = 0.7617 \times 25$$
$$C = (0.7617 \times 25) \div 0.2383$$

Therefore: original consumption $= 79.91 \, \text{tonne/day}$

## A. E2.66:

| Width | SM | Product | Lever | Product | Lever | Product |
|-------|-----|---------|-------|---------|-------|---------|
| 6.0 | 1 | 6.0 | 0 | — | 0 | — |
| 6.0 | 4 | 24.0 | 1 | 24.0 | 1 | 24.0 |
| 5.3 | 2 | 10.6 | 2 | 21.2 | 2 | 42.4 |
| 3.6 | 4 | 14.4 | 3 | 43.2 | 3 | 129.6 |
| 0.6 | 1 | 0.6 | 4 | 2.4 | 4 | 9.6 |
| | | | | 90.8 | | 205.6 |

$$\text{1st moment} = (h^2 \div 3)\Sigma_m$$
$$= (3^2 \div 3) \times 90.8$$
$$= 272.4 \text{ m}^3$$
$$\text{Load on bulkhead} = \rho\,g \times \text{1st moment}$$
$$= 0.80 \times 9.81 \times 272.4$$
$$= 2138 \text{ kN}$$
$$= 2.138 \text{ MN}$$
$$\text{Centre of pressure} = \text{2nd moment} \div \text{1st moment}$$
$$= (h\,\Sigma_i) \div \Sigma_m$$
$$= 3 \times 205.6 \div 90.8$$
$$= 6.79 \text{ m from top of bulkhead}$$

**A. E2.67:**

| V (knots) | 12.50 | 13.25 | 14.00 |
|-----------|-------|-------|-------|
| ep (kW)   | 1440  | 1800  | 2230  |
| QPC       | 0.705 | 0.713 | 0.708 |
| dp (KW)   | 2043  | 2525  | 3150  |

$$dp = ep \div QPC$$

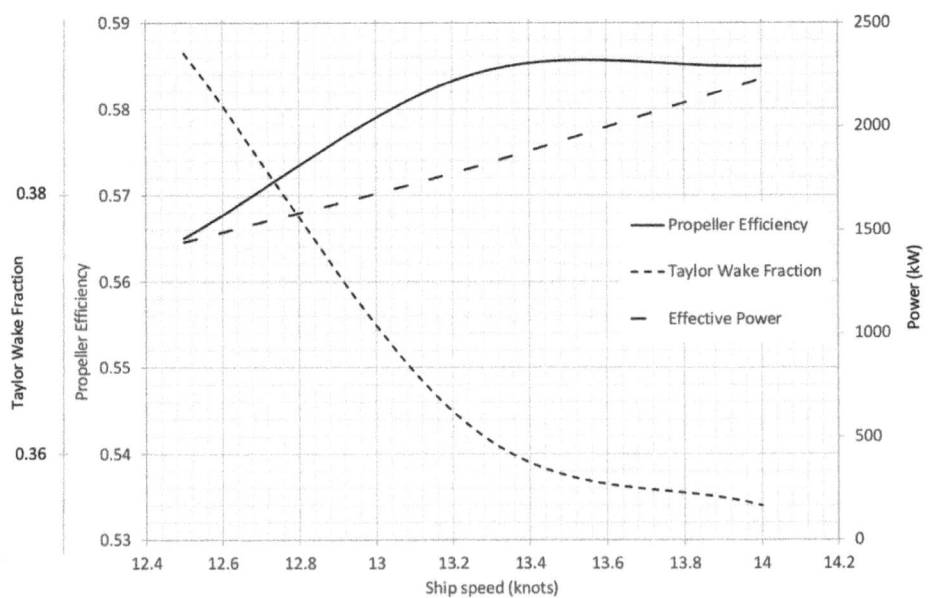

▲ **Figure A. E2.22**

| Ship speed (knots) | 12.50 | 13.25 | 14.00 |
|---|---|---|---|
| Effective power (kW) | 1440 | 1800 | 2230 |
| QPC | 0.705 | 0.713 | 0.708 |
| Propeller efficiency | 0.565 | 0.584 | 0.585 |
| Taylor wake fraction | 0.391 | 0.362 | 0.356 |

From graph, when dp is 2385 kW, ship speed = 13.06 knots

$$At\ this\ speed, w_t = 0.365\ and\ propeller\ efficiency = 0.581$$
$$Thrust\ power\ tp = 2385 \times 0.581$$
$$= 1385\ kW$$
$$Ship\ speed = 13.06 \times 0.5144$$
$$Ship\ speed = 6.72\ m/s$$
$$0.365 = (6.72 - V_a) \div V_a$$
$$V_a = 6.72 \times (1 - 0.365)$$
$$= 4.267\ m/s$$
$$tp = T \times V_a$$
$$Propeller\ thrust\ T = 1385 \div 4.266$$
$$= 325\ kN$$

**A. E2.68:**

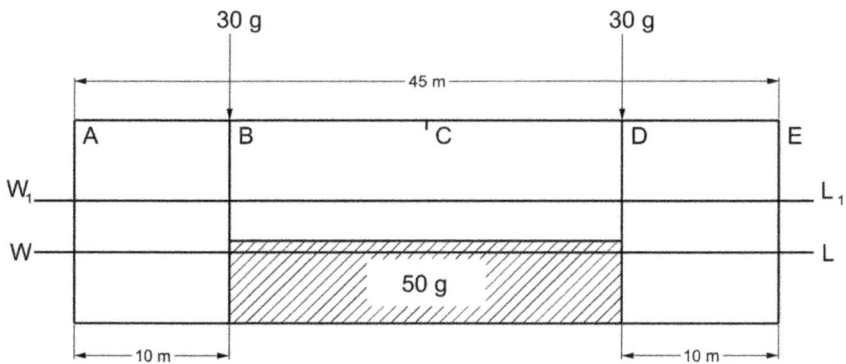

▲ **Figure A. E2.23**

Since initially the load is uniformly distributed along the vessel's length, there will be no shearing force. After the addition of the masses there will be shearing forces due to the difference in loading along the length of the vessel.

$$\text{Uniformly distributed load, B to D} = 50\,g \div 25$$
$$= 2g\ \text{kN/m}$$
$$\text{Additional buoyancy required} = (30\,g + 30\,g + 50\,g) \div 45$$
$$= 2.444\,g$$
$$\text{Shearing force at A} = 0$$
$$\text{Shearing force at left hand of B} = 2.444\,g \times 10$$
$$= 24.44\,g$$

Shearing force at right hand of B

$$= 24.44\,g - 30g$$
$$= -5.56\,g$$
$$\text{Shearing force at C} = 2.444\,g \times 22.5 - 30\,g - 2\,g \times 12.5$$
$$= 55\,g - 30\,g - 25\,g$$
$$= 0$$

Since the vessel is symmetrically loaded, these values will be repeated, but of opposite sign.

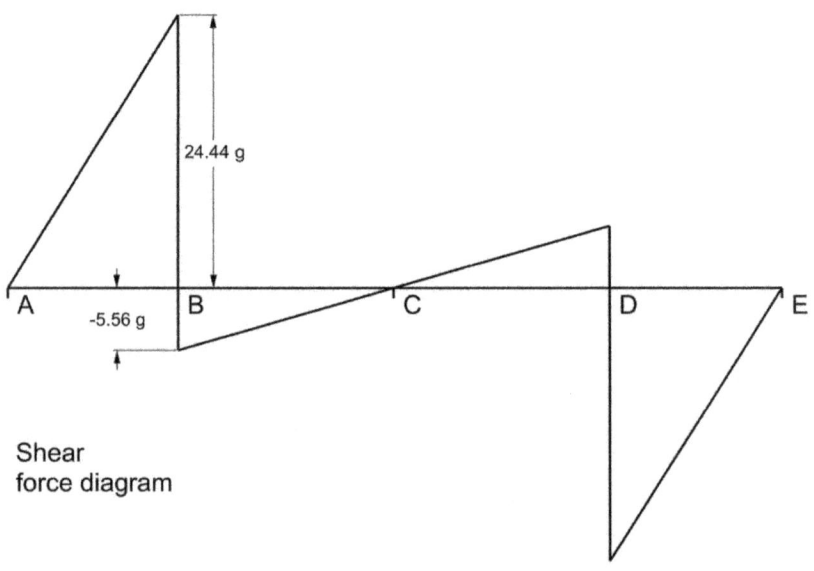

Shear
force diagram

▲ Figure A. E2.24

The maximum shearing force occurs at B and D.

$$\text{Maximum shearing force} = 24.44\,g$$
$$= 239.8\ \text{kN}$$

**A. E2.69:**

$$R_t = T(1-t)$$
$$= 378(1-0.24)$$
$$= 287.3 \, \text{kN}$$
$$\text{Ship speed} = 15.5 \times 0.5144$$
$$= 7.974 \, \text{m/s}$$

**a)**      Effective power $ep = 287.3 \times 7.974$
$$= 2291 \, \text{kW}$$
$$V_t = 4.87 \times 1.58$$
$$= 7.695 \, \text{m/s}$$
$$0.30 = (7.695 - V_a) \div 7.695$$
$$V_a = 7.695(1-0.30) = 5.386 \, \text{m/s}$$

**b)**      Taylor wake fraction $w_t = (7.974 - 5.386) \div 7.974$
$$= 0.325$$
$$\text{Shaft power } sp = 3540 \div 0.97$$
$$= 3649.5 \, \text{kW}$$

**c)**      Propulsive coefficient $= ep \div sp$
$$= 2291 \div 3649.5$$
$$= 0.628$$
$$ep = ep_n + \text{appendage allowance}$$
$$ep_n = 2291 \div 1.15 = 1992 \, \text{kW}$$

**d)**      Quasi-propulsive coefficient $QPC = ep_n \div dp$
$$= 1992 \div 3540$$
$$= 0.563$$

$$\text{MCT 1 cm} = (14000 \times 120) \div (100 \times 125)$$
$$= 134.4 \, \text{tonne m}$$
$$\text{Let } m = \text{mass added}$$
$$d = \text{distance of mass F}$$
$$\text{Change in trim required} = (8.5 - 7.9) \times 100$$
$$= 60 \, \text{cm}$$

Since the after draught remains constant, the change in trim aft must be equal to the bodily sinkage.

$$\text{Change in trim aft} = (60 \div 125) \times (125 \div 2 + 3)$$
$$= 31.44 \text{ cm}$$
$$\text{Bodily sinkage} = m \div 19$$
$$m \div 19 = 31.44$$
$$m = 31.44 \times 19$$
$$= 597.36 \text{ tonne}$$
$$\text{But change in trim } 60 = (m \times d) \div \text{MCT 1cm}$$
$$d = (60 \times 134.4) \div 597.36$$
$$= 13.5 \text{ m forward}$$

Thus: 597.36 tonne must be added 13.5 m forward of midships.

**A. E2.71:**

When a mass is suspended from a derrick head, its centre of gravity may be taken at the derrick head.

**a)** $\quad GG_1 = (50 \times 18) \div 4000$
$$= 0.225 \text{ m up}$$

**b)** The mass has been moved the same distance:

ie $GG_1 = 0.225$ m up

**c)** $\quad GG_1 = 50 \times 12 \div 4000$
$$= 0.15 \text{ m up}$$

**d)** $\quad GG_1 = 50 \times 14 \div 4000$
$$= 0.175 \text{ m outboard}$$

**A. E2.72:**

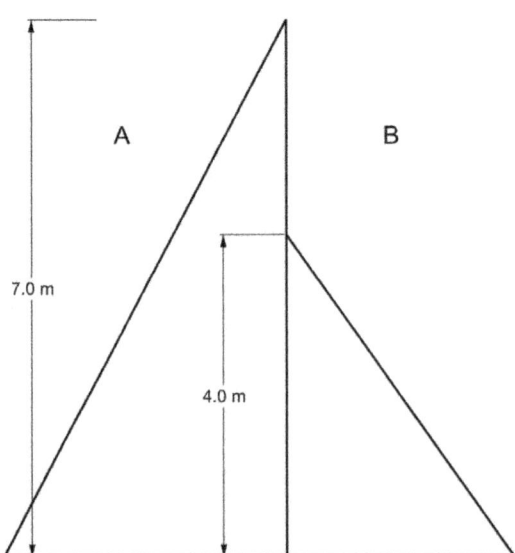

▲ Figure A. E2.25

Load on side A = $\rho g A H$

$$= 1.00 \times 9.81 \times 7 \times 8 \times (7 \div 2)$$
$$= 1922.76 \text{ kN}$$

Centre of pressure from bottom = $(7 \div 3)$m

Load on side B = $0.850 \times 9.81 \times 4 \times 8 \times (4 \div 2)$
$$= 533.66 \text{ kN}$$

Centre of pressure from bottom = $(4 \div 3)$ m

a)      Resultant load = $1922.76 - 533.66$
$$= 1389.1 \text{kN}$$

b) Take moments about the bottom of the bulkhead:

Resultant centre of pressure = $(1922.76 \times (7 \div 3) - 533.66 \times (4 \div 3)) \div 1389.1$
$$= (4486.44 - 711.55) \div 1389.1$$
$$= 2.718 \text{ m from bottom}$$

**A. E2.73:**

| Section | $\frac{1}{2}$ area | SM | Product for volume | Lever | Product for moment |
|---------|------|-----|--------------------|-------|--------------------|
| AP | 0.4 | 1 | 0.4 | +5 | +2.0 |
| 1 | 7.6 | 4 | 30.4 | +4 | +121.6 |
| 2 | 21.4 | 2 | 42.8 | +3 | +128.4 |
| 3 | 33.5 | 4 | 134.0 | +2 | +268.0 |
| 4 | 40.8 | 2 | 81.6 | +1 | +81.6 |
| 5 | 45.5 | 4 | 182.0 | 0 | +601.6 |
| 6 | 48.4 | 2 | 96.8 | −1 | −96.8 |
| 7 | 52.0 | 4 | 208.0 | −2 | −416.0 |
| 8 | 51.1 | 2 | 102.2 | −3 | −306.6 |
| 9 | 34.4 | 4 | 137.6 | −4 | −550.4 |
| FP | 0 | 1 | — | −5 | — |
| | | | 1015.8 | | −1369.8 |

a) New displacement $= \frac{2}{3} \times 9.15 \times 1015.8 \times 1.025$

$$= 6351.3 \text{ tonne}$$

Original displacement $= 5750$ tonne

Therefore: mass of water added $= 601.3$ tonne

b) Moment of buoyancy about midships $= \frac{2}{3} \times 9.15^2 \times (-1369.8 + 601.6) \times 1.025$

$$= -43949 \text{ tonne m } (-\text{ve implies forward})$$

Original moment of buoyancy $= 5750 \times 4.6$

$$= 26450 \text{ tonne m forward}$$

Therefore: moment of ballast about midships $= 43949 - 26450$

$$= 17499 \text{ tonne m forward}$$

Centre of gravity of ballast from midships $= 17499 \div 601.3$

$$= 29.10 \text{ m forward}$$

Mass of water pumped out $= 14 \times 12 \times 0.6 \times 10.25$

$$= 103.3 \text{ tonne}$$

## A. E2.74:

The centre of gravity of this water is 0.9 m above the keel. Taking moments about the keel:

New $KG = (7500 \times 6.7 - 103.3 \times 0.9) \div (7500 - 103.3)$

$$= (50250 - 93) \div (7396.7)$$

$$= 6.781 \text{m}$$

Free surface effect $= i \div \nabla$

$$= (14 \times 12^3 \times 1.025) \div (12 \times 7396.7)$$

$$= 0.279 \text{ m}$$

Original $KM = 6.7 + 0.45$

$$= 7.15 \text{ m}$$

New $GM = 7.15 - 6.781 - 0.279$

$$= 0.09 \text{ m}$$

## A. E2.75:

Change in mean draught $= (\Delta \times 100 \div A_w) \times (\rho_s - \rho_r) \div (\rho_s \rho_r) \text{ cm}$

$$= (9100 \times 100 \times 1.024) \div (16.8 \times 100) \times (1.024 - 1.000) \div$$
$$(1.024 \times 1.000) \text{ cm}$$

$$= 9100 \times 0.024 \div 16.8$$

$$= 13 \text{ cm reduction}$$

Shift in centre of buoyancy

$$= (\rho_s - \rho_r) \div \rho_r$$
$$= (1.024 - 1.000) \div 1.000 \times (0.6 + 2.25)$$
$$= 0.0684 \text{ m}$$
$$\text{Change in trim} = 9100 \times 0.0684 \div 122$$
$$= 5.1 \text{ cm by the stern}$$
$$\text{Change forward} = -(5.10 \div 120) \times (120 \div 2 + 0.6)$$
$$= -2.6 \text{ cm}$$
$$\text{Change aft} = (5.10 \div 120) \times (120 \div 2 - 0.6)$$
$$= +2.5 \text{ cm}$$
$$\text{New draught forward} = 6.7 - 0.13 - 0.026$$
$$= 6.544 \text{ m}$$
$$\text{New draught aft} = 6.7 - 0.13 + 0.025$$
$$= 6.595 \text{ m}$$

**A. E2.76:**

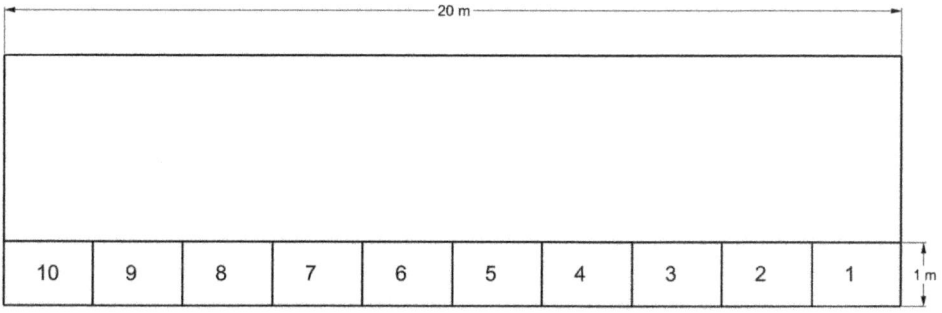

▲ Figure A. E2.26

$$\text{Mass added/tank} = 2 \times 10 \times 1 \times 1.025$$
$$= 20.5 \text{ tonne}$$
$$\text{Total mass added} = 41 \text{ tonne}$$
$$\text{mass/m} = 20.5 \div 2$$
$$= 10.25 \text{ tonne}$$
$$\text{Weight/m} = 10.25 \, g \text{ kN}$$
$$\text{Buoyancy required/m} = 41 \, g \div 20$$
$$= 2.05 \, g$$

Hence, in way of ballast,

$$\text{Excess load/m} = 10.25 \, g - 2.05 \, g$$
$$= 8.2 \, g \text{ kN}$$

**a)** With No 1 and No 10 tanks filled:

$$\text{SF at aft end of vessel} = 0$$
$$\text{SF at fore end of No.10} = -8.2\,g \times 2$$
$$= -16.40\,g \text{ kN}$$
$$\text{SF at midships} = -16.40\,g + 2.05\,g \times 8$$
$$= 0$$
$$\text{SF at aft end of No 1} = +2.05\,g \times 8$$
$$= +16.4\,g$$
$$\text{SF at fore end of vessel} = +16.4\,g - 8.2\,g \times 2$$
$$= 0$$

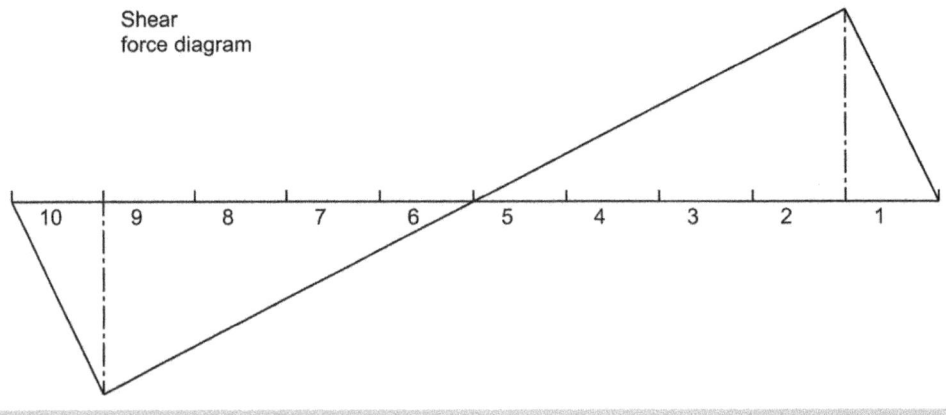

Shear force diagram

▲ **Figure A. E2.27**

**b)** With No 3 and No 8 tanks filled:

$$\text{SF at aft end of vessel} = 0$$
$$\text{SF at aft end of No 8} = +2.05\,g \times 4$$
$$= +8.2\,g$$
$$\text{SF at fore end of No 8} = +8.2\,g - 8.2\,g \times 2$$
$$= -8.2\,g$$
$$\text{SF at midships} = -8.2\,g + 2.05\,g \times 4$$
$$= 0$$
$$\text{SF at aft end of No 3} = +2.05\,g \times 4$$
$$= +8.2\,g$$
$$\text{SF at fore end of No 3} = +8.2\,g - 8.2\,g \times 2$$
$$= -8.2\,g$$
$$\text{SF at fore end of vessel} = 0$$

Shear
force diagram

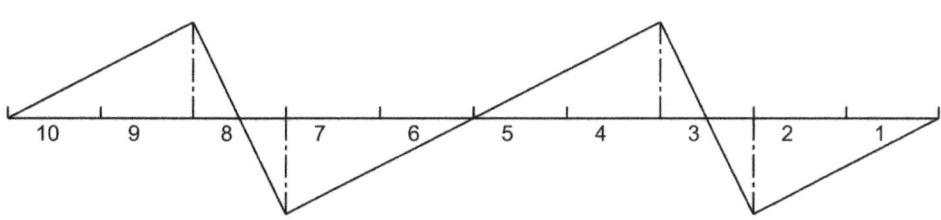

▲ **Figure A. E2.28**

c)  With No 5 and No 6 tanks filled:

$$SF \text{ at aft end of vessel} = 0$$
$$SF \text{ at aft end of No } 6 = +2.05\,g \times 8$$
$$= +16.4\,g$$
$$SF \text{ at fore end of No } 5 = +16.4\,g - 8.2\,g \times 4$$
$$= -16.4\,g$$
$$SF \text{ at fore end of vessel} = 0$$

Shear
force diagram

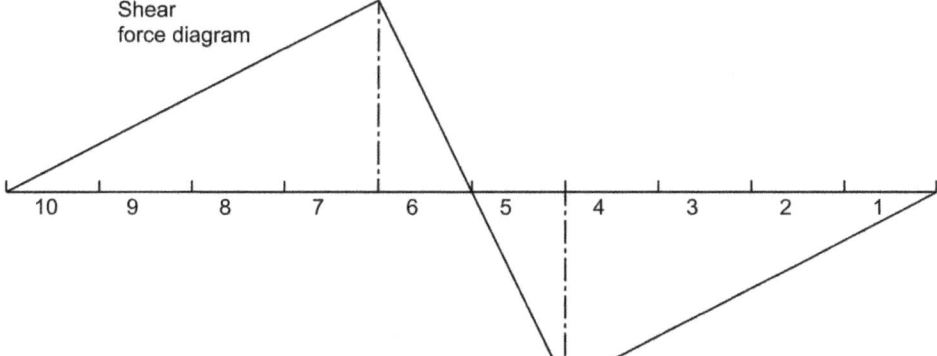

▲ **Figure A. E2.29**

The maximum shearing force in case b) is half of the maximum values in cases a) and c).
Thus b) is the best loaded condition.

**A. E2.77:**

$$\text{Final displacement} = 216 + 81 + 54$$
$$= 351\,\text{tonne}$$
$$= L \times B \times d \times \rho$$
$$\text{New draught} = 351 \div (32 \times 5.5 \times 1.025)$$
$$= 1.946\,\text{m}$$
$$KB = 1.946 \div 2$$
$$= 0.973\,\text{m}$$
$$BM = (32 \times 5.5^3) \div (12 \times 32 \times 5.5 \times 1.946)$$
$$= 1.295\,\text{m}$$
$$KM = 0.973 + 1.295$$
$$= 2.268\,\text{m}$$
$$\text{Final } GM = 0.13\,\text{m}$$
$$\text{Therefore: final } KG = 2.138\,\text{m}$$
$$\text{Let } x = Kg \text{ of machinery}$$
$$351 \times 2.138 = 216 \times 1.8 + 81x + 54 \times 0.15$$
$$750.44 = 388.8 + 81x + 8.1$$
$$81x = 353.54$$
$$Kg \text{ of machinery } x = 4.365\,\text{m}$$

**A. E2.78:**

If a ship moves from sea water of 1.025 t/m³ into fresh water of 1.000 t/m³, change in mean draught $= \Delta \div (40\,\text{TPC})$ cm

$$\text{Therefore}: 23 = \Delta \div (40 \times 23)$$
$$\Delta = 23 \times 40 \times 23$$
$$= 21160\,\text{tonne}$$

The draught in the river water is the same as the allowable draught in the fresh water, but the displacement is 202 tonne greater, ie 21362 tonne.

$$\text{Let } \rho_R = \text{density of river water}$$
$$\text{Volume of displacement in fresh water} = 21160 \div 1.000\,\text{m}^3$$
$$\text{Volume of displacement in river water} = 21362 \div \rho_R \text{m}^3$$
$$\text{But volume in fresh water} = \text{volume in river water}$$
$$21160 \div 1.000 = 21362 \div \rho_R$$
$$\rho_R = 21362 \div 21160$$
$$\text{Density of the river water} = 1.010\,\text{t/m}^3$$

**A. E2.79:**

| Speed of advance | (m/s) | 1.22 | 1.46 | 1.70 | 1.94 |
|---|---|---|---|---|---|
| Thrust | (N) | 93.7 | 72.3 | 49.7 | 24.3 |
| Thrust power | (W) | 114.3 | 105.6 | 84.5 | 47.1 |
| Torque | (Nm) | 3.90 | 3.23 | 2.50 | 1.61 |
| Delivered power | (W) | 196.1 | 162.4 | 125.7 | 80.9 |

$$\text{Thrust power} = \text{thrust} \times \text{speed of advance}$$
$$\text{Delivered power} = \text{torque} \times 2\pi \times \text{rev/s}$$

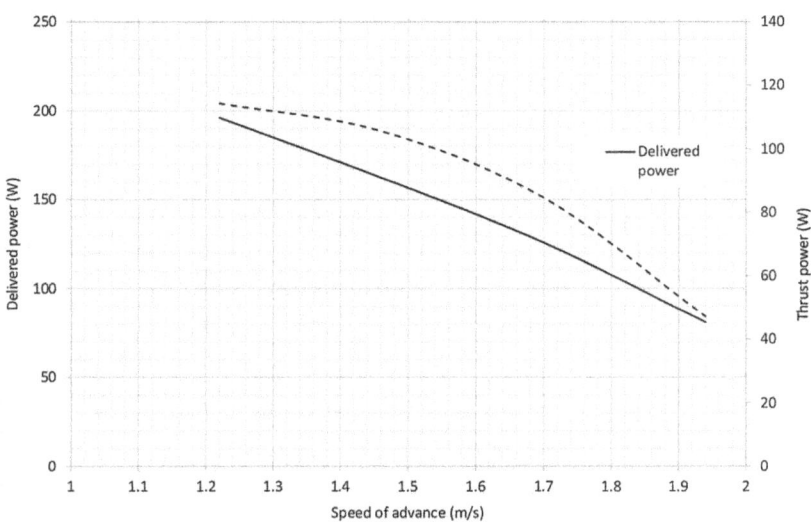

▲ **Figure A. E2.30**

For ship dp = 3000 kW in sea water

Equivalent for model dp = $3000(0.3 \div 4.8)^{3.5} \times (1.000 \div 1.025)$

$= 178.6$ W

From graph at this dp: $V_a = 4.348$ m/s

and at this speed: tp = 111.1W

a) For ship tp = $111.1 \times (4.8 \div 0.3)^{3.5} \times (1.025 \div 1.000)$

$= 1866$ kW

b) Propeller efficiency = $1866 \div 3000 \times 100$

$= 62.2\%$

**A. E2.80:**

| Draught | $C_w$ | SM | Product for volume | (1) Lever | Product for vert. moment | LCF | Product for (2) long. moment |
|---------|-------|-----|--------------------|-----------|--------------------------|-----------|------------------------------|
| 1.22 | 0.78 | 1 | 0.78 | 1 | 0.78 | + 1.30 | + 1.014 |
| 2.44 | 0.82 | 4 | 3.28 | 2 | 6.56 | + 1.21 | + 3.969 |
| 3.66 | 0.85 | 2 | 1.70 | 3 | 5.10 | + 0.93 | + 1.581 |
| 4.88 | 0.88 | 4 | 3.52 | 4 | 14.08 | + 0.50 | + 1.760 |
| 6.10 | 0.90 | 1 | 0.90 | 5 | 4.50 | − 0.06 | − 0.054 |
| | | | 10.18 | | 31.02 | | 8.270 |

**(1)** Levers taken from the keel

**(2)** Product of volume column and LCF

**a)**  Displacement of layer $= (1.22 \div 3) \times 10.18 \times 128 \times 16.75 \times 1.025$
$= 9097.8$ tonne
Longitudinal moment $= (1.22 \div 3) \times 8.27 \times 128 \times 16.75$

**b)**
**i)**  LCB of layer from midships

$= ((1.22 \div 3) \times 8.27 \times 128 \times 16.75) \div ((1.22 \div 3) \times 10.18 \times 128 \times 16.75)$
$= 8.27 \div 10.18$
$= 0.812 \, \text{m forward}$

**ii)**  Vertical moment $= (1.22^2 \div 3) \times 31.02 \times 128 \times 16.75$

VCB of layer from keel

$= ((1.22^2 \div 3) \times 31.02 \times 128 \times 16.75) \div ((1.22 \div 3) \times 10.18 \times 128 \times 16.75)$
$= (1.22 \times 31.02) \div 10.18$
$= 3.717 \, \text{m}$

**A. E2.81:**
**a)**  Second moment of area about LCF

$$I_F = I_m - A\,x^2$$
$$= 2326048 - 2110 \times 4.6^2$$
$$= 2281400 \text{ m}^4$$
$$BM_L = I_F \div \nabla$$
$$= 2281400 \times 1.025 \div 14000$$
$$= 167.03 \text{ m}$$
$$GM_L = KB + BM_L - KG$$
$$= 4.27 + 167.03 - 8.54$$
$$= 162.76 \text{ m}$$
$$\text{MCT 1 cm} = \Delta \times GM_L \div (100L)$$
$$= 14000 \times 162.76 \div (100 \times 140)$$
$$= 162.76 \text{ tm}$$

**b)**     Distance from LCF to aft end $= (140 \div 2) - 4.6$
$$= 65.4 \text{ m}$$
Change in trim over this distance $= 7.45 - 7.0$
$$= 0.45 \text{ m}$$
Change in trim over 140 m $= (0.45 \div 65.4) \times 140$
$$= 0.963 \text{ m}$$
Change in trim forward $= 0.963 - 0.45$
$$= 0.513 \text{ m}$$
New draught forward $= 7.0 - 0.513$
$$= 6.487 \text{ m}$$

**c)**     Longitudinal shift in centre of gravity $= (\text{change in trim} \times \text{MCT 1 cm}) \div \Delta$
$$= (0.963 \times 100 \times 162.76) \div 14000$$
$$= 1.12 \text{ m aft}$$
New position of centre of gravity $= 0.88 + 1.12$
$$= 2 \text{ m aft of midships}$$
**d)**     Shift in centre of gravity $= m \times d \div \Delta$
$$m = 1.12 \times 14000 \div 112$$
$$= 140 \text{ tonne}$$

**A. E2.82:**

a) 
$$ep = R_t \times V$$
$$R_f = f\, SV^a$$
$$= 1.432 \times 5946 \times V^{1.825}$$
$$= 8514.7V^{1.825}\ (N)$$
$$Rt = R_f \div 0.74$$
$$= 8514.7V^{1.825} \div 0.74$$
$$= 11.506V^{1.825}\ (kN)$$
$$11250 = 11.506V^{1.825}V$$
$$V^{2.825} = 11250 \div 11.506$$
$$V = 11.44\ m/s$$
$$V = 22.24\ knots$$

b) 
$$sp = ep \div pc$$
$$= 11250 \div 0.60$$
$$= 18750\ kW$$
$$\text{Fuel consumption/day} = 0.22 \times 18750 \times 24 \times 10^{-3}$$
$$= 99\ tonne$$

fc is proportional to $V^3$

Assuming that the specific consumption remains unchanged

$$\text{Cons/day at reduced speed} = 99 \times (0.9V \div V)$$
$$= 72.17\ tonne$$
$$\text{Actual cons/day} = 83\ tonne$$
$$\text{Increase in cons/days} = 10.83\ tonne$$
$$\text{Percentage increase} = (10.83 \div 72.17) \times 100$$
$$= 15\%$$

**A. E2.83:**

a)

|              | Mass | Kg  | Moment |
|--------------|------|-----|--------|
| Light barge  | 300  | 2.6 | 780    |
| Lower hold   | 1000 | 4.7 | 4700   |
| 'tween deck  | 500  | 6.1 | 3050   |
| Deep tank    | 200  | 3.4 | 680    |
| Displacement | 2000 |     | 9210   |

$$KG = 9210 \div 2000$$
$$= 4.605 \, \text{m}$$
$$\text{Draught} = 2000 \div (60 \times 10.5 \times 1.025)$$
$$= 3.097 \, \text{m}$$
$$KB = 3.097 \div 2$$
$$= 1.549 \, \text{m}$$
$$BM = I \div \nabla$$
$$= B^2 \div (12d)$$
$$= 10.5^2 \div (12 \times 3.097)$$
$$= 2.967 \, \text{m}$$

Metacentric height $GM = KB + BM - KG$

**i)**
$$= 1.548 + 2.966 - 4.605$$
$$= -0.091 \, \text{m}$$

**ii) For an unstable, wall-sided vessel**

$$\tan^2 \theta = -2GM \div BM$$
$$= 0.180 \div 2.966$$

Angle of heel $\theta = 13.84°$

**b)** Required $GM = +0.15 \, \text{m}$

Change in $KG = 0.15 + 0.09$
$$= 0.24 \, \text{m}$$
$$= (m \times d) \div \Delta$$
$$m = (0.24 \times 2000) \div (6.1 - 3.4)$$

Mass transferred $= 177.8 \, \text{tonne}$

**A. E2.84:**

**a)** Let $l =$ length of centre compartment

Volume of lost buoyancy $= l \times 12 \times 7$

Area of intact waterplane $= (80 - l) \times 12$

Increase in draught $3 = (l \times 12 \times 7) \div ((80 - l) \times 12)$
$$240 - 3l = 7l$$
$$10l = 240$$

Length of centre compartment $l = 24 \, \text{m}$

New displacement $= 80 \times 12 \times 7 \times 1.025$
$$= 6888 \, \text{tonne}$$

Cargo added $= 6888 - 888$

**b)** $= 6000 \, \text{tonne}$

Length of end compartment $= (80 - 24) \div 2$
$$= 28 \, \text{m}$$

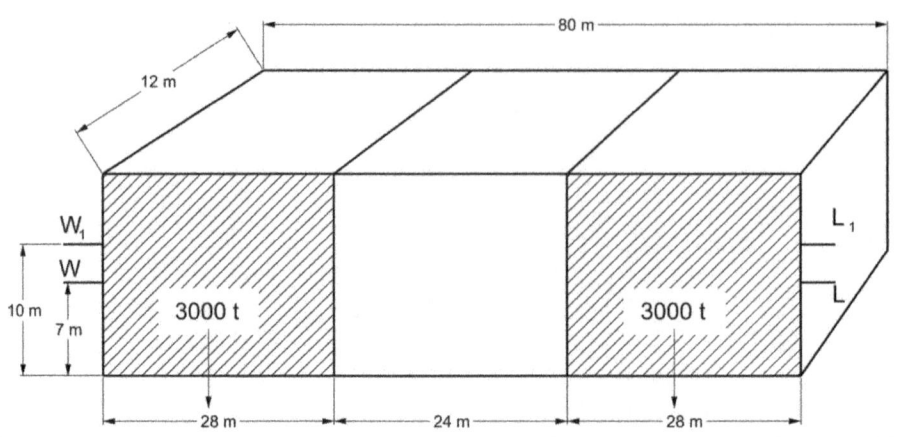

▲ Figure A. E2.31

i) Moment of buoyancy about midships $= (6888 \div 2) \times 20 \times g$
$$= 68880\,g$$
Moment of weight about midships $= (888 \div 2) \times 20 \times g + 3000 \times (14 + 12)g$
$$= 8880\,g + 78000\,g$$
$$= 86880\,g$$
Bending moment at midships $= 86880\,g - 68880\,g$
$$= 18000\,g\ \text{kN m hog}$$
Moment of buoyancy about midships $= (6888 \div 2) \times 26 \times g$
$$= 89544\,g$$
ii) Bending moment at midships $= 86880\,g - 89544\,g$
$$= -2664\,g\ \text{kNm}$$
$$= 2664\,g\ \text{kN m sag}$$

**A. E2.85:**

**a)** Block coefficient $C_b = 355190 \div (325 \times 56 \times 22.4 \times 1.025)$
$$= 0.85$$
$$w_t = 0.5 \times 0.85 - 0.05$$
$$= 0.375$$
$$\text{Pitch } P = 7.4 \times 0.85$$
$$= 6.29 \text{ m}$$
$$V_T = 6.29 \times 1.5$$
$$= 9.435 \text{ m/s}$$
$$\text{Real slip} = (V_T - V_a) \div V_T$$
$$V_a = 9.435(1 - 0.4888)$$
$$= 4.823 \text{ m/s}$$
$$\text{Taylor wake fraction } w_t = (V - V_a) \div V$$
$$\text{Ship speed} = 4.823 \div (1 - 0.375)$$
$$= 7.717 \text{ m/s}$$
$$= 7.717 \div 0.5144$$
$$= 15 \text{ knots}$$

**b)** Voyage consumption is proportional to $V^2$

$$(VC_1 \div VC_2) = (V_1 \div V_2)^2$$
$$(VC_1 \div 0.5VC_1) = (15 \div V_2)^2$$
$$V_2 = 15\sqrt{0.5}$$
$$\text{Reduced speed } V_2 = 10.61 \text{ knots}$$

**c)** Let distance travelled $= D$
At service speed, time taken $= D \div (15 \times 24)$ days
At reduced speed, time taken $= D \div (10.61 \times 24)$ days
$$= D \div (15 \times 24) + 6$$

Therefore:

$$D \div (10.61 \times 24) = D \div (15 \times 24) + 6$$
$$0.00393D - 0.0028D = 6$$
$$D = 5215 \text{ n.m.}$$

**d)** At service speed, fuel used $= 165 \times 5215 \div (15 \times 24)$
$$= 2390 \text{ tonne}$$

At reduced speed, fuel required

$$= 0.5 \times 2390$$
$$= 1195 \text{ tonne}$$

**A. E2.86:**

$$\text{Wetted surface area } S = 2.58\sqrt{(13716 \times 137)}$$
$$= 3537\,\text{m}^2$$
$$\text{Effective power} = 4847 \times 0.67$$
$$= 3247\,\text{kW}$$
$$\text{Ship speed} = 15\,\text{knots}$$
$$= 15 \times 0.5144$$
$$= 7.716\,\text{m/s}$$
$$\text{Total resistance } R_t = \text{ep} \div V$$
$$= 3247 \div 7.716$$
$$= 420.8\,\text{kN}$$
$$\text{At } L = 137\,\text{(by interpolation)}$$
$$f = 1.4163$$
$$R_f = fsV^2$$
$$= 1.4163 \times 3537 \times 7.716^{1.825}$$
$$= 208.6\,\text{kN}$$
$$R_r = 420.8 - 208.6$$
$$= 212.2\,\text{kN}$$
$$\text{New ship } \Delta = 18288\,\text{tonne}$$
$$R_r \text{ is proportional to } L^3$$
$$R_r \text{ is proportional to } \Delta$$

Therefore:

$$R_r = 212.2 \times 18288 \div 13716$$
$$= 282.9\,\text{kN}$$

$\Delta$ is proportional to $L^3$

Therefore:

$$L_1 = 137 \times (18288 \div 13716)^{\frac{1}{3}}$$
$$= 150.8\,\text{m}$$

$S$ is proportional to $L^2$

$$S = 3537(150.8 \div 137)^2$$
$$= 4285\,\text{m}^2$$

At corresponding speeds $V$ is proportional to $\sqrt{L}$

$$V_1 = 7.716\sqrt{(150.8 \div 137)}$$
$$= 8.095\,\text{m/s}$$
$$\text{At } L_1 = 150.8 \text{ (by interpolation)}$$
$$f = 1.4149$$
$$R_f = 1.4149 \times 4285 \times 8.095^{1.825}$$
$$= 275.5\,\text{kN}$$
$$R_t = 275.5 + 282.9$$
$$= 558.4\,\text{kN}$$
$$\text{ep} = 558.4 \times 8.095$$
$$= 4520\,\text{kW}$$
$$\text{sp} = 4521 \div 0.67$$
$$\text{Shaft power sp} = 6748\,\text{kW}$$

# ACKNOWLEDGEMENTS

I would like to thank Duncan Mitchell for his technical advice and Jenny Clark from Bloomsbury for her support in the preparation of this book. I would particularly like to thank Jenni Davis, for her attention to detail while editing this volume. I would also like to thank Lucy, Alice and Miles Pemberton for their patience.

# INDEX

# REEDS MARINE ENGINEERING AND TECHNOLOGY SERIES

**Vol. 1 Mathematics for Marine Engineers**
Kevin Corner, Leslie Jackson and William Embleton
ISBN 9781472974037

**Vol. 2 Applied Mechanics for Marine Engineers**
Paul A Russell, Leslie Jackson and William Embleton
ISBN 9781472988188

**Vol. 3 Applied Thermodynamics for Marine Engineers**
Leslie Jackson, William Embleton and Paul A Russell
ISBN 9781472993403

**Vol. 4 Naval Architecture for Marine Engineers**
Richard Pemberton, E A Stokoe
ISBN 9781399410120

**Vol. 5 Ship Construction for Marine Engineers**
Paul A Russell, E A Stokoe
ISBN 9781472989208

**Vol. 6 Basic Electrotechnology for Marine Engineers**
Christopher Lavers, Edmund G R Kraal and Stanley Buyers
ISBN 9781472963833

**Vol. 7 Advanced Electrotechnology for Marine Engineers**
Christopher Lavers and Edmund G R Kraal
ISBN 9781408176030